Reliability and Risk Assessment

Reliability and Risk Assessment

by

J D Andrews and T R Moss

Professional Engineering Publishing Limited
London and Bury St Edmunds, UK

First published 2002

ISBN 1 86058 290 7

A CIP catalogue record for this book is available from the British Library.

Printed and bound in Great Britain by Antony Rowe Ltd., Chippenham, Wiltshire, UK.

Related Titles

The Reliability of Mechanical Systems	J Davidson (Ed.) C Hunsley (Technical Ed.)	0 85298 881 8
A Practical Guide to Engineering Failure Investigation	C Matthews	1 86058 086 6
IMechE Engineers' Data Book (Second Edition)	C Matthews	1 86058 248 6
Engineering System Safety	G J Terry	0 85298 781 1
Improving Maintainability and Reliability through Design	G Thompson	1 86058 135 8
Process Machinery – Safety and Reliability	W Wong (Ed.)	1 86058 046 7

For the full range of titles published by
Professional Engineering Publishing contact:

Marketing Department, Professional Engineering Publishing Limited,
Northgate Avenue, Bury St Edmunds, Suffolk, IP32 6BW, UK
Tel: +44 (0)1284 724384
Fax: +44 (0)1284 718692
Website: www.pepublishing.com

About the Authors

John Andrews graduated in Mathematics from Birmingham University in 1978 and later gained his doctorate in Mathematical Engineering. His initial employment was as a Research Scientist for British Gas at their Midlands Research Station. After 10 years industrial experience he moved to academia taking up his current position in the Department of Mathematical Sciences at Loughborough University in 1989. In 2001 John was awarded a chair in Mathematical Engineering.

John Andrews' main research interests are in risk and reliability modelling methods and he currently has over 80 publications on this topic. He is also a member of the Editorial Boards of *Quality and Reliability Engineering International* and the Institution of Mechanical Engineers Proceedings, *Journal of Process Mechanical Engineering*. Much of the research conducted by his group at Loughborough University is sponsored by industry. Support has come from MOD, Rolls-Royce Aero-engines, Bectel, ExxonMobil, DaimlerChrysler, and Dyson.

Bob Moss is a consulting engineer with over 30 years experience in safety, reliability, and risk analysis. Prior to forming his own company in 1979 he was a Principal Engineer with the UKAEA Safety and Reliability Directorate with responsibility for the group that developed methods for the availability assessment of large, potentially hazardous chemical plant. Since 1979 he has been responsible for a wide range of projects concerned with nuclear installations, petrochemical plant, and offshore and onshore oil and gas platforms. He was one of the principal architects of the OREDA (Offshore Reliability Data) project and supervised the collection and analysis of reliability data from the UK sector of the North Sea.

Bob is a past Vice Chairman of the Institution of Mechanical Engineers Process Industries Division, past Chairman of its

Mechanical Reliability Committee and an Honorary Member of the European Safety Reliability and Data Association. For 8 years he was a Visiting Research Fellow in the Department of Mathematical Sciences at Loughborough University and was awarded a PhD for research into rotating machinery reliability. He has lectured extensively worldwide and has published over 50 papers on reliability and risk assessment. His current interests focus on the application of reliability and risk assessment methods to maintenance and asset management.

Contents

Preface

Over recent years major accidents such as Flixborough, Bhopal, Chernobyl, and Piper Alpha have taken a sad toll of lives and increased public perception of the risks associated with operating large process plant. After such accidents the reaction is always to say 'it must never happen again'; however, it is clearly impossible to eliminate all risks. For the public and Government Regulators in a modern society there is, therefore, the need for compromise to resolve the apparent paradox of obtaining the benefits of modern technology without incurring the problems that such technology can bring.

We need to eliminate hazards as far as possible and reduce the risks from the plant so that the remaining hazards can be seen to make only a very small addition to the inherent background risks of everyday life. This can never be achieved by the age-old method of learning from past experience. Each new plant is different from any previous one and therefore it needs to be assessed to identify, evaluate, and control the particular hazards associated with it.

Reliability and risk analysis techniques are the methods advocated by many regulatory bodies to assess the safety of modern, complex process plant and their protective systems. The term quantified risk assessment (which includes both reliability and risk analysis) is now incorporated into the requirements for safety cases in the nuclear, chemical/petrochemical, and offshore industries. The methods have also been adopted in the defence, marine, and automotive industries.

Reliability and risk analysis techniques also find applications in production and maintenance studies. The trend towards large, single-stream process plant means that outages due to failure and maintenance can have a significant effect on profitability. Reliability and risk assessment can be applied to identify the systems, equipment, and components in the plant which are likely to have a major impact on product availability. Attention to these areas by the introduction of redundancy and improved maintainability for these availability-critical items during the design phase of new plant can lead to dramatic improvements in profitability during the plant lifetime. The technique can, of course, also be applied to existing plant although with less dramatic effect since the basic plant design is generally less easily modified.

The objective of this book is to provide the reader with a comprehensive description of the main probabilistic methods employed in reliability and risk assessment, particularly fault tree analysis and failure mode and effect analysis and their derivatives. Each chapter is self-contained and is illustrated with worked examples. The final chapter features four major case studies from recent projects to show how the methods have been applied to identify and quantify the risk and reliability characteristics of major plant against specific performance requirements.

This book focuses on the risk and reliability assessment of process plant. It will provide design, operation, and regulatory engineers involved directly with the safety or availability assessment of potentially hazardous process plant with a comprehensive reference to these powerful analysis techniques.

Since this book is based on a programme of lectures provided for university undergraduates and post-graduates and additional specialist courses for industry, the authors hope it will also be found valuable as a teaching aid.

Acknowledgements

We acknowledge with thanks the permission given by John Wiley and Sons to reproduce Case Study 1 in Chapter 14, which was originally published in *Quality and Reliability International*, Vol. 7, 1991.

We would also like to thank Henri Procaccia and Andre Lannoy for permission to include the in-service inspection case study in Chapter 14.

Notation

A	Availability
A_s	System availability
$A(t)$	Availability function
C	Number of censored items; consequence of an incident
$[\mathbf{C}]$	Connectivity matrix of a network
C_i	Minimal cut set i
C_m	Failure mode criticality number
C_r	System criticality number
D	Demand probability
dt	Infinitesimally small interval
du	Infinitesimally small interval
F	Cumulative failures; probability that an item has failed
Fc	Common cause failure probability
Fs	Probability that a system is in a failed state
$F(t)$	Probability of failure before time t; cumulative distribution function
f/Mh	Failure per million hours
F/year	Failures per year
FAR	Fatal accident rate
FDT	Fractional dead time
$f(t)$	Failure probability density function
$f(u)$	Probability density function
$G_i(q)$	Criticality function for component i (Birnbaum's measure of importance)
$G(t)$	Cumulative repair distribution

$g(t)$	Repair function p.d.f.
$H(t)$	Cumulative hazard function
$h(t)$	Hazard rate
I_i	Importance measure for component i
i	Failure number
K_i	Stress factor for stress i
k	Total number of failures
LCC	Life cycle costs
MO_i	Mean order number
MTBF	Mean time between failures
MTTF	Mean time to failure
MTTR	Mean time to repair
m	Expected number of failures
$m(t)$	Conditional repair rate (repair process only)
N	Number of items in a population
N_f	Cumulative number of failed components
N_p	Number of minimal path sets
N_s	Number of surviving components
$N(t)$	Cumulative failures before time t
$N(T)$	Total number of failures
n	Number of items in a sample; number of units up
n_C	Number of minimal cut sets
$n(t)$	Number of units which fail in each time period
p.d.f.	Probability density function
P	Probability; probability of an event
$[\mathbf{P}]$	Transition probability matrix
$[P]$	Minimal path set matrix of a network
P_F	Failure probability
P_S	Probability that a system is working
$P_w(t)$	Probability that a component or system is working at time t
P_μ	Estimate of percentage failed at mean life
Q	System unavailability
Q_{AV}	Average unavailability

Q_{MCSU}	Minimum cut set upper bond approximation
Q_S	System failure probability
$Q(t)$	Unavailability function
q	Component reliability
R	Risk
$R(t)$	Reliability function; probability of survival to time t
$R_s(t)$	System reliability function
r	Number of units operating; component reliability; severity class (FMECA)
s	Number of survivors
s_i	Survival time
t	Time; aggregate system life
t_0	Initial time; start time
t_1	Smallest ordered age at failure
t_2	Second smallest ordered age at failure
t_i	ith ordered age at failure; interval between $(i-1)$th failure and the ith failure; times to failure
t_k	Largest ordered age at failure
tr_i	ith repair times
U_S	System unavailability
$V(t_0,t_1)$	Expected number of repairs
$v(t)$	Unconditional repair intensity
$W(t_0,t_1)$	Expected number of failures
$w(t)$	Unconditional failure intensity
x	Number of failures
$x(t)$	Binary variable to indicate component state
$x_F(t)$	Number of component failures to time t
$x_R(t)$	Number of component repairs to time t
y_i	System age at the ith failure
z	Normal $N(0,1)$ random variable

α	Proportion of failures in specific failure mode (FMEA)
β	Conditional probability of a consequence resulting from a potential cause; ratio of common cause failure rate to the total failure rate; Weibull shape parameter
γ	Weibull location constant
η	Weibull scale parameter (characteristic life)
θ	Test or inspection interval
λ	Failure rate or hazard rate
λ_b	Base failure rate
λ_{cc}	Common cause component failure rate
λ_i	Independent component failure rate
λ_p	Part failure rate
$\lambda(t)$	Conditional failure intensity (failure and repair process)
μ	Mean time to failure; distribution mean
$\mu(t)$	Conditional repair intensity (failure and repair process)
ν	Constant repair rate
σ	Mean deviation of a random variable or of a population
τ	Mean time to repair
ϕ	Network incidence function

Chapter 1

An Introduction to Reliability and Risk Assessment

1.1 Introduction

Reliability and risk assessment methods have developed over a number of years from a variety of different initiatives. The pioneer work in Germany on missile systems during the Second World War, the development of reliability methods for defence equipment by the US Department of Defense, and in the UK the contributions on hazard analysis by Trevor Kletz of ICI Ltd and on reliability methods by Green and Bourne of the UKAEA are all worthy of note. Milestones also exist such as the reports of the public enquiries following the accidents at Windscale, Flixborough, Piper Alpha, the well-known WASH 1400 report on nuclear safety, the Canvey Island risk assessment report and many others. All of these initiatives have contributed to the increased awareness of the general public to potential hazards and highlighted the need for better methods for ensuring the reliability of complex systems and improving the safety of hazardous plant.

Every engineer and scientist involved in reliability and risk assessment will have developed an understanding of the techniques in a different way. In the case of the authors, a general interest in reliability was the initial trigger with risk assessment following some time later. Now we see the various methods as integrated techniques which can be used over a wide range of applications as illustrated by the case studies in Chapter 14.

It is certain that a number of reliability and risk assessment methods will not be covered in this book. Our aim has been to concentrate on the techniques that we have employed extensively to good effect over many years for the assessment of reliability and risk in process plant.

1.2 Quantified reliability

After the First World War, reliability became of major interest to the aircraft industry. Comparisons were made between the reliability of single- and multi-engined aircraft although at that time little attempt was made to express reliability in quantitative terms. Most attempts at reliability improvement were based on trial and error. When something failed, its replacement would be designed and manufactured using any improvements in technology together with experience gained from investigating the failure. As time went by information was gradually collected on system failures which led naturally to the concept of expressing reliability in terms of the failure rate for a particular type of aircraft or system. In the 1940s the reliability requirement was given for maximum permissible failure rates of 1 per 100 000 flying hours. It was also estimated that there was a fatal accident risk of 1 in 10^6 landings prior to automatic landing systems. When automatic landing systems were introduced in the 1960s the reliability requirement of the system was specified in terms of the fatal accident risk being not greater than 1 in 10^7 landings. Therefore, quantification gradually became part of the design specification.

The missile and space industry is a product of the Second World War and is the area where quantitative reliability became formalized. After the initial poor reliability of the V1 missile, Lusser, a German mathematician, queried the original assumption that the reliability of a chain of components was determined by the strength of its weakest link. He showed that a large number of fairly 'strong' links can be inherently more unreliable than a single 'weak' link because of the variability of component strengths and operational loading. Quality assurance was therefore applied to increase the reliability of all components resulting in a dramatic improvement of V1 reliability which ultimately achieved a success rate of over 60 per cent. Because of the increase in complexity of modern systems the reliability of such devices is still significantly below 100 per cent.

The American armed forces took an increasing interest in reliability and its measurement in the Second World War because the unreliability of vital equipment and systems was causing significant problems. In the Korean War the unreliability of electronic equipment was reported as costing $2/year to maintain each dollar's worth of equipment. In the UK at this time it was also noted that the lack of reliability and maintainability of equipment was forcing the armed services to spend over half of its resources on maintenance rather than operations.

Arising from this appreciation of the importance of reliability and maintainability a series of US Department of Defense Standards (MIL-STDs) were introduced and implemented. Subsequently the UK Ministry of Defence also introduced similar standards. By and large their introduction has significantly improved the reliability and maintainability of military equipment and systems.

Reliability methods have also been introduced into the process industries. Here the objectives are to improve the safety and availability of new and existing plant. The technology of reliability has benefited significantly from this interest and the concepts of unrevealed failures and techniques, such as fault trees, Markov and human reliability analysis, have found many applications in process plant assessment.

1.3 Reliability terminology

In this section some of the terminology and expressions used in reliability assessment are defined. A more detailed discussion of these concepts is given in Chapter 5.

1.3.1 Reliability

Quantitative reliability can be defined as:

> 'the probability that an item (component, equipment, or system) will operate without failure for a stated period of time under specified conditions'.

Reliability is therefore a measure of the probability of successful performance of the system over a period of time. Reliability assessment is initially carried out for systems of components which are assumed to have settled down into steady-state or useful-life phase. The reliability characteristics of most component families will follow the so-called 'reliability bath-tub' curve shown in Fig. 1.1.

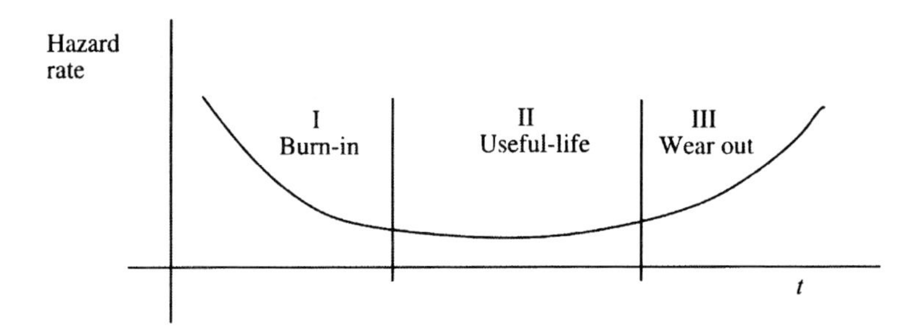

Fig. 1.1 Reliability bath-tub curve

In phase I in the figure the hazard rate (failure rate) will reduce as weak components are eliminated; in phase II (the useful-life phase) it will remain approximately constant; and in phase III components will start to wear out and the hazard rate will increase.

For constant hazard rate the reliability of the system can be represented by the simple expression

$$R(t) = \mathrm{e}^{-\lambda t} \tag{1.1}$$

where

$R(t)$ = probability of successful operation for a period of time t
λ = (constant) hazard or failure rate

When the hazard rate is constant it is frequently referred to as the failure rate and is represented by λ.

1.3.2 Availability

When system failure can be tolerated and repair can be instigated, an important measure of the system performance is the system availability represented by the expression

$$A = \frac{\text{MTTF}}{\text{MTTF} + \text{MTTR}} \tag{1.2}$$

The mean time to failure (MTTF) is the reciprocal of the (constant) failure rate, i.e. MTTF = $1/\lambda$. The mean time to repair (MTTR) is the average time taken from the failure of the system to its start-up.

Availability can be defined as:

> 'the fraction of the total time that a device or system is able to perform its required function'.

Availability is represented by the symbol A and its complement – unavailability – by the symbol Q. Since availability is probabilistic:

$$A + Q = 1 \tag{1.3}$$

1.3.3 Unrevealed failures

For systems residing in the dormant state, such as automatic protection systems, it is evident that failure will only be revealed when a demand is made on the system. The demand may be in the form of a measured-parameter deviation or a test. The probability of the system being in the failed state at any time can be represented by the **unavailability** or **fractional dead time** (FDT) of the system. For a single-channel system

$$\text{FDT} = \lambda\left(\frac{\theta}{2} + \tau\right) \tag{1.4}$$

where

λ = unrevealed failure rate of the system
θ = test interval
τ = mean repair time

The mean repair time τ is frequently much shorter than the test interval and the fractional dead time is therefore often approximated by the expression

$$\text{FDT} = \frac{\lambda\theta}{2} \tag{1.5}$$

The probability of a single-channel protection system failing to successfully protect against a hazard is the product of the fractional dead time and the demand probability, *D*:

$$P_f = \text{FDT} \times D \tag{1.6}$$

These are the expressions on which the majority of the methods in process system reliability studies are based.

1.4 Reliability programmes

Reliability assessment should be carried out during every stage of a project. During the initial design stage, system boundaries will be defined and targets set for the required reliability, maintainability, and availability performance. A broad-brush assessment is subsequently carried out during the concept design phase to predict the reliability characteristics of the overall plant and its systems for comparison with the target values and to identify the reliability-critical areas. More detailed assessments will be carried out in these critical areas during the detail design phase to ensure that robust, failure-tolerant designs are produced which can be adequately maintained. Reliability and quality assurance specifications will be produced at this stage.

A formal reliability programme is essential on all projects of any size or importance. An effective reliability programme will be based on the responsibility and authority of one person generally termed the Reliability Programme Manager. This responsibility must relate to a defined objective which may be a maximum warranty cost figure, an MTTF to be demonstrated, or a minimum predicted availability.

The reliability programme should begin at the earliest stage in a project and must be defined in outline before the concept design phase starts. It is at this stage that fundamental decisions involving trade-offs between reliability, complexity, performance, and price are made. The reliability engineer will be involved in the assessment of these trade-offs and the generation of specific reliability objectives.

As development proceeds from initial study to detail design, the reliability will be assured by formal, documented procedures for the review of the design and the imposition of design rules relating to component, material and process selection, derating policy, tolerancing, etc. The objectives will be to ensure that known, good engineering practice is applied, that deviations from the reliability specification are detected and corrected, and that areas of uncertainty are highlighted for further action. Data collection and analysis should be carried out during manufacture and testing and should continue during the operational phase. Throughout the product life-cycle, therefore, the reliability should be assessed, at first by initial predictions based on past experience to determine feasibility and to set objectives, and subsequently by refining the predictions as the detail design proceeds. Finally reliability performance should be reported and analysed during the test, manufacture, and operational phases. This performance data must be fed back to generate corrective action and to provide data and experience to aid future predictions.

The elements of a reliability programme are outlined in MIL-STD-785 and BS 5760. Figure 1.2, adapted from BS 5760, shows the nature of the activities involved in a typical reliability programme.

The implementation of a reliability programme will always ensure an increased understanding of the system and improved communications within the organization. Reliability assessment will identify the critical areas and the uncertainties associated with the safety, maintenance, and operation of the plant. Its systematic application in an interdisciplinary manner provides an independent evaluation of the design which ensures that robust, cost-effective designs are produced. Experience shows that the application of reliability techniques throughout all phases of system design, manufacture, and operation ensures that problems are minimized during installation, commissioning, and operation. This is not to say that all problems will be eliminated but the reduction in the problem areas will clearly make for lower system failure rates, better maintainability, and a significant improvement in the operability of the system.

1.5 Quantified risk assessment

Reliability and risk assessment methods are both employed in safety studies to identify the various combinations of faults which can lead to reduced safety. Here risk evaluation and risk reduction are briefly discussed and examples are given of the benefits which can be obtained from reliability and risk assessment in safety studies.

SYSTEM LIFE–CYCLE				
SYSTEM DEFINITION PHASE	CONCEPT DESIGN PHASE	DETAIL DESIGN PHASE	MANUFACTURING PHASE	OPERATING PHASE
Establish reliability requirements Set provisional reliability/ availability targets Prepare outline reliability specification	Perform global safety/ availability assessments (various scenarios) Identify critical areas and components Confirm/review targets	FMEA/FTA of critical systems and components Review reliability database Carry out detailed system reliability assessment Prepare safety case	Prepare and implement reliability specifications Review reliability demonstrations	Audit reliability performance Collect and analyse reliability, test and maintenance data Assess reliability impact of modifications

Fig. 1.2 Outline reliability programme for process plant

1.5.1 Background

The hazards associated with process plant involve significant risks. Risks can be lessened by spending money but it is not possible to avoid risks entirely. In the end, society has to judge how much money it is worth spending to save each additional life. The difficulty faced by safety assessors lies in convincing regulators and decision-makers that at some point a process system is 'safe enough'.

For major hazard assessments risk is generally defined as the probability of a specific undesired event (say, explosion or toxic/radioactive release). Alternatively, risk or 'expected loss' can be defined quantitatively as the product of the consequences of a specific incident and the probability over a time period or frequency of its occurrence:

$$R = C \times P \tag{1.7}$$

The risk can be reduced by reducing the consequences of the incident (C) or by reducing the probability or frequency of its occurrence (P). Methods for reducing consequential losses by the use of simpler plants or inherently safe systems, etc. are covered elsewhere (**1**). Here we are concerned with evaluating the frequencies of incidents which have major safety implications by the use of probabilistic methods.

When risks have been identified and evaluated we can judge if these are 'acceptable' or if we need to make modifications to the plant to improve its safety and reliability. It is obvious that we must make this judgement against realistic targets.

In risk assessment studies there are generally three main types of risk to be considered:

1. Occupational risks – that is, risks to the workforce of the plant.
2. Community risks – risks to people living nearby and the environment.
3. Economic risks – the financial penalties arising from loss of capital assets, production and compensation.

Estimates of the risks existing in particular plant must be compared against specific criteria of acceptability. These criteria will need to take account of a whole range of factors, depending on whether the hazards are confined to the plant or if they can affect people or the environment outside the plant. A target probability must be derived for specific hazards in each case.

Economic risks are generally covered by insurance.

1.5.2 *Occupational risks*

In the United Kingdom a criterion of acceptability has been adopted based on experience from industry. For the UK chemical industry, Kletz (2) has shown that the average fatal accident rate (FAR) pre-Flixborough was four fatalities in 10^8 working hours (i.e. four fatalities in approximately 10 000 man years). Table 1.1, from a paper by Kletz, shows typical figures for other UK industries and occupations, compared with some non-occupational activities expressed in a similar manner. Note these are the risk to an individual.

Table 1.1 Fatal accident rates for UK occupations and activities

	Individual risk per 10^8 hours (FAR)
Occupational:	
British industry (premises covered by Factories Act)	4
Clothing and footwear	0.15
Vehicles	1.3
Timber, furniture, etc.	3
Metal manufacturing, shipbuilding	8
Agriculture	10
Coal mining	16
Railway linesmen	45
Construction erectors	67
Non-occupational:	
Staying at home (men 16–65)	1
Travelling by train	5
Travelling by car	57

Kletz notes that the chemical industry average FAR of four is made up of common background risk (e.g. falling off a ladder or being involved in a road accident) and special risks peculiar to each sector. He suggests that (for the UK chemical industry) an acceptable target is probably a FAR of 0.4 in 10^8 h – a factor of 10 reduction on the average for the industry – that is 0.4 deaths per 10 000 men per year, or a risk per man of 4×10^{-5}/year.

This is a realistic approach to setting targets against which to compare occupational risks and it has been extensively adopted in UK industry.

1.5.3 Community risks

For risks involving the community or the environment, the criterion must be based on the probability of releases from the plant and their consequences since pollution and damage outside the boundaries of the plant are involved. One of the pioneers in developing an environmental risk acceptance criterion was Farmer of the UKAEA (**3**). The basis of his argument was that no plant or structure could be considered entirely risk free and that so-called 'incredible' accidents were often made up of combinations of very ordinary events. Hence, as the consequences of a potential hazard increased, so the occurrence probability should decrease. Figure 1.3 shows an example of the Farmer curve, where probabilities above the line reflect high-risk situations and those below the line reflect 'acceptable' or low-risk situations.

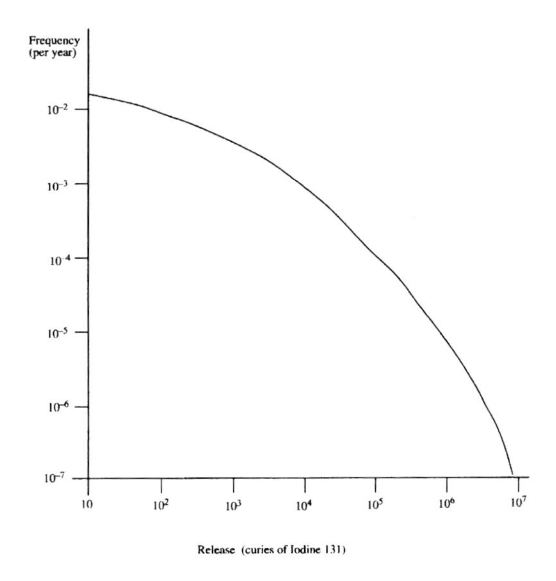

Fig. 1.3 Farmer curve: accident release frequency limit line

The Farmer approach was subsequently applied in the USA. For example, in 1975, a comprehensive assessment of accidental risks in a US commercial nuclear power plant was published (**4**). This study was a milestone in probabilistic safety studies and contributed a great deal to the application of quantitative risk assessment for large-scale hazardous situations.

In the UK, after the explosion at the Flixborough chemical plant in 1974, the Health and Safety Commission set up an Advisory Committee on Major Hazards which issued its first report in 1976 (**5**). Among the many recommendations advocated was the registration of plant whose inventory of toxic, flammable, or explosive materials exceeded prescribed limits. These plants became known as Notifiable Installations. With regard to accident frequencies, it suggested that a 10^{-4} chance of a serious accident per year might perhaps be on the borderline of acceptability. There has been much speculation as to the definition of the term 'serious accident' and we have normally assumed this to involve ten or more fatalities, although one could argue for any figure between, say, 1 and 30.

Subsequently the Health and Safety Executive (HSE) commissioned a major study of toxic and flammable hazards in the Canvey Island/Thurrock area. This study (**6**) was the first example of a quantified risk assessment published by a regulatory authority. However, a number of other major studies have since been completed notably for LNG/LPG terminals, pipelines, ethylene storage, nuclear, and offshore installations in addition to a wide variety of process plant.

In many of these studies, a careful distinction is made between individuals at risk and the risk to the community at large – the so-called 'societal' risk. Cox (**7**) has suggested that in the general community, risks imposed on individuals can be related to the background level experienced as part of everyday life, as shown by mortality statistics. Table 1.2 from his paper shows mortality statistics for different types of accident.

From Table 1.2, it is possible to identify several sub-groups as follows:

1. Risks of an everyday kind without any compensating benefits, which are accepted principally because they are fundamentally unavoidable and immutable, such as falling, fire, natural and environmental risks (including lightning strikes), and poisoning.
2. Risks which are in principle avoidable, but provide direct compensating benefits and in practice are unavoidable for people who wish to be part of modern society, i.e. all transport accidents and poisoning by medication.
3. Risks which are truly avoidable in the sense that people who are exposed to them do so of their own free will in order to gain some other benefit, i.e. most of the drowning cases, hang-gliding, etc.

Table 1.2 Mortality statistics for several Western countries*

Cause of death	Rate per person per year
All causes	0.0110
Accidents:	
All	4.6×10^{-4}
Railway	4.3×10^{-6}
Air	6.5×10^{-6}
Water transport	5.3×10^{-6}
Motor car	2.4×10^{-4}
Poisoning by medical drugs, etc.	3.4×10^{-6}
Poisoning by other substances	1.5×10^{-5}
Falling	1.3×10^{-4}
Fire	2.8×10^{-5}
Natural and environmental factors	3.8×10^{-7}
Lightning	8.7×10^{-7}
Drowning	2.4×10^{-7}

* USA, UK, France, Belgium, Netherlands.

For populations living near to process plants, the risk will be perceived as one which is essentially unavoidable (involuntary) and from which they do not receive a direct benefit. It is therefore comparable with the risk in sub-group 1, except that the plant is a man-made hazard rather than a natural one and it is therefore reasonable to expect the man-made risk to be sufficiently low that it does not make a significant difference to the pre-existing comparable natural risk. In the context of the sub-group 1 risks in Table 1.2, excluding falls (which affect almost exclusively the elderly and are therefore not comparable), a man-made risk of level 1×10^{-6}/year would make about 3 per cent difference to the total sub-group 1 risk, and about 0.22 per cent difference to the total risk of accidental death of all kinds, and again about 0.01 per cent difference to the total risk of death from all causes. This would therefore appear to represent an appropriate level of acceptable risk for the siting of potentially hazardous plant.

These principles were incorporated into the general requirements of the Control of Industrial Major Accident Hazards (CIMAH) Regulations 1984 (now updated to the COMAH regulations (**8**) to conform to the EU Seveso II directive). For COMAH, acceptable criteria are based on the HSE-recommended three-zone approach. At the high end (incidents above the Farmer curve in Fig. 1.3) the risks are considered unacceptable; the probability of occurrence or the consequences must be reduced to bring the risks into an acceptable region. At the low end the risk are negligible and thus acceptable. In the intermediate region the risks are required to be reduced to As Low As Reasonably Practical (ALARP). This may involve, for example, reducing the stored quantities of hazardous chemicals or introducing additional protective systems to reduce the probability of an incident. To determine which region a specific hazard belongs in commonly requires a quantitative risk assessment.

1.6 Risk assessment studies

Quantitative risk assessment (QRA) involves four basic stages:

(a) the identification of the potential safety hazards
(b) the estimation of the consequences of each hazard
(c) the estimation of the probability of occurrence of each hazard
(d) a comparison of the results of the analysis against the acceptability criteria.

The identification of safety-related hazards can be carried out by using checklists based on accident statistics for the industry or from company records. Preliminary hazard analysis (PHA), which employs criticality ranking of the effect of various hazardous conditions, can also be employed effectively to identify the sub-systems which are likely to pose the major hazards in operation. An example of a PHA for an offshore system is shown in Fig. 1.4.

Hazard and operability (HAZOP) studies are also employed extensively in industry, both onshore and offshore, for safety studies. They consist of rigorous examination by a team of engineers, including operators and maintenance and reliability engineers, of all possible variations of operating conditions on each item of plant through the use of 'guide words'. Thus, examination of a storage tank would include consideration of HIGH level, LOW level, HIGH flow, LOW flow, REVERSE flow, HIGH temperature, LOW temperature, contents OTHER THAN normal and any other deviations from the design intention.

The HAZOP team will then examine the consequence of such deviations and where these give cause for concern they are considered in more detail. The more detailed analysis may involve a reliability assessment using one or more of the techniques discussed in the next section, or consideration of a variety of possible release scenarios,

SUB-SYSTEM OR FUNCTION	ITEM NO.	HAZARDOUS ELEMENT	EVENT CAUSING HAZARD	HAZARDOUS CONDITION	EVENT CAUSING HAZARD	POTENTIAL ACCIDENT	EFFECT	HAZARD CLASS	PREVENTATIVE MEASURES		
									HARDWARE	PROCED-URES	PERSON-NEL
Gas metering	M12	Gas pressure	Leak rupture Damage to equipment Instrument failure	Gas released to module	Spark flame Static electricity	Fire Explosion	Injury to personnel Damage to equipment	III or IV	ESD system Fire suppression system		
Flare KO drum	PV4	Gas pressure Oil	Relief system in-operative due to ESD 2, 3, or 4 fault Instrument or relief valve fault	Leak rupture High noise level	Excess pressure damage to equipment Excess liquid carry-over to drum or boom flare-back	Fire Explosion debris and missiles	Injury to personnel Damage to equipment Damage to structure	II or III			
Produced water separator	PV8	Salt water at high temperature	Leak rupture	Hot salt water released in module	Personnel in vicinity of leak Panic reaction by personnel	Steam/water burns Water damage to equipment Electrical faults	Injury to personnel Damage to equipment	II			

Fig 1.4 Example of preliminary hazard analysis – gas compression system

frequently involving the use of computer models to simulate, say, flammable gas cloud build-up and dispersion (and possible ignition), under a wide range of atmospheric conditions.

Although the use of HAZOP studies is principally carried out to identify major hazards on a plant, it is worth noting that it also provides a valuable insight into operation of the system under normal and fault conditions. These frequently result in recommendations for modifications to improve the safety and operability of the plant.

1.7 Reliability in risk assessment

New plant should always receive consideration from the reliability point of view during the various design stages so that safety aspects can be evaluated and optimized. It is always more difficult and expensive – particularly in the layout of plant – to rectify faults at a later stage. At conceptual design a quantitative reliability assessment may not be feasible, but an independent review based on a study of the effect of a range of failures is invaluable as a design check. As the design progresses, more detail will become available and it is then possible to study failure causes in a more detailed and systematic manner. Increasingly the reliability engineer becomes involved in helping the designer to solve particular reliability problems.

Reliability assessment encompasses a wide range of methods but special techniques are extensively applied in safety analysis and therefore warrant particular consideration. In this book we concentrate particularly on fault tree analysis and failure mode and effect analysis.

1.7.1 Fault trees

Fault trees are used extensively in safety studies to identify combinations of events which could cause a major incident. The fault tree has an established form and is built from the top, undesired event, downward using logical AND/OR gates to combine the causal events. The probability of the top event can then be calculated given the probability of the events terminating the branches of the tree. Many different fault trees may be required to evaluate the risks associated with a particular type of hazard in a plant. The method is well proven and computer programs are available to calculate top-event probabilities of a constructed fault tree. Fault trees have also proved valuable as diagnostic aids. A small fault tree example is shown in Fig. 1.5. This method is described extensively in Chapter 7.

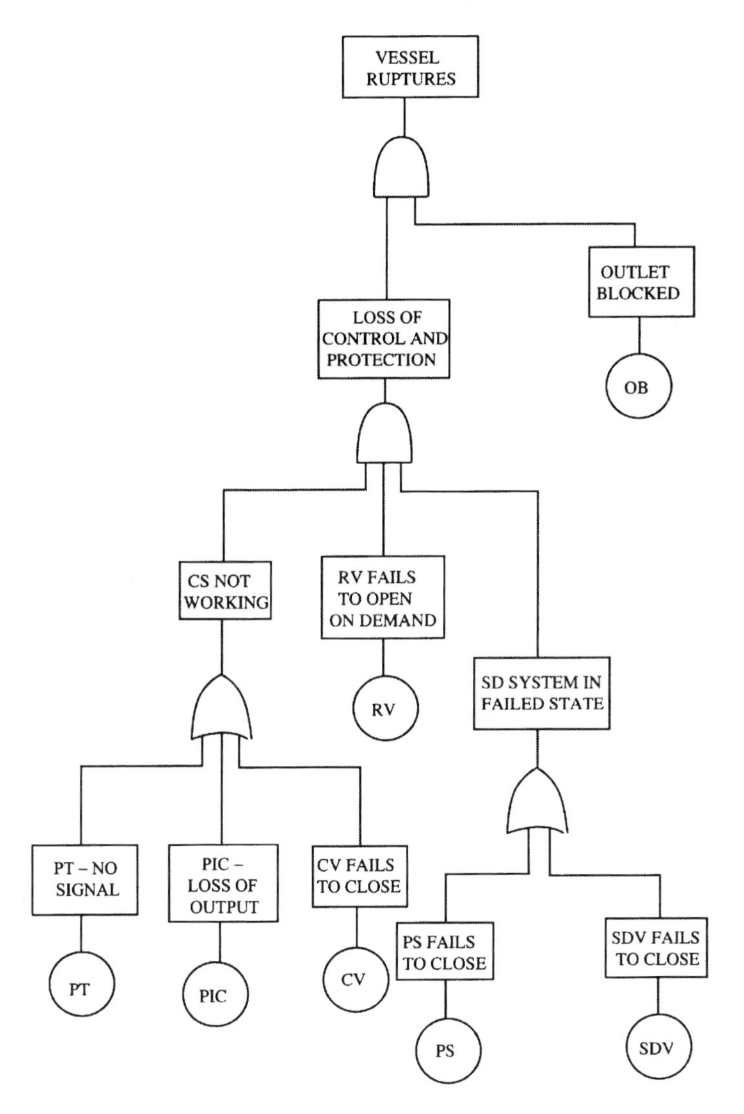

Fig. 1.5 Fault tree example

1.7.2 Failure mode and effect analysis (FMEA)

FMEA is another powerful tool used by the reliability engineer for analysing equipment and systems. It is frequently referred to as a 'bottom-up' approach (fault trees are 'top-down') since it takes a 'what happens if' approach. The technique basically involves:

(a) breaking down the equipment into component or sub-assembly blocks
(b) examining each block for its modes of failure
(c) classifying each mode of failure according to the effect it has on the system.
(d) applying appropriately stressed failure rates for each failure mode where quantification is required.

The process is fully described in MIL-STD-1629A (**9**). The analysis can be carried out in terms of the hardware and the actual failure modes which occur. Alternatively an approach based on components or sub-assemblies, and their effect on the function of the equipment as a whole, can be employed. In the early stage of the design where the hardware is not completely defined, a functional approach often provides the best solution. In cases where specific failure modes in specific pieces of hardware are concerned, the more detailed approach using physical failure modes is generally preferred. One sheet of a typical FMEA is shown in Fig. 1.6. Chapter 4 provides a more comprehensive account of this important technique.

1.8 Risk ranking

The risks associated with different areas, equipment, operations, or functions can be ranked to show the main items affecting safety. Those which give unfavourable indications with respect to the target criteria need action in the form of decisions on how to reduce risks. This can be achieved by reducing the consequences or by increasing the reliability of the plant and its protective systems. The choice should always consider the relative economic penalties involved.

High cost penalties may be associated with increasing personnel safety and reducing environmental risks. Hence equipment procurement, repair, and replacement policies all need to be considered as well as the failure process. The methods employed to determine these aspects involve the use of failure and repair data to model the reliability and availability characteristics of the system. Financial penalties associated with replacement, testing, maintenance, loss of production, etc. can then be applied to the model to quantify the expected economic losses and identify the solutions giving the greatest cost-benefit for a specific reduction in the overall risk.

400xW WATER CHILLER UNIT

INDENTURE LEVEL: 2
Sheet no. 6 of 20
Mission Phase 1 - OPERATING

MANUFACTURING ORDER 08/018
Study by:
Prepared by:
Approved by:
Date:

SYSTEM Ref-Description-Function	FAILURE		POSSIBLE CAUSES	SYMPTOM DETECTED BY	EFFECT OF FAILURE		Compensating provision against failure	Severity class	REMARKS
	Entry code	Mode			Local	On next level			
1.4.0 Lubrication system Provide cooled lube oil at pressure to bearings and gearing. Auxiliary oil pump from starting. Jet pump and low- and high-speed centrifugal pumps whilst running. Duplicate filters and single tube oil cooler	1401	Leakage	Loose connectors. Auxiliary oil pump fault	Observation – gas in air monitors. Fall of sump level	Slow leaks have no effect	Eventual shut-down if uncorrected by loss of oil pressure. Performance loss if air ingress	2 hourly inspections. Automatice shutdown on low oil pressure	2	Sump contains 30 litres, so a loss of, say, 25 l, acceptable to unit, should be readily observed
	1402	Oil too hot	Restriction of oil flow. Restriction of water cooling. Fouling of heat exchanger	High oil temperature warning	Inadequate lubrication due to low oil viscosity and loss of bearing cooling	Eventual shutdown	Unit shutdown due to high lube oil temperature	2	Slow rise of oil temperature would be found by study of readings
	1403	Oil pressure low	Blockage. Wear. Jet pump fails	Low at pressure warning	Inadequate lubrication due to reduction of oil flow	Eventual shutdown	Unit shutdown due to low lube oil pressure	2	Oil pressure is one of the most important parameters monitored. Will be watched to decide on state of oil filter
	1404	Oil contaminated	Filter not cleaning oil. Water in oil	Daily observ-ation of oil condition	Discoloration and/or emulsi-fication of oil	Eventual damage if not corrected	Oil condition is judged daily. Oil is changed annually	2	Condition of 'replaced' oil filters is examined for evidence of any abnormal state
	1405	Oil level in sump low	System leak. Oil transfer to refrigerant circuit	Observed in sight glass	Only sudden, severe leak would have effect	Sudden, severe leak would lead to shutdown on low lube oil pressure	Pre-start-up checks and daily checks. Sudden severe leak gives shutdown or low oil pressure	2	Oil can be topped up, a standard routine. When sufficient oil has migrated to refrigerant it can be returned slowly
	1406	Oil level in sump high	Refrigerant in oil	Observed in sight glass. Foaming	Reduced lub-rication due to gas separating out from oil	Eventually no output. Shutdown or low lube oil pressure	Pre-start-up checks and daily check-up. Pre-start requirement to keep oil hot for 12 h	3	Gas separating out from oil would give low lubricating oil pressure shutdown

Fig. 1.6 FMEA example

1.9 Summary

This introduction has reviewed the general approach taken in reliability and risk assessment studies and identified the important role associated with reliability. With the current complexity and size of modern plant, the consequences of equipment failure provide an increased contribution to the overall risk. Reliability methods, as discussed here, provide the methodology for assessing the probability of failure and the optimum mix of redundant equipment, testing, and maintenance which are required to meet the requisite safety and economic standards.

Overall decisions are required and can best be made on the basis of quantitative assessments of the risks (in terms of loss of life, pollution, or cost) which can be compared with acceptance criteria. The criteria will be of two basic types: the individual/societal risks and the economic constraints which apply to any particular situation.

1.10 References

(1) **Kletz, T. A.** (1970) *Cheaper, Safer Plants*. Institute of Chemical Engineers.

(2) **Kletz, T. A.** (1971) *Hazard Analysis: A Quantitative Approach to Safety*. Institute of Chemical Engineers Symposium Series no. 34.

(3) **Farmer, F. R.** (1967) Siting criteria: a new approach. In IAEA Symposium on *The Containment of Nuclear Power Reactors*, Vienna, April.

(4) USAEC (1974) An assessment of accidental risk in US commercial nuclear power plants, WASH, 1400, August.

(5) Health and Safety Commission (1976) First report: Advisory Committee on Major Hazards.

(6) Health and Safety Executive (1978) CANVEY – An investigation of potential hazards from operations in the Canvey Island/Thurrock area.

(7) **Cox, R. A.** (1981) Improving risk assessment methods. In Proceedings of the 2nd National Conference on *Engineering Hazards*, London, January.

(8) *The Control of Major Accident Hazard Regulations 1999 (COMAH)*. HSE Books.

(9) MIL-STD-1629A (1977) *US Department of Defense* Procedure for performing a failure mode and effects analysis.

Chapter 2

Reliability Mathematics

2.1 Probability theory

Assessing the reliability performance of systems involves dealing with events whose occurrence or non-occurrence at any time cannot be predicted. We are unable to tell exactly when a working component or system will fail. If a group of identical components, manufactured in the same batch, installed by the same engineer and used under identical operating and environmental conditions were monitored, all components would fail at different times. Handling events whose occurrence is non-deterministic is a problem commonly experienced in many branches of engineering. Its solution requires some means by which the likelihood of an event can be expressed in a quantitative manner. This enables comparisons to be made as to which of several possibilities is the most likely to happen. The branch of statistics used to overcome these difficulties is **probability theory**.

The probability of an event occurring is a 'scientific measure' of chance which quantitatively expresses its likelihood. Probability is a number placed anywhere on a scale of zero to one. If an event is said to have a probability of zero it is an impossible event. At the other extreme, events which are certain to happen have a probability of one. Most events of interest do not have probabilities of either zero or one; their probabilities are somewhere between these two numbers.

In general terms, probability is a measure of the likelihood that a particular **event** will occur in any one **trial** or **experiment** carried out in prescribed conditions. Each separate possible result from the trial is an **outcome**. For example, if we throw a die to determine which number will be uppermost when the die comes to rest, we are performing a random experiment since the result cannot be predicted. If the die is loaded such that it shows a six on every throw, we are not performing a random experiment because we can always determine the result. A random experiment may be characterized by itemizing its possible outcomes. In the case of throwing an unloaded die this is easy: the possible outcomes are '1', '2', '3', '4', '5', or '6'. The itemization of all possible outcomes produces a list known as the **sample space**. To summarize we have:

Random experiment:	the throw of a die.
Object of experiment:	to determine which number is uppermost when the die comes to rest.
Sample space:	$\{1, 2, 3, 4, 5, 6\}$.

In this case the number of outcomes is finite; however, other experiments may have a very large, perhaps infinite, number of outcomes. For instance, a more appropriate experiment for reliability assessment may be:

Random experiment:	the operation of a particular component in a system.
Object of experiment:	to determine the time of failure, t, of the component to the nearest hour.
Sample space:	can be denoted as $\{t_1, t_2, \ldots\}$.

The determination of the probability of any event can be approached in two ways: (a) an empirical or experimental approach, or (b) a theoretical approach.

2.1.1 Empirical or experimental probability

This is based on known experimental results and is calculated as a relative frequency. For a random experiment with sample space $\{E_1, E_2, E_3, \ldots, E_n\}$, the experiment is repeated N times and N_i the number of times each outcome E_i occurs is noted. The ratio N_i/N then represents the relative frequency of occurrence of outcome E_i in exactly N repetitions of this random experiment. If this ratio approaches some definite limit as N approaches infinity, this limit is said to be the probability associated with event E_i, i.e.

$$P(E_i) = \lim_{N \to \infty} \left(\frac{N_i}{N} \right) \tag{2.1}$$

This means that if the experiment is performed very often, the relative frequency of event E_i is approximately equal to the probability of E_i in the experiment. Therefore, $P(E_i)$ has the following properties:

$$0 \leq P(E_i) \leq 1, \ i = 1, 2, \ldots, n$$

and if $P(E_i) = 1$, E_i is certain to occur, and if $P(E_i) = 0$, E_i is impossible.

2.1.2 Sample size

The size of the sample from which the probability is established affects the reliability of the result. Probabilities derived from small samples will be unlikely to reflect the probability of the whole population. The larger the sample size the better the estimate.

2.1.3 Theoretical probability

The theoretical or classical approach to probability is obtained by considering the number of ways in which it is theoretically possible for an event to occur. Therefore for event A

$$P(A) = \frac{\text{number of ways in which } A \text{ occurs}}{\text{total number of possible outcomes}} \tag{2.2}$$

This can only be used providing each of the possible outcomes to the experiment is equally likely.

Consider the information contained in Table 2.1, below, which records the number of times a coin is tossed and how many times it lands with the 'head' side uppermost. Results are recorded in batches of ten. So of the first ten coins tossed, eight came down showing 'heads'. An estimate of the probability of this event based on ten trials is therefore 0.8. The second batch of ten trials produced five which landed on 'heads' and five on 'tails', a revised estimate of the probability of a 'heads' occurring based on 20 trials is 0.65. This process was continued for 200 experiments and the results are shown in Fig. 2.1. After 200 experiments $P(\text{H})$ is estimated on a relative frequency basis as 0.525.

Table 2.1 Results recorded for tossing a die (groups of 10 trials)

	Group									
	1	2	3	4	5	6	7	8	9	10
N(H)	8	5	8	6	5	7	6	6	5	5
P(H)	0.8	0.65	0.7	0.68	0.64	0.65	0.64	0.64	0.62	0.61
	Group									
	11	12	13	14	15	16	17	18	19	20
N(H)	6	6	6	3	4	4	5	4	2	4
P(H)	0.61	0.61	0.61	0.58	0.57	0.56	0.56	0.55	0.53	0.525

Using the classical approach to estimate probabilities we would obtain $P(\mathrm{H}) = 0.5$ (one out of a possible two equally likely outcomes for the experiment). Figure 2.1 shows that the probability obtained from the experimental approach may well be near to the theoretical value after more experiments have been performed. This emphasizes that probabilities estimated using empirical or experimental approaches do need to be based on a large number of trials before confidence can be placed in their accuracy.

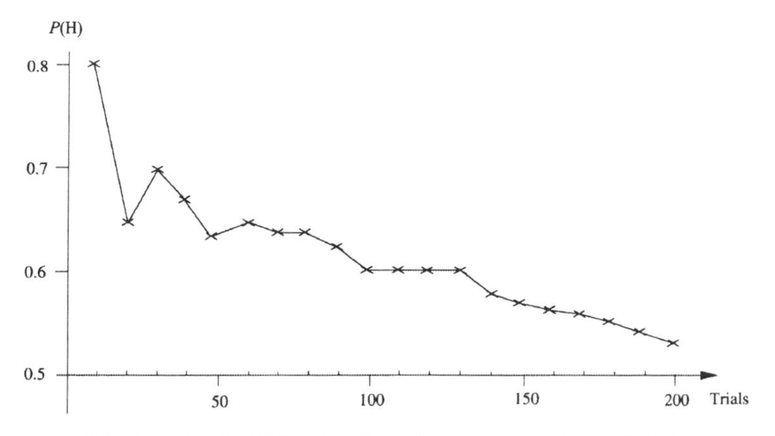

Fig. 2.1 Probability of a 'head' versus number of trials

Probabilities of complex events can, in certain circumstances, be obtained by combining the probabilities of occurrence of simpler events. If events are **mutually exclusive** then the addition law of probability can be used to obtain the probability of one or more of the group of events occurring.

2.1.4 *Mutually exclusive events*

Mutually exclusive events are events which cannot occur together. For example when rolling a die, the six different outcomes, '1', '2', '3', '4', '5', and '6', are mutually exclusive events as they cannot occur at the same time. Similarly each possible result obtained by drawing a card from a pack of cards is mutually exclusive.

2.1.5 *Non-mutually exclusive events*

Non-mutually exclusive events are events which can occur simultaneously. For example when rolling a die the events 'obtaining a result which is a multiple of 3' and 'obtaining a result which is a multiple of 2' can occur at the same time (if a six is thrown).

2.1.6 *The addition law of probability*

If events A and B are mutually exclusive then

$$P(A \text{ or } B) = P(A) + P(B) \tag{2.3}$$

Example

If there are n possible outcomes to a trial of which x give event A and y give event B, then if A and B are mutually exclusive

$$\begin{aligned} P(A \text{ or } B) &= P(A) + P(B) \\ &= \frac{x}{n} + \frac{y}{n} = \frac{x+y}{n} \end{aligned}$$

The addition rule applies for any number of mutually exclusive events. So if $E_1, E_2, \ldots, E_n$ are all mutually exclusive then

$$P(E_1 \text{ or } E_2 \text{ or} \ldots \text{or } E_n) = \sum_{i=1}^{n} P(E_i) \tag{2.4}$$

If $E_1, E_2, \ldots, E_n$ are also exhaustive (i.e. include all possible outcomes) then

$$P(E_1 \text{ or } E_2 \text{ or} \ldots \text{or } E_n) = \sum_{i=1}^{n} P(E_i) = 1$$

If two events A and B are non-exclusive then the rule must be modified to take account of events occurring at the same time:

$$P(A \text{ or } B) = P(A) + P(B) - P(A \text{ and } B) \tag{2.5}$$

For example when a die is thrown, if event A denotes obtaining a score which is a multiple of 3 and event B denotes obtaining a score which is a multiple of 2 then

$$A = \{3, 6\}$$
$$B = \{2, 4, 6\}$$

These are not mutually exclusive since 6 occurs in both.

Therefore

$$\begin{aligned} P(A \text{ or } B) &= P(A) + P(B) - P(A \text{ and } B) \\ &= \frac{2}{6} + \frac{3}{6} - \frac{1}{6} \\ &= \frac{4}{6} = \frac{2}{3} \end{aligned}$$

This result we can confirm since A or B = {2, 3, 4, 6}.

This can also be extended to any number of events. For n events $E_1, E_2, E_3, \ldots, E_n$, the addition rule becomes

$$\begin{aligned} P(E_1 \text{ or } E_2 \text{ or} \ldots \text{or } E_n) = {} & \sum_{i=1}^{n} P(E_i) - \sum_{i=1}^{n-1} \sum_{j=i+1}^{n} P(E_i \text{ and } E_j) \\ & + \sum_{i=1}^{n-2} \sum_{j=i+1}^{n-1} \sum_{k=j+1}^{n} P(E_i \text{ and } E_j \text{ and } E_k) + \cdots + \\ & + (-1)^{n+1} P(E_1 \text{ and } E_2 \text{ and } E_3 \text{ and} \ldots \text{and } E_n) \end{aligned} \tag{2.6}$$

so for $n = 3$

$$P(E_1 \text{ or } E_2 \text{ or } E_3) = \{P(E_1)+P(E_2)+P(E_3)\} \\ -\{P(E_1)P(E_2)+P(E_2)P(E_3)+P(E_3)P(E_1)\} \\ +P(E_1)P(E_2)P(E_3) \tag{2.7}$$

When the probability of individual events is small this expression can be approximated by the first term in the expansion, i.e.

$$P(E_1 \text{ or } E_2 \text{ or } E_3) = P(E_1)+P(E_2)+P(E_3)$$

2.1.7 *Independent events*

Two events are independent when the occurrence of one does not affect the probability of the occurrence of the second event. For example rolling a die on two occasions, the outcome of the first throw will not affect the outcome of the second, so the two throws have independent outcomes.

2.1.8 *Dependent events*

Events are dependent when one event does affect the probability of the occurrence of the second. An example of this would be drawing a second card from a pack of cards without replacing the first. The outcome of the second trial is then dependent upon the outcome of the first.

2.1.9 *Multiplication law of probability*

Let event A = throwing a six when rolling a die and event B = drawing an ace from a pack of cards. These are two independent events, the occurrence of one can have no bearing on the likelihood of occurrence of the second:

$$P(A) = \frac{1}{6}$$

and

$$P(B) = \frac{4}{52} = \frac{1}{13}$$

If the trial is then to throw the die and draw a card, then to get both a six and an ace there are 6 × 52 possible outcomes and only four will combine a six and an ace, so

$$P(A \text{ and } B) = P(A) \times P(B)$$

$$= \frac{1}{6} \times \frac{4}{52} = \frac{1}{78}$$

This rule also applies to any number of independent events $E_1, E_2, \ldots, E_n$.

$$P(E_1 \text{ and } E_2 \text{ and} \ldots \text{and } E_n) = \prod_{i=1}^{n} P(E_i) \qquad (2.8)$$

2.1.10 Conditional probability

In reliability analysis many events will not be independent. Failure of one component can increase the stress on a second component and increase its likelihood of failure. In this instance the failure of the second component is **conditional** on the failure of the first component. Also, in quality checks when sampling components from a batch, the probability of the second component sampled from a batch being defective will depend on whether or not the first component sampled was replaced or not. Hence replacing the first selected component results in independent probabilities of selecting a defective component on the second sample. Non-replacement does not.

Example

Consider the situation where there is a batch of 25 components of which five are known to be defective. If one is selected at random its probability of being defective is 5/25 = 1/5. If a second component is then selected from the batch without replacing the first, the probability that this is defective is dependent upon the result of our first trial. If the first item was a working component then there are still five defectives and the sample size is now reduced to 24. So the probability of choosing a defective item on the second selection given that a working component was chosen at the first selection is 5/24. Had the first selection produced a defective component the probability of then choosing a second defective component would be 4/24.

$P(A/B)$ is used to denote the probability that event A occurs given that the event B already has. Therefore

$$P(A \text{ and } B) = P(A)\, P(B|A) = P(B)\, P(A|B)$$

or

$$P(A|B) = \frac{P(A \text{ and } B)}{P(B)} \text{ and } P(B|A) = \frac{P(A \text{ and } B)}{P(A)} \tag{2.9}$$

From the equations above, we can see that if A and B are independent then $P(A \text{ and } B) = P(A)\, P(B)$ and this gives $P(A|B) = P(A)$.

Example

Again consider a batch of 25 components of which five are known to be defective. It is required to find the probability of selecting one working and one defective component if components are selected without replacement.

Let

D_1 = first component drawn is defective

W_1 = first component drawn is working

D_2 = second component drawn is defective

W_2 = second component drawn is working

P (one defective and one working) = $P(D_1\, W_2 \text{ or } W_1\, D_2)$

$$P(D_1W_2) = P(D_1)\, P(W_2|D_1) = \frac{5}{25} \times \frac{20}{24} = \frac{1}{6}$$

$$P(W_1D_2) = P(W_1)\, P(D_2|W_1) = \frac{20}{25} \times \frac{5}{24} = \frac{1}{6}$$

Since D_1W_2 and W_1D_2 are mutually exclusive:

$$\begin{aligned} P(D_1W_2 \text{ or } W_1D_2) &= P(D_1W_2) + P(W_1D_2) \\ &= \frac{1}{6} + \frac{1}{6} \\ &= \frac{1}{3} \end{aligned}$$

A diagram of this process, known as a probability tree, is shown in Fig. 2.2.

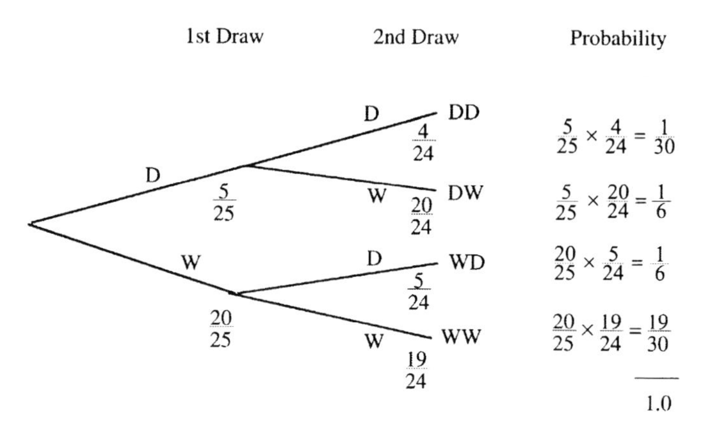

Fig. 2.2 Probability tree

When the first component selection is made, two outcomes can result, either a defective is chosen or a working component is chosen with probabilities of 5/25 and 20/25 respectively. These two options are exhaustive and their probabilities sum to one. These options are represented by two branches starting from the initial point on the left of the diagram. When the second draw is carried out, the same two outcomes are possible but the probabilities depend on the result of the first draw. Treating the outcome of the first draw as branch points gives two further paths added to each point, one representing the selection of a defective, the other the selection of a working component with appropriate probabilities. Since only two draws were carried out in the experiment the tree is terminated at this point. Each terminating branch point now represents one of the four possible outcomes of the experiment and the probability of each outcome can be obtained by multiplying the probabilities along the path leading to that point. Since all branch points represent outcomes which are mutually exclusive and exhaustive their probabilities sum to one.

The problem required the probability of obtaining one defective and one working component. The termination of the second and third branches, DW and WD, results in this option. Since branch points represent mutually exclusive events:

$$P(\text{DW or WD}) = P(\text{DW}) + P(\text{WD})$$
$$= \frac{1}{6} + \frac{1}{6} = \frac{1}{3}$$

as obtained using conditional probabilities.

2.1.11 *Binomial distribution*

The probability tree illustrated in Fig. 2.3 refers to the situation where three trials are conducted and two outcomes are possible for each trial, labelled success (S) and failure (F). Eight possible results can occur at the termination of the three trials as indicated giving all combinations of success and failure over each trial. For each result the number of 'successes' has been listed on the far right of the tree.

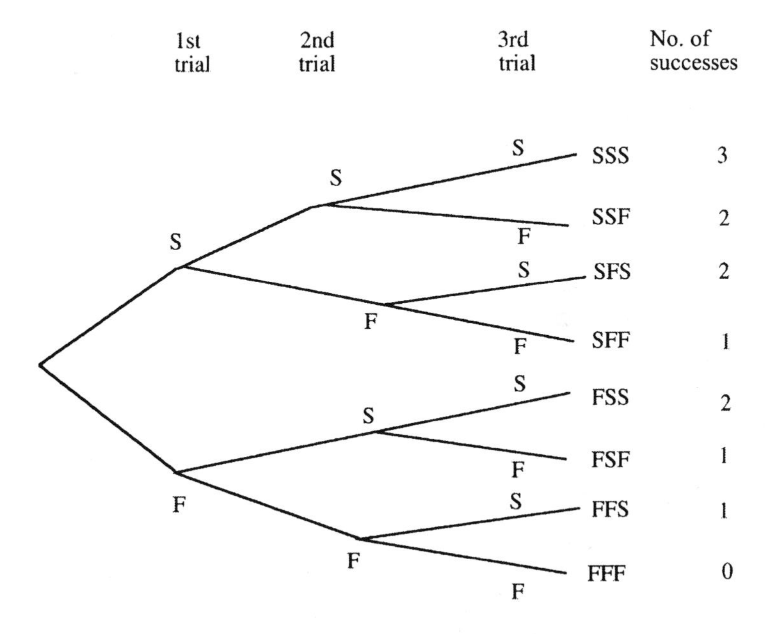

Fig. 2.3 Probability tree for three trials each with two outcomes

Providing that the probability of success $P(\text{S}) = p$ and failure $P(\text{F}) = q$ $(p + q = 1)$ remain constant at each trial we can construct the following table to obtain the probability of 0, 1, 2, or 3 successes in the three trials of this experiment.

Successes	0	1	2	3
Probability	q^3	pq^2	p^2q	p^3
No. of occurrences	1	3	3	1
Total probability =	q^3	$3pq^2$	$3p^2q$	p^3

The result line for the total probability can be compared with $(p + q)^3 = q^3 + 3pq^2 + 3p^2q + p^3$. Had we conducted n trials of the above experiment the total probability could have been obtained by expanding $(p + q)^n$. Probabilities of obtaining r successes then becomes a matter of selecting the correct term in the resulting binomial expansion. This can be generalized as follows.

When an experiment has only two outcomes identified as 'success' and 'failure', with probabilities p and q respectively ($p + q = 1$), and is repeated n times independently then the probability distribution for r successes is given by

$$P(r) = \binom{n}{r} p^r (1-p)^{n-r}, \qquad r = 0, 1, \ldots, n \tag{2.10}$$

where

$$\binom{n}{r} = {}^nC_r = \frac{n!}{(n-r)!r!}$$

(the number of ways of choosing r items from n).

This distribution is known as the **binomial distribution** and has mean $\mu = np$ and standard deviation $\sigma = \sqrt{(npq)}$. It is a two-parameter distribution requiring only the number of trials n and the probability of success p at each trial to completely specify the distribution.

Example

Twenty per cent of items produced on a machine are outside stated tolerances. Determine the probability distribution of the number of defectives in a pack of five items.

Therefore $n = 5$, $p = 0.2$, and $q = 1 - p = 0.8$ (note by success we mean defective component).

$$r=0 \quad {}^5C_0=1 \quad P(r=0)=q^5 \quad =(0.8)^5 \quad =0.3277$$

$$r=1 \quad {}^5C_1=5 \quad P(r=1)=5pq^4 \quad =5(0.2)(0.8)^4 \quad =0.4096$$

$$r=2 \quad {}^5C_2=10 \quad P(r=2)=10p^2q^3 \quad =10(0.2)^2(0.8)^3 \quad =0.2048$$

$$r=3 \quad {}^5C_3=10 \quad P(r=3)=10p^3q^2 \quad =10(0.2)^3(0.8)^2 \quad =0.0512$$

$$r=4 \quad {}^5C_4=5 \quad P(r=4)=5p^4q \quad =5(0.2)^4(0.8) \quad =0.0064$$

$$r=5 \quad {}^5C_5=1 \quad P(r=5)=p^5 \quad =(0.2)^5 \quad =0.0003$$

$$\left[(p+q)^5=q^5+5pq^4+10p^2q^3+10p^3q^2+5p^4q+p^5\right]$$

Since r can take only integer values between 0 and n the resulting discrete probability function can be represented pictorially, as shown in Fig. 2.4.

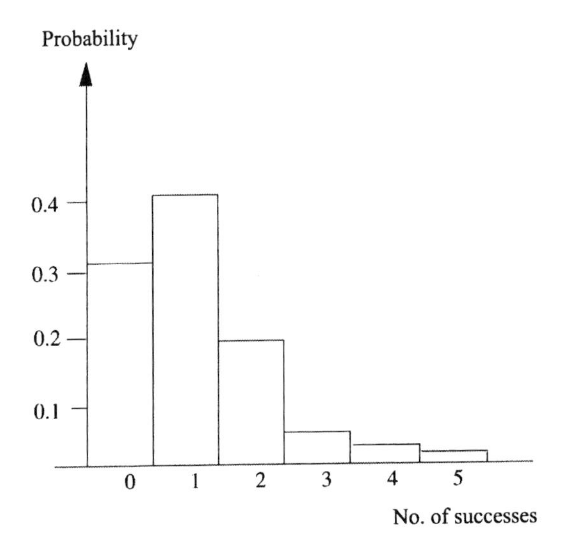

Fig. 2.4 Binomial distribution for n = 5, p = 0.2

(It may be argued that in selecting the first of the items for our pack of five that since we do not replace this item the probability of obtaining a defective for the second selection will be affected and conditional probabilities should be used. This is strictly true. However, in selecting from a very large sample the change in the probability of obtaining a defective is so small that this probability can accurately be assumed constant.)

2.1.12 Poisson distribution

The Poisson distribution, like the binomial, is a discrete distribution. It is appropriate to use this distribution to predict probabilities for the number of times an event will occur during a set period of time given the mean occurrence rate. The difference between this and the binomial distribution is that the non-occurrence of the event is not considered. The binomial deals with a fixed number of trials so each result can be classified as either an event occurrence or non-occurrence. There are many applications where event non-occurrence either has no meaning or cannot be calculated, for example, incoming calls at a telephone exchange per minute or the number of casualties arriving at a hospital per hour. Average rates of arrival can be monitored and the Poisson distribution used to predict the distribution of the number of event occurrences over the period. Non-occurrence of a telephone call or a casualty cannot be measured since it really has no meaning.

The Poisson distribution is:

$$P(r) = \frac{e^{-\mu} \mu^r}{r!} \quad \text{for} \quad r = 0, 1, 2, \ldots \tag{2.11}$$

where μ is the mean arrival rate over the time period. Remember $0! = 1$.

Example

A component has a failure rate of 0.3 failures per 500 h operating time. Over a 10 000 h period what is the probability of 0, 1, 2, 3, and more than 3 failures?

$$\text{Mean number of failures per hour} = \frac{0.3}{500}$$

$$\text{Mean number of failures in 10 000 h} = \frac{0.3}{500} \times 10\,000 = 6$$

i.e. $\mu = 6$.

$$P(r)=\frac{e^{-\mu}\mu^{r}}{r!}$$

0 failures $\quad P(0)=\frac{e^{-6}\times 6^{0}}{0!}=e^{-6}\quad =0.0025$

1 failure $\quad P(1)=\frac{e^{-6}\times 6^{1}}{1!}=\quad =0.0149$

2 failures $\quad P(2)=\frac{e^{-6}\times 6^{2}}{2!}=\quad =0.0446$

3 failures $\quad P(3)=\frac{e^{-6}\times 6^{3}}{3!}=\quad =0.0892$

$$P(\text{more than 3 failures})=1-\left[P(0)+P(1)+P(2)+P(3)\right]=0.8488$$

The Poisson distribution is a one-parameter distribution which can be used, as shown, in its own right. It can also be used to approximate the binomial distribution. When applying the binomial distribution to situations when $n \geq 50$ and $p \leq 1/10$, then an accurate approximation which is easier to calculate can be obtained by using the Poisson distribution. The parameter μ is obtained by equating the distribution means, so

$$\mu = np$$

2.1.13 Continuous probability distributions

The binomial and Poisson distributions refer to discrete events, e.g. the number of successes resulting in a sequence of trials. This has applications in reliability engineering, such as in counting the number of failed or defective components in manufacturing batches or the number of machines breaking down over a time interval. What is also of interest is the time to failure of a component or system and its time to repair. Time is a continuous variable and as such these distributions of times to failure or repair are represented by a continuous probability distribution. A cumulative distribution function $F(t)$ is defined as the probability that T, the random variable, assumes a value less than or equal to the specific value t.

$$F(t)=P(T \leq t) \tag{2.12}$$

(In reliability applications $F(t)$ is the failure function or unreliability function, i.e. the probability that a component/system fails to survive beyond a set time t.)

Since $F(t)$ is a probability we know that

$$0 \leq F(t) \leq 1 \tag{2.13}$$

Also since $F(t)$ is a cumulative distribution it can never decrease as t increases so $F(t)$ is a non-decreasing function. From $F(t)$ we can derive the probability density function, p.d.f., represented by $f(t)$

$$f(t) = \frac{\mathrm{d}}{\mathrm{d}t} F(t) \tag{2.14}$$

So

$$F(t) = \int_{-\infty}^{t} f(u)\,\mathrm{d}u$$

As $f(t)$ is the slope of a non-decreasing function then

$$f(t) \geq 0$$

and since its integral is a probability, integration over the entire range must result in unity:

$$\int_{-\infty}^{\infty} f(u)\,\mathrm{d}u = 1 \tag{2.15}$$

To obtain the probability of the random variable being between two limits we can find the area under $f(t)$ between these limits and so

$$P[t_1 \leq T \leq t_2] = \int_{t_1}^{t_2} f(u)\,\mathrm{d}u \tag{2.16}$$

For a continuous variable, probabilities must be expressed in terms of intervals. This is because there are an infinite number of values of T in any range so that the probability of some specific value t is always zero. Hence, $f(t)\,\mathrm{d}t$ is taken as the probability that the point of interest will lie in between t and $t + \mathrm{d}t$.

2.1.14 Normal distribution

The normal distribution is one of the most commonly used probability distributions in engineering applications. Its probability density function is symmetric about the mean as shown in Fig. 2.5. Lying within one standard deviation of the mean are 68.26 per cent of a normal population, 34.13 per cent distributed either side of the mean. Between one and two standard deviations from the mean are a further 27.18 per cent of the population; 4.3 per cent then lie between two and three standard deviations from the mean. In total this means 99.74 per cent of the population of a normally distributed random variable are within three standard deviations of the mean. This distribution of the population will be true of any normally distributed random variable, whatever the values taken by the mean and standard deviation.

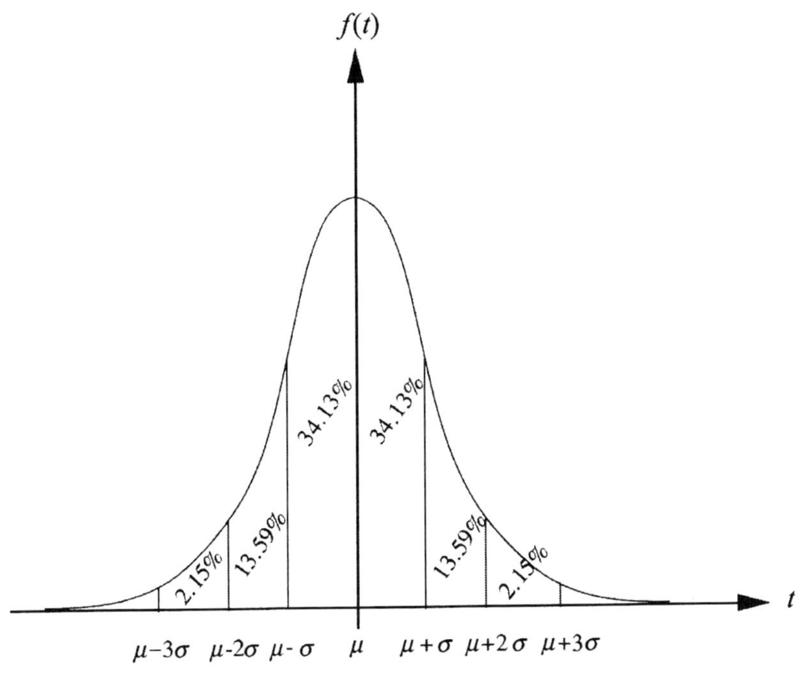

Fig. 2.5 The normal or Gaussian probability distribution

The normal distribution $N(\mu, \sigma)$, defined by two parameters, the mean μ and the standard deviation σ, has a probability density function given by

$$f(t) = \frac{1}{\sigma\sqrt{(2\pi)}} e^{-(t-\mu)^2/(2\sigma^2)} \qquad -\infty \le t \le \infty \tag{2.17}$$

The cumulative probability function can then be obtained by integrating $f(t)$. Unfortunately this cannot be integrated to produce a formula which could be used for any values of μ and σ. Integration has to be performed using numerical methods and the results tabulated for different limits of the integration. It would be very inconvenient if this required a different table for every possible combination of parameters μ and σ. This can be avoided by transforming any normally distributed variable with mean μ and standard deviation σ into a random variable with mean 0 and standard deviation 1, i.e. $N(0, 1)$. Centring the distribution on a mean of zero is achieved by translating the whole distribution, i.e. subtracting μ from the random variable. The dispersion of the distribution about its mean is then expanded or contracted to produce a standard deviation of unity by dividing by σ. This produces a new variable z where

$$z = \frac{t - \mu}{\sigma} \tag{2.18}$$

Since z is $N(0, 1)$ only one table of values for the cumulative probability is needed.

Table 2.2 contains values for the area under a $N(0, 1)$ distribution between the mean and any value of the random variable z. Using the symmetry of the distribution and the fact that 50 per cent of the population lie either side of the mean, the area between any two values z_1 and z_2 can be determined.

Example

A type of bearing has an average life of 1500 h and a standard deviation of 40 h. Assuming a normal distribution determine the number of bearings in a batch of 1200 likely to fail before 1600 h.

The bearing lifetime distribution density function will therefore be as shown in Fig. 2.6.

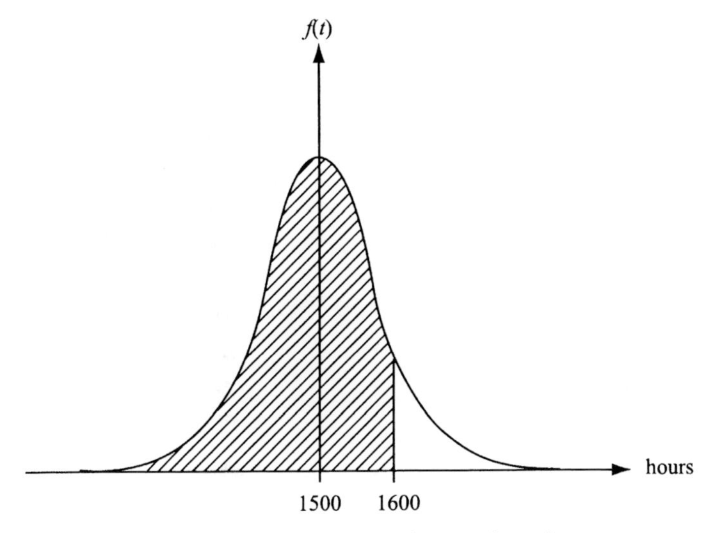

Fig. 2.6 Bearing failure density function

It is required to find first of all the probability that an individual bearing will fail before 1600 h lifetime. This is equivalent to finding the area under the $N(1500, 40)$ distribution as shaded above. The values tabulated in Table 2.2 refer to areas between the mean and the value z for the $N(0, 1)$ distribution. So the upper limit of 1600 h needs to be transformed. Therefore

$$z = \frac{t-\mu}{\sigma} = \frac{1600-1500}{40}$$
$$= 2.5$$

The area we want is the left-hand half of the distribution (area = 0.5), plus the area between the mean and 2.5 standard deviations above the mean. The second part of the area can be obtained from looking up the value of z = 2.5 in Table 2.2. This gives a value of 0.4938, i.e. $\Phi(2.5) = 0.4938$. Therefore

$$P(T \leq 1600) = 0.5 + 0.4938$$
$$= 0.9938$$

Table 2.2 Normal probability distribution

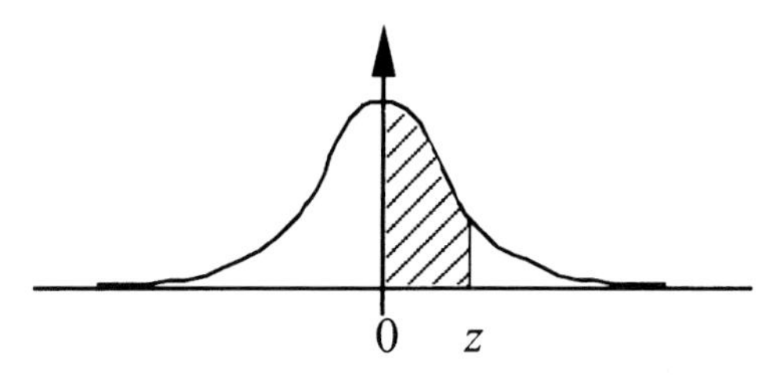

Partial areas under the standardized curve

$z = \frac{x - \bar{x}}{\sigma}$	0	1	2	3	4	5	6	7	8	9
0.0	0.0000	0.0040	0.0080	0.0120	0.0159	0.0199	0.0239	0.0279	0.0319	0.0359
0.1	0.0398	0.0438	0.0478	0.0517	0.0557	0.0596	0.0636	0.0678	0.0714	0.0753
0.2	0.0793	0.0832	0.0871	0.0910	0.0948	0.0987	0.1026	0.1064	0.1103	0.1141
0.3	0.1179	0.1217	0.1255	0.1293	0.1331	0.1388	0.1406	0.1443	0.1480	0.1517
0.4	0.1554	0.1891	0.1628	0.1664	0.1700	0.1736	0.1772	0.1808	0.1844	0.1879
0.5	0.1915	0.1950	0.1985	0.2019	0.2054	0.2086	0.2123	0.2157	0.2190	0.2224
0.6	0.2257	0.2291	0.2324	0.2357	0.2389	0.2422	0.2454	0.2486	0.2517	0.2549
0.7	0.2580	0.2611	0.2642	0.2673	0.2704	0.2734	0.2760	0.2794	0.2823	0.2852
0.8	0.2881	0.2910	0.2939	0.2967	0.2995	0.3023	0.3051	0.3078	0.3106	0.3133
0.9	0.3159	0.3186	0.3212	0.3238	0.3264	0.3289	0.3215	0.3340	0.3365	0.3389
1.0	0.3413	0.3438	0.3451	0.3485	0.3508	0.3531	0.3554	0.3577	0.3599	0.3621
1.1	0.3643	0.3665	0.3686	0.3708	0.3729	0.3749	0.3770	0.3790	0.3810	0.3830
1.2	0.3849	0.3869	0.3888	0.3907	0.3925	0.3944	0.3962	0.3980	0.3997	0.4015
1.3	0.4032	0.4049	0.4066	0.4082	0.4099	0.4115	0.4131	0.4147	0.4162	0.4177
1.4	0.4192	0.4207	0.4222	0.4236	0.4251	0.4265	0.4279	0.4292	0.4306	0.4319
1.5	0.4332	0.4345	0.4357	0.4370	0.4382	0.4394	0.4406	0.4418	0.4430	0.4441
1.6	0.4452	0.4463	0.4474	0.4484	0.4495	0.4505	0.4515	0.4525	0.4535	0.4545
1.7	0.4554	0.4564	0.4573	0.4582	0.4591	0.4599	0.4608	0.4616	0.4625	0.4633
1.8	0.4641	0.4649	0.4656	0.4664	0.4671	0.4678	0.4686	0.4693	0.4699	0.4706
1.9	0.4713	0.4719	0.4726	0.4732	0.4738	0.4744	0.4750	0.4756	0.4762	0.4767

Table 2.2 Continued

$z=\frac{x-\bar{x}}{\sigma}$	0	1	2	3	4	5	6	7	8	9
2.0	0.4772	0.4778	0.4783	0.4785	0.4793	0.4798	0.4803	0.4808	0.4812	0.4817
2.1	0.4821	0.4826	0.4830	0.4834	0.4838	0.4842	0.4846	0.4850	0.4854	0.4857
2.2	0.4861	0.4864	0.4868	0.4871	0.4875	0.4878	0.4881	0.4884	0.4882	0.4890
2.3	0.4893	0.4896	0.4898	0.4901	0.4904	0.4906	0.4909	0.4911	0.4913	0.4916
2.4	0.4918	0.4920	0.4922	0.4925	0.4927	0.4929	0.4931	0.4932	0.4934	0.4936
2.5	0.4938	0.4940	0.4941	0.4943	0.4945	0.4946	0.4948	0.4949	0.4951	0.4952
2.6	0.4953	0.4955	0.4956	0.4957	0.4959	0.4960	0.4961	0.4962	0.4963	0.4964
2.7	0.4965	0.4966	0.4967	0.4968	0.4969	0.4970	0.4971	0.4972	0.4973	0.4974
2.8	0.4974	0.4975	0.4976	0.4977	0.4977	0.4978	0.4979	0.4980	0.4980	0.4981
2.9	0.4981	0.4982	0.4982	0.4983	0.4984	0.4984	0.4985	0.4985	0.4986	0.4986
3.0	0.4987	0.4987	0.4987	0.4988	0.4988	0.4989	0.4989	0.4989	0.4990	0.4990
3.1	0.4990	0.4991	0.4991	0.4991	0.4992	0.4992	0.4992	0.4992	0.4993	0.4993
3.2	0.4993	0.4993	0.4994	0.4994	0.4994	0.4994	0.4994	0.4995	0.4995	0.4995
3.3	0.4995	0.4995	0.4995	0.4996	0.4996	0.4996	0.4996	0.4996	0.4996	0.4997
3.4	0.4997	0.4997	0.4997	0.4997	0.4997	0.4997	0.4997	0.4997	0.4997	0.4998
3.5	0.4998	0.4998	0.4998	0.4998	0.4998	0.4998	0.4998	0.4998	0.4998	0.4998
3.6	0.4998	0.4998	0.4999	0.4999	0.4999	0.4999	0.4999	0.4999	0.4999	0.4999
3.7	0.4999	0.4999	0.4999	0.4999	0.4999	0.4999	0.4999	0.4999	0.4999	0.4999
3.8	0.4999	0.4999	0.4999	0.4999	0.4999	0.4999	0.4999	0.4999	0.4999	0.4999
3.9	0.5000	0.5000	0.5000	0.5000	0.5000	0.5000	0.5000	0.5000	0.5000	0.5000

The probability of an individual bearing taken from this sample failing prior to 1600 h life is 0.9938. From a sample of 1200 we can expect ($1200 \times 0.9938 = 1192.56$), i.e. 1192, to fail.

2.1.15 Log-normal distribution

The normal distribution described in the previous section can be used to model lifetimes of components or repair times. An unfortunate property of this distribution is that it allows negative lifetime/repair time with a non-zero probability. More frequently, particularly with repair times, the normal distribution is used for the log of the life-time or repair times. This assumes that the times are log-normally distributed. The log-normal density function is given by

$$f(t) = \frac{1}{\sigma t \sqrt{(2\pi)}} \exp\left\{-(\log t - \mu)^2 / 2\sigma^2\right\} \qquad t > 0 \tag{2.19}$$

This distribution has mean

$$\exp\left(\mu + \frac{1}{2}\sigma^2\right)$$

and variance

$$\exp\left(2\mu + \sigma^2\right)\left\{\exp\sigma^2 - 1\right\}$$

2.1.16 Negative exponential distribution

The negative exponential distribution is one of the most important probability distributions in its applicability to reliability engineering. Its role is discussed in detail in Chapter 5 which deals with component failure distributions. At this point in the book, discussions on the distribution will be limited to defining its mathematical form and deriving the distribution mean which will be required later in the text.

The negative exponential distribution is applicable when the hazard rate λ is constant. In this situation the reliability distribution $R(t)$ takes the form of the negative exponential function:

$$R(t) = e^{-\lambda t} \tag{2.20}$$

therefore the corresponding density function is $-\lambda e^{-\lambda t}$. Since $R(t) + F(t) = 1$, the unreliability $F(t)$ is $1 - e^{-\lambda t}$ and the corresponding density function for this is $f(t) = \lambda e^{-\lambda t}$ as shown in Fig. 2.7.

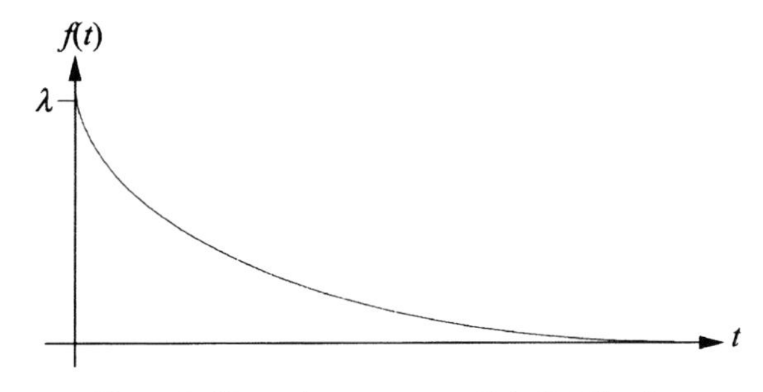

Fig. 2.7 Negative exponential distribution

By taking the expectation of this distribution (see Appendix A) we can obtain the mean time to failure:

$$E(t) = \int_0^\infty tf(t)\,\mathrm{d}t$$
$$= \int_0^\infty \lambda t \mathrm{e}^{-\lambda t}\,\mathrm{d}t$$

Integrating by parts gives

$$\left[-t\ \mathrm{e}^{-\lambda t}\right]_0^\infty + \int_0^\infty \mathrm{e}^{-\lambda t}\ \mathrm{d}t = \left[\frac{-\mathrm{e}^{-\lambda t}}{\lambda}\right]_0^\infty = \frac{1}{\lambda}$$

i.e. the mean time to failure is the reciprocal of the failure rate.

2.1.17 Weibull distribution

The Weibull distribution, like the exponential, is of considerable importance in reliability studies and is dealt with in detail later in the book (Chapter 5). As such, it is only defined at this point. One feature of this distribution which makes it so versatile is that it has no characteristic shape; by varying its parameters different ranges of shapes can be achieved. This makes it particularly suitable for fitting to component failure data where a decreasing, constant or increasing hazard rate is applicable.

The failure density function of the Weibull, two-parameter distribution is defined as

$$f(t) = \frac{\beta t^{\beta-1}}{\eta^\beta} \exp\left[-\left(\frac{t}{\eta}\right)^\beta\right], \quad t \geq 0, \quad \beta > 0, \quad \text{and} \quad \eta > 0 \qquad (2.21)$$

and the cumulative failure distribution is

$$F(t) = \int_0^t f(u)\ \mathrm{d}u = 1 - \exp\left[-\left(\frac{t}{\eta}\right)^\beta\right]$$

It can be seen from the equation above that setting $\beta = 1$ gives a negative exponential distribution with hazard rate $1/\eta$. The negative exponential distribution is a special case of the Weibull distribution.

2.2 Set theory

The preceding section dealt with ways in which the probability of events can be calculated. Predicting the likelihood of particular outcomes from a random experiment involves distinguishing those outcomes from the experiment which are of concern or interest from those which are not. In this way there is a strong link between probability theory and set theory. Set theory is a means of grouping or organizing events into defined categories. A **set** is simply a collection of items having some recognizable feature in common which distinguishes them from items excluded from the set. For example, a set could be made up of even numbers, integers, cars, all possible different states of a component, or combinations of components whose failure will cause a system to fail in a dangerous way. The first three of these examples are just general illustrations of possible groupings of events into sets. The last two examples would be the organization of particular events which may be of interest in a reliability study. Probabilities of occurrence of sets can then be calculated from the probabilities of their components using the probability laws stated in the previous section. A very brief description of the basic elements of set theory and its relationship to probability theory follows.

2.2.1 Notation

Sets are usually denoted by single capital letters, and the simplest way to express or define a set is to list, either explicitly or implicitly, all its elements. For example:

$A = \{2, 4, 6, 8, 10\}$ or
$A = \{\text{even numbers greater than 1 and less than 12}\}$
$B = \{1, 2, 3, 4, 5\}$
$C = \{\text{Fiesta, Escort, Sierra, Capri}\}$
$D = \{\text{Capri, Escort, Fiesta, Sierra}\}$
$E = \{2, 4\}$

If A is a set and x is an **element** or **member** of A then this is written $x \in A$ or x belongs to A.

So

$2 \in A$
$5 \in B$
Escort $\in C$

The negation of this, x is not a member of A, is written $x \notin A$. So Escort $\in C$ but Escort $\notin B$.

Two sets X and Y are **equal** (written $X = Y$) if and only if they have the same elements, so that every element in X is an element of Y, and every element of Y is an element of X. (Therefore, in the sets listed above, $C = D$.) Also a set X is a **sub-set** of Y if every element of X is an element of Y. This is written $X \subset Y$. X is **contained** in Y (so $E \subset B$ and also $C \subset D$ and $D \subset C$).

In any application of the theory of sets, all sets under investigation are sub-sets of a fixed set. We call this the **universal set** and this can be denoted by U, 1, or E. It is also necessary to introduce the concept of the **empty** or **null set** denoted by $\varnothing$ or { }. For any set X, $\varnothing \subset X \subset U$.

The universal set is often implicitly used in set theory calculations as it enables the formulation of the **complementary** set of X, denoted $\overline{X}$ or X'. $\overline{X}$ contains all the elements that do not belong to X.

2.2.2 Venn diagrams

A Venn diagram provides a very useful pictorial means of representing sets. The universal set is represented by some geometrical shape (usually a rectangle) and events and sub-sets are illustrated inside.

Example

U = set of integers between 0 and 11
A = {even numbers}
B = {1, 3, 5, 6, 8, 10}

This can be represented by using a Venn diagram. Consider Fig. 2.8: the outer rectangle is the universal set and contains all events of interest, i.e. the integers between 0 and 11. Of these events all those which are contained in set A, the even numbers, are placed within the circular boundary labelled A. Similarly events placed in set B are placed within the rectangle labelled B. Events {6, 8, 10} occur in both A and B and hence appear on the diagram where the circle and rectangle representing these sets overlap.

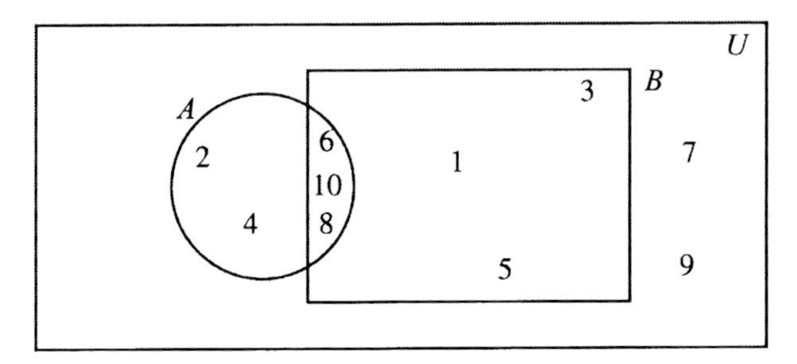

Fig. 2.8 Venn diagram

2.2.3 Operations on a set

Once sets have been formulated operations can be performed on their members. There are three basic operations: 'union', 'intersection', and 'complementation'. These are defined below.

Union

The union of two sets A and B is the set which contains all elements that are either in A or B or both. This is written $A \cup B$, and is illustrated by the Venn diagram in Fig. 2.9.

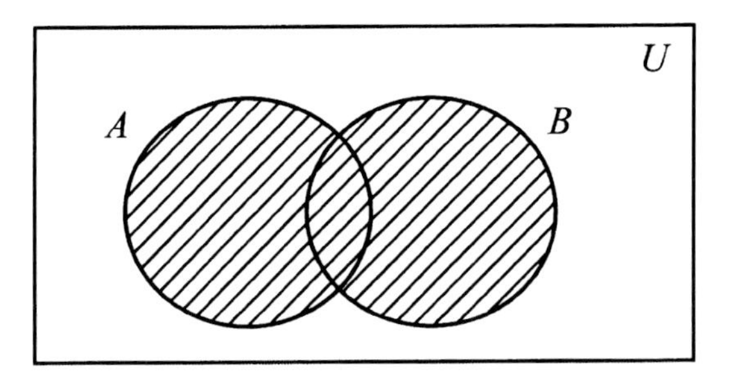

Fig. 2.9 Union

Intersection

The intersection of two sets A and B is the set which contains all elements that are common to both A and B. This is written $A \cap B$, and its Venn diagram is shown in Fig. 2.10.

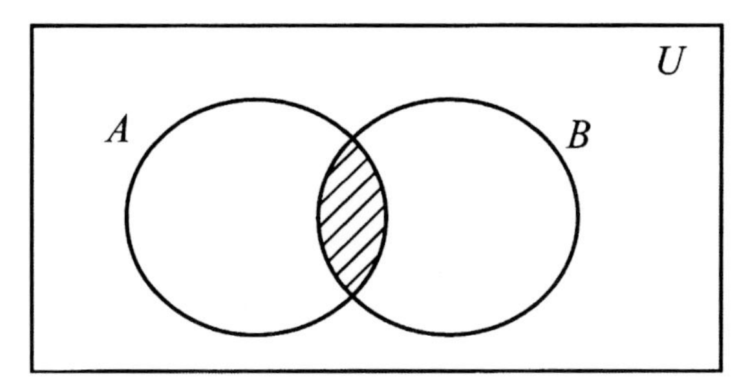

Fig. 2.10 Intersection

Complementation

The complement of a set X is the set which contains all the elements that are not in X. This is written $\overline{X}$. On the Venn diagram in Fig. 2.11 it is everything outside of set X.

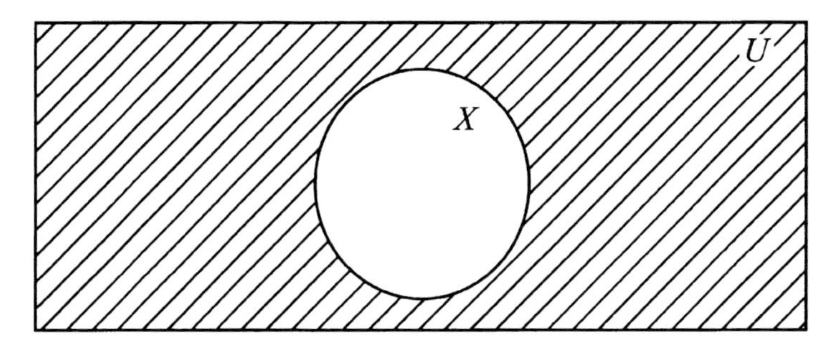

Fig. 2.11 Complementation

Example

Outcomes from the toss of a die:

A = the result is an even number = $\{2, 4, 6\}$
B = the result is less than 4 = $\{1, 2, 3\}$
C = the result is divisible by 7 = $\varnothing$

The Venn diagram which represents these sets is shown in Fig. 2.12.

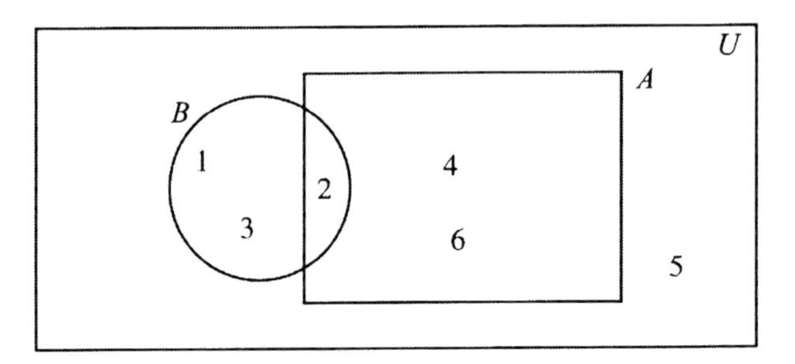

Fig. 2.12 Example Venn diagram

From this we can see that

$$2 \in B, \quad 2 \in A$$
$$1 \in B, \quad 1 \notin A$$

Also the union and intersection of the two sets A and B give

$$A \cup B = \{1, 2, 3, 4, 6\}$$
$$A \cap B = \{2\}$$

Only element 5 lies outside the union of sets A and B; therefore

$$\left(\overline{A \cup B}\right) = 5$$

2.2.4 Probability and Venn diagrams

The Venn diagram can also be used to represent the probability of occurrence of each set. This is achieved by making the area assigned to each set equal to its probability of occurrence. Thus the total area of the universal set is 1. The probability of compound events can then be found from the area representing the set, for example:

Union of mutually exclusive or disjoint events

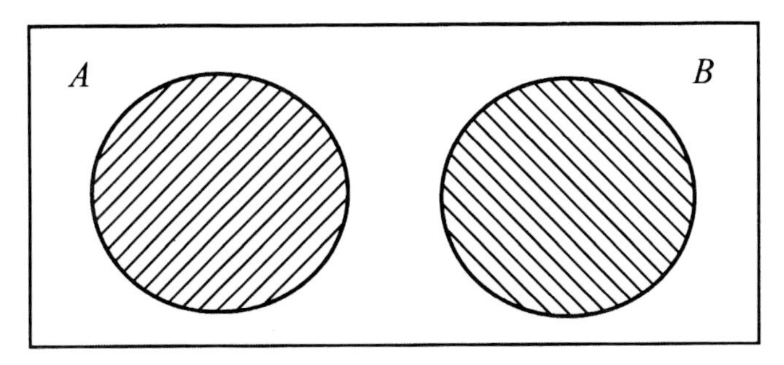

Fig. 2.13 Mutually exclusive events

If an event is in the union of sets A and B then it is contained in the area shaded in the Venn diagram above. A and B are disjoint (there is no overlap) and therefore

$$P(A \cup B) = P(A) + P(B)$$

This is the addition law of probabilities given in the previous section [equation (2.3)].

Union of independent events

When the same element is contained in two sets they will overlap on the Venn diagram. To find the area contained within the two sets now requires us to modify the above equation. If we add the area of A to the area of B we have included the area of the intersection (darker shaded area in Fig. 2.14) twice. We therefore take the area of the intersection away from the sum of the two areas, giving the following equation, relevant for two independent but not necessarily mutually exclusive sets:

$$P(A \cup B) = P(A) + P(B) - P(A \cap B)$$

If the two sets are mutually exclusive then $P(A \cap B) = 0$ and this reduces to the previous equation.

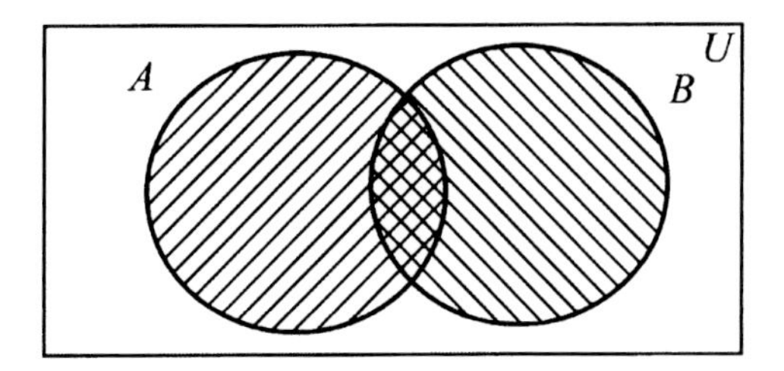

Fig. 2.14 Non-mutually exclusive events

Intersection of independent events

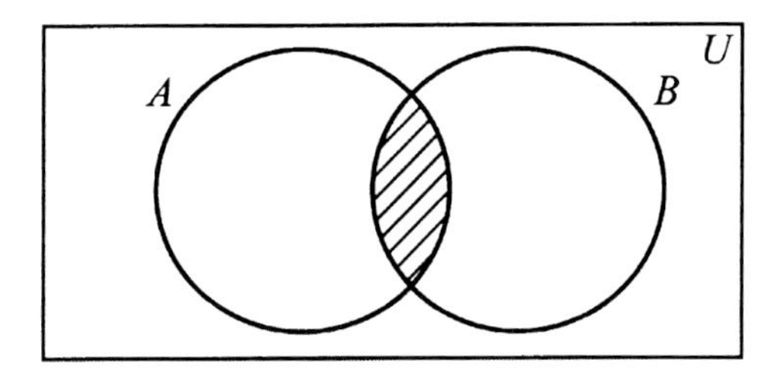

Fig. 2.15 Intersection of events

To calculate the area of the intersection of two sets A and B then from equation (2.8) we can multiply the areas of the two sets:

$$P(A \cup B) = P(A)\, P(B)$$

Complement of an event

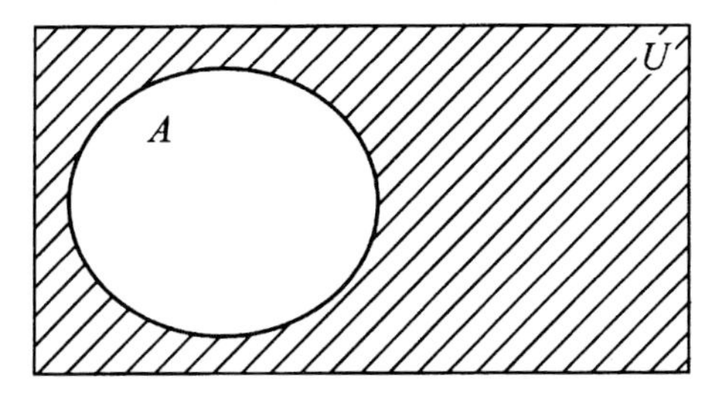

Fig. 2.16 Complementation

Since the area of the universal set (Fig. 2.16) is 1 we get

$$P(\bar{A}) = 1 - P(A)$$

(Note: the notation of $\cup$ for union and $\cap$ for intersection is cumbersome and is frequently modified in reliability work, particularly by fault tree analysts, to + and . respectively.)

Example

Assume a system consists of three components. Let us denote the 'success' or functioning of these components by $\bar{A}$, $\bar{B}$, and $\bar{C}$ and use A, B, and C to represent the failure of these components. The system is designed such that it will fail if two or more of its components fail.

As there are three components which can be assumed to exist in one of two states, there are $2^3 = 8$ combinations of system states, i.e.

$$U = \{ABC, AB\bar{C}, A\bar{B}C, \bar{A}BC, A\bar{B}\bar{C}, \bar{A}B\bar{C}, \bar{A}\bar{B}C, \bar{A}\bar{B}\bar{C}\}$$

The sets of events corresponding to system failure (F) and success (S) are

$$F = \{ABC, AB\bar{C}, A\bar{B}C, \bar{A}BC\}$$

$$S = \{A\bar{B}\bar{C}, \bar{A}B\bar{C}, \bar{A}\bar{B}C, \bar{A}\bar{B}\bar{C}\}$$

2.3 Boolean algebra

The final mathematical topic covered in this chapter which is of importance to system reliability evaluation techniques is that of Boolean algebra. This particular branch of mathematics is very important in situations involving items which can exist in one of two states; for example, switches being open or closed in an electrical circuit, valves used to isolate flow in a pipe being open or closed. In reliability assessments components of a system are commonly considered to exist in either the normal working state, or in the failed state. Techniques such as reliability networks or fault tree analysis then represent the system failure in terms of component failures. Hence the Boolean techniques described below have immediate relevance in those methods where the system failure can be translated into an equivalent set of Boolean equations.

In Boolean algebra, binary states 1 and 0 are used to represent the two states of each event, i.e. occurrence and non-occurrence. So any particular event has an associated Boolean variable, say A; then

$$A = \begin{cases} 1 & \text{event occurs} \\ 0 & \text{event does not occur} \end{cases} \tag{2.22}$$

Boolean algebra is concerned with events, as are probability theory and set theory already covered. Consider the simple experiment which involves throwing a die where the event of interest is that the throw results in a '4'. This event we will call A; then

$$A = \{4\}$$

and

$$\bar{A} = \{1, 2, 3, 5, 6\}$$

If a '4' is achieved then we let the Boolean variable $A = 1$; if not, then $A = 0$.

Manipulating several events to provide information on more complex situations can be achieved by applying algebra to the events' Boolean variables. Events are combined using the three logical operators OR, AND, and NOT (these are equivalent to union, intersection, and complementation in set theory). The following truth tables demonstrate the result of combining the two Boolean variables A and B by these operations for all possible values of these variables.

2.3.1 *A* OR *B*

Table 2.3 Logical OR

A	*B*	*A* OR *B* ($A + B$)
0	0	0
1	0	1
0	1	1
1	1	1

In Table 2.3 the two left-hand columns headed A and B list all possible combinations for the occurrence and non-occurrence of events A and B. The OR operator used to combine these events means that the outcome event will occur whenever A or B or both events occur. From the truth table it can be seen that only the first line, where neither A nor B occurs, results in the non-occurrence of the output event.

2.3.2 *A AND B*

For the result of this operation to be true (to occur), then both events A and B must occur as illustrated in the truth table in Table 2.4.

Table 2.4 Logical AND

A	B	A AND B (A . B)
0	0	0
1	0	0
0	1	0
1	1	1

2.3.3 *NOT A*

The NOT logical operator has the same effect as complementation and applied to a single event A produces the truth in Table 2.5:

Table 2.5 Logical NOT

A	$\bar{A}$
0	1
1	0

The initial truth tables given above provide the outcome for all possible combinations of two events using AND, OR, and NOT operators. Larger tables can be constructed when more variables are encountered and a larger logical expression results. Unfortunately in reliability-type work the large number of Boolean variables in the expressions which need manipulation preclude the use of tables on the grounds of size. Manipulation and simplification of larger expressions can be accomplished by directly using the laws of Boolean algebra. These laws are listed below: '+' is used to denote OR and '.' to signify AND. (Other symbols sometimes used in other texts are $\vee$ for OR and $\wedge$ for AND.)

2.3.4 Rules of Boolean algebra

1. Commutative laws:

$$A + B = B + A \qquad A \,.\, B = B \,.\, A \tag{2.23}$$

2. Associative laws:

$$(A + B) + C = A + (B + C) \qquad (A \,.\, B) \,.\, C = A \,.\, (B \,.\, C) \tag{2.24}$$

3. Distributive laws:

$$A + (B \,.\, C) = (A + B) \,.\, (A + C) \tag{2.25a}$$

$$A \,.\, (B + C) = A \,.\, B + A \,.\, C \tag{2.25b}$$

   [Note: although equation (2.25b) is analogous to the distributive law in ordinary algebra, equation (2.25a) is not.]

4. Identities:

$$\begin{matrix} A + 0 = A & A + 1 = 1 \\ A \,.\, 0 = 0 & A \,.\, 1 = A \end{matrix} \tag{2.26}$$

5. Idempotent law:

$$A + A = A \qquad A \,.\, A = A \tag{2.27}$$

6. Absorption law:

$$A + A \,.\, B = A \qquad A \,.\, (A + B) = A \tag{2.28}$$

   Laws 5 and 6 enable the removal of redundancies in expressions.

7. Complementation:

$$\begin{matrix} A + \overline{A} = 1 \\ A \,.\, \overline{A} = 0 \\ \overline{\left(\overline{A}\right)} = A \end{matrix} \tag{2.29}$$

8. De Morgan's laws:

$$\overline{(A+B)} = \overline{A}.\overline{B}$$
$$\overline{(A.B)} = \overline{A}+\overline{B} \qquad (2.30)$$

Frequently in qualitative reliability work Boolean expressions need to be manipulated. Sometimes simplification, i.e. removal of redundant elements, is required, at other times a special form of the expression is needed such as the sum-of-products form required for fault tree analysis as will be shown in Chapter 7. Three examples of simplifying Boolean expressions are now given.

Examples

1. Simplify $A\,.\,(B + C\,.\,A)$

$= A\,.\,B + A\,.\,C\,.\,A$	(distributive law)
$= A\,.\,B + A\,.\,A\,.\,C$	(commutative law)
$= A\,.\,B + A\,.\,C$	(as $A\,.\,A = A$)
$= A\,.\,(B + C)$	(distributive law)

2. Simplify $\overline{A}\,.\,B\,.\,\left(C\,.\,\overline{B}+\overline{A}\,.\,B\,.\,C\right)$

$= \overline{A}\,.\,B\,.\,C\,.\,\overline{B}+\overline{A}\,.\,B\,.\,\overline{A}\,.\,B\,.\,C$	(distributive law)
$= 0+\overline{A}\,.\,B\,.\,C$	(as $B\,.\,\overline{B}=0$)
$= \overline{A}.B.C$	

3. Simplify $A.\overline{B}.\overline{C}.\overline{D}+\overline{A}.B.\overline{C}+\overline{A}.B.C+A.\overline{B}.\overline{C}.D$

$= A.\overline{BC}.\left(\overline{D}+D\right)+\overline{A}B\left(\overline{C}+C\right)$	(grouping together first and fourth terms, and second and third terms, and then factoring)
$= A\overline{BC}+\overline{A}B$	$\left(\overline{D}+D=1,\quad \overline{C}+C=1\right)$

While not providing a mathematical proof, Venn diagrams can be used to demonstrate the equivalence of logical expressions. For example, take the first of De Morgan's laws:

$$\overline{(A+B)} = \bar{A}.\bar{B}$$

The Venn diagram which represents the left-hand side of this equation is given in Fig. 2.17.

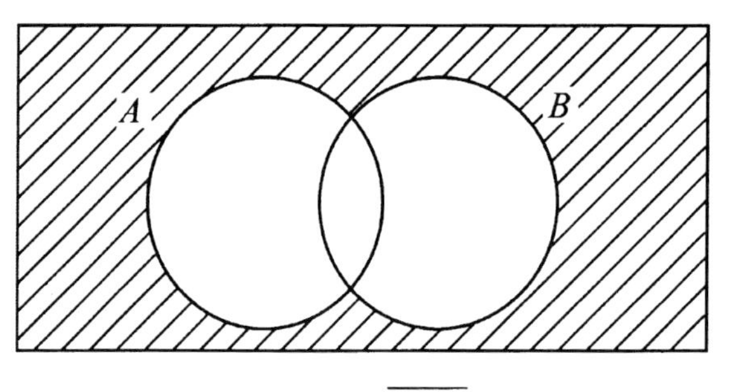

Fig. 2.17 $\overline{A+B}$

Since this is the complement of the union of A OR B, the Venn diagram for the right-hand side is then as given in Fig 2.18.

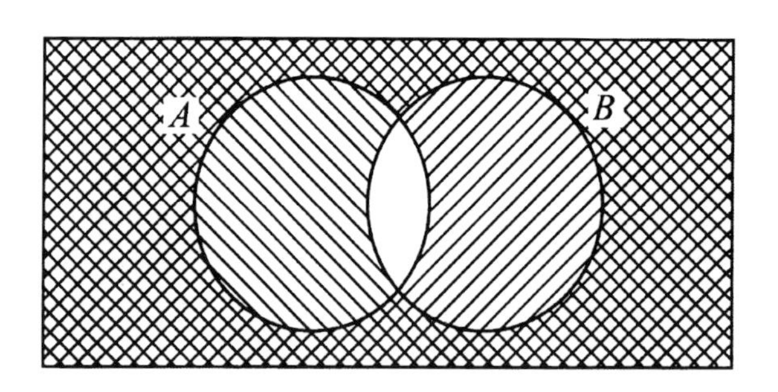

$\bar{A}$ =

$\bar{B}$ =

Fig. 2.18 $\bar{A}.\bar{B}$

The intersection of $\overline{A}$ and $\overline{B}$ is the area where both types of shading appear which is the same as that shaded for the left-hand-side Venn diagram (Fig. 2.17). The equality is demonstrated using a truth table (Table 2.6) to show that De Morgan's rule is true for any values of A and B.

Table 2.6 De Morgan's law

A	B	$A + B$	$\overline{A+B}$	$\overline{A}$	$\overline{B}$	$\overline{A} . \overline{B}$
0	0	0	1	1	1	1
0	1	1	0	1	0	0
1	0	1	0	0	1	0
1	1	1	0	0	0	0

The first two columns give all possible combinations of values taken by the variables A and B. Column 3 provides an intermediate step in evaluating the left-hand side of the equation which is shown in column 4. Similarly the right-hand side expression is then evaluated in steps resulting in column 7. Since column 4 has identical entries to those in column 7 then $\overline{(A+B)} = \overline{A} . \overline{B}$ for all values of A and B.

2.4 Summary

This chapter has presented the basic mathematical tools necessary to understand and apply the reliability techniques which will be developed in succeeding chapters. Reliability theory is concerned with the occurrence and non-occurrence of events. Probability theory enables us to determine how likely events such as hardware failure, human error, and software failure are to occur. Using Boolean algebra we can determine how basic failure events combine to cause the system to fail. From the laws of probability we now have the tools to express the system failure likelihood in terms of component failure likelihoods. The means which determine how this manipulation is achieved both qualitatively and quantitatively will now be the subject of the remainder of the book.

2.5 Bibliography

Craft, A., Davison, R., and **Hargreaves, M.** (2001) *Engineering Mathematics* (3rd Edition). Prentice Hall.

Green, J. A. (1965) *Sets and Groups.* Routledge and Kegan Paul Limited.

Grosh, D. L. (1989) *A Primer of Reliability Theory.* John Wiley.

Kreysig, E. (1999) *Advanced Engineering Mathematics* (8th Edition). John Wiley.

Montgomery, D. C. and **Runger, G. C.** (1999) *Applied Statistics and Probability for Engineers* (2nd Edition). John Wiley.

Spiegel, M. (1975) *Probability and Statistics.* McGraw-Hill.

Stroud, K. A. (1987) *Engineering Mathematics.* Macmillan.

Triola, M. F. (1995) *Elementary Statistics*. Addison-Wesley.

Chapter 3

Qualitative Methods

3.1 Introduction

Qualitative methods can provide an important input to reliability and risk assessment studies. They are frequently applied during feasibility and conceptual phases when equipment selection and system configurations are not fully developed. The objective of qualitative studies is to identify and rank by importance the potential hazards, plant areas, equipment types, or operating/maintenance procedures which may critically affect the safety or availability of the plant.

The application of qualitative methods can lead to significant reductions in risks, and improved reliability. Qualitative techniques can also provide important input to more detailed quantitative studies employing event trees, fault trees, or other reliability/availability models. The qualitative methods discussed in this chapter are particularly relevant to process plant studies. They include the use of hazard analysis, including hazard and operability studies, hazard checklists, and a number of related techniques.

Failure mode and effect analysis is also an important qualitative technique and this is discussed separately in Chapter 4 with specific emphasis on its application to process plant reliability studies.

3.2 Hazard analysis

Hazard analysis has been defined as the identification of undesired events which lead to the materialization of a hazard, the analysis of mechanisms by which these undesired events could occur, and usually the estimation of the extent (magnitude) and likelihood of any harmful effects.

The identification of safety-related hazards is carried out using checklists based on accident statistics, experience in the industry and techniques such as hazard and operability studies. Hazards can be generated from causes external to the process or they may be process related. A checklist of external hazards noted by the Norwegian Petroleum Directorate (**1**) for offshore installations includes the following:

- blowouts
- fires and explosions
- falling objects
- ship and helicopter collisions
- earthquakes
- other possible relevant types of accident
- extreme weather conditions
- relevant combinations of these accidents.

These hazards clearly need consideration in any risk assessment of offshore operations. Similar checklists are published for other industries.

Process-related hazards are best addressed by hazard and operability studies, which are discussed later, but other less exhaustive techniques can be employed effectively in certain situations. The main techniques used in hazard analysis are discussed in the following sections.

3.3 Checklists

Hazard identification checklists have been produced for many potentially hazardous industries. They are used to support a systematic evaluation of the plant and process to identify the potential for material releases which could prove hazardous to personnel working on the plant, members of the public resident in the area surrounding the plant, or the ecology.

The first stage is to consider and list the events which could cause a hazardous release or explosion. The main effects and possible secondary effects of the hazard are then evaluated together with consideration of the likely causes of the event. Finally an estimate is made of the expected frequency of each event and the demands that may be made on protective devices and unprotected systems. A typical general-purpose checklist for process plant (**2**) is shown in Fig. 3.1. The objective is to identify the set of undesired events that need further study.

Hazard Identification Checklist

---

1. What serious EVENT could occur – start with the WORST CONCEIVABLE, e.g.

Toxic	)	
Explosive	)	Release
Flammable	)	
Aggressive (chemically/thermally)	)	

Internal explosion
Offensive emission
Anything else?

2. What EFFECT has this on

Plant fabric	)	e.g.	blast, missile damage, flame,
Operators	)		radiation, toxic effects, chemical
Property	)		attack, corrosion; offensive noise,
Public	)		smell; effluent
Business	)		

Any secondary events (domino effects)?

3. What would CAUSE this?

Materials	–	fuels: ignition sources, air ingress, etc.
		any reaction shift, side reactions, etc.
Processes	–	batch or catalysed – hangfire reactions
Deviations	–	pressure, vacuum, temperature, flow, etc.
Human intervention	–	maloperation, error, nullified safeguards, etc.
External events	–	mechanical damage, collapse, stress, vibration, services failure
Equipment	–	adapted/redundant units, changes in duty

Other causes?

4. With what FREQUENCY would it happen?

- Demands on unprotected systems
- Demands on failed protective devices

Having assessed the severity and frequency of this event, now return to 1 and consider other events of lesser and also differing nature.

Fig. 3.1 General-purpose hazard checklist

3.4 Hazard and operability studies

Hazard and operability (HAZOP) studies were introduced to the chemical industry by ICI Ltd as a method by which plants could be assessed to determine the potential hazards they present to operators and the general public. HAZOP is defined by the British Chemical Industry Safety Council as:

> 'The application of a formal systematic critical examination of the process and engineering intentions of the new facilities to assess the hazard potential of maloperation or malfunction of individual items of equipment and the consequential effects on the facility as a whole.'

The technique aims to stimulate the imagination of designers and operators in a systematic manner so that they can identify the cause of potential hazards in a design. It is a flexible methodology which can be applied to a wide range of industrial installations.

The technique can be used by small as well as large organizations. HAZOP studies can be performed at any time during a plant's life; however, the most effective time is when the design is frozen. At this stage sufficient detail is normally available and the eradication of identified hazards need not necessarily incur considerable expense. There is no reason why HAZOP studies should not be performed on existing plants, although at this stage modifications may be very expensive.

A comprehensive guide to HAZOP is published by the Chemical Industries Association (**3**). The following sections concentrate on the practical aspects of applying the technique.

3.4.1 HAZOP methodology

The basic concept of a HAZOP study is to identify hazards which may arise within a specific system or as a result of system interactions with an industrial process. This requires the expertise of a number of specialists familiar with the design and operation of the plant.

The team of experts systematically consider each item of plant, applying a set of guide words to determine the consequence of operating conditions outside the design intentions. Because of the structured form of a HAZOP it is necessary that a number of terms are clearly defined.

- *Intentions*: defines how the part is expected to function.
- *Deviations*: departures from the design intention which are discovered by systematic application of the guide words.

- *Causes*: the reasons why deviations might occur. Causes can be classified as realistic or unrealistic. Deviations due to the latter can be rejected.
- *Consequences*: the results of the deviations.
- *Hazards*: consequences which can cause damage, injury or loss.
- *Guide words*: simple words which are used to qualify the intention and hence deviations. The list of guide words are:

NO/NOT	No flow, no pressure, etc.
MORE	High flow, high pressure, etc.
LESS	Low flow, low pressure, etc.
AS WELL AS	Material in addition to the normal process fluids.
PART OF	Process only part of the fluid.
REVERSE	Reverse flow of process fluids.

3.4.2 The HAZOP team

HAZOP studies are normally carried out by multi-disciplinary teams, the members of the team providing a technical contribution or supporting role. A typical team composition would be of engineers from design, process, instrumentation, and structural disciplines with a reliability engineer undertaking the role of Chairman.

Members of the HAZOP team need to have knowledge of the plant design and operating procedures. This team should between them be able to answer the majority of questions generated by the application of the guide words. The team should be restricted to about five or six members. If the study seems to require a larger team it may be advantageous to break it down into convenient parts which can be considered separately during the appropriate sessions.

During HAZOP sessions it is necessary to have someone who controls the discussion. This person is known as the Study Leader or Chairman. In addition a Secretary who notes the hazards and who is responsible for the actions from the meeting can significantly increase the output of a HAZOP study. These two people form the support team and are not expected to make a major technical contribution to the study. However, it is important that the Study Leader at least should have sufficient technical knowledge of the process and experience in applying the HAZOP technique to enable him to understand and control the team discussions.

3.4.3 The HAZOP study

To perform a HAZOP study the team will need to be in possession of plant details commensurate with the study objective. This will generally be in the form of flow sheets, line diagrams, operating instructions, piping and instrumentation diagrams (P and IDs), and material inventories.

The Study Leader is responsible for preparatory work such as the scheduling of the study and defining the operator and equipment interfaces. These preparations will necessarily involve project engineers and manufacturers and may require considerable effort.

Once preparations are complete the examination sessions can begin. These sessions are highly structured and the Study Leader will make sure they follow the predetermined plan. As hazards are detected the Study Leader should ensure everyone fully understands them. Solutions may be agreed in the sessions and design modifications implemented (provided the team members have the authority to do so). Alternatively, the appropriate modifications may be determined through discussions outside the examination sessions.

It is very important that the study sessions are accurately recorded, particularly with respect to the hazards identified, the solutions proposed, and those responsible for follow-up actions. This is often best achieved by including a Secretary in the support team and by structuring the study sessions. Recording the study sessions can be useful as a back-up to the Secretary's note-taking and can facilitate any subsequent, more detailed evaluation of the hazard. One recommended approach is to run the HAZOP sessions in the mornings. The afternoons are then spent by the Secretary completing the HAZOP records and progressing actions while the Study Leader prepares for the following morning's session. This can involve the Study Leader in tasks such as preparing and reviewing the relevant documentation, defining system boundaries, and colour coding the different product streams on the P and IDs. Examples of a typical HAZOP record sheet and a HAZOP action sheet are shown in Figs 3.2 and 3.3 respectively.

Guide word	Deviation	Possible cause	Consequences	Safeguards/ interlocks	Action required
System 23 process vent					
MORE OF	High flow	Failure to close off the UMV and WV with subsequent opening of EVV Erosion of restriction orifice	Possible overpressure of vent systems. Orifice RO 022 sized to hold back wellhead pressure upstream and maintain pressure downstream at or below design pressure of vent system	Detection of flow is via TSL 503 set at −10 °C	Procedures should allow for inspection of orifice plate after the use of the blowdown system
LESS OF	Low flow (i) Manual venting of test separator and liquid measurement drum	Blockage, especially of flame arrestor, especially by hydrates during venting	Vent pipework will see full system pressure. This will overpressure the vent system		The removal of the flame arrestor should be considered because of the hazard potential of the blocked arrestor
	(ii) Emergency venting via EVV	As above but also the possibility that the valve does not open fully	Takes longer to depressurize, and thus may increase inventory		HAZ1 – The whole of the ventilation system needs to be reviewed with respect to safe dispersion of released gases
	(iii) Relief valves	Failure of heat tracing downstream of PSV and a leaking PSV will lead to hydrate blockage. Subsequent relief requirement may not be met. Relief valve failure to lift	Possible overpressure of vessel in first case. Note that relief valves will not protect the vessel in the case of a jet fire		Procedures to include periodic check of all heat tracing
NO/NONE	No flow	EVV fails to lift. As for low flow	No hazard because personnel available to correct the situation		

Fig. 3.2 HAZOP record sheet example

HAZOP (II) ACTION SHEET	Action No.	HAZ 1

Meeting no. 4 Date:9/12 Tape Ref: 3D3 Mod Sheet:

Drawing no. PID X 123

Action: R Bloggs To be cleared by: 31/12

The whole of the venting system needs to be reviewed with respect to dispersion of the released gases. Procedures for the selection of vent direction with respect to wind speed and direction need to be modified to take account of any change that may prove hazardous.

Reply:

Accepted: All venting operations and utilization of platform venting systems have been reviewed. It is concluded that the Operations Manual adequately addresses venting issues. For each venting operation prevailing weather conditions will be considered and venting will be delayed if conditions are unsuitable.

Reviewed 14/1 and agreed satisfactory

P J Holding

Fig. 3.3 A HAZOP action sheet

3.5 Rapid ranking

Rapid ranking methods were also developed in ICI (**4**) for identifying and ranking hazards on existing plant where post-operation HAZOP studies are considered too time-consuming and demanding. It is a screening technique which can be applied to a wide range of chemical plant to quickly reveal areas where significant risk seems to exist. Resources are then concentrated in these areas.

Useful features of the method are:

(a) It is quick to apply – typically 4 half-days of study per plant.
(b) Plant and plant areas are ordered in priority of study.
(c) The ordering has a rational basis and is not subjective.

The basis of the method is a table which categorizes the severity of incidents – starting with minor consequences and progressing to the more serious consequences involving deaths and major outages of the plant.

The categories are arranged to cover severity in terms of orders of magnitude, hence a logarithmic scale is used. A typical classification system is shown in Fig. 3.4. The table uses terms which enable differing kinds of incidents – fires, toxic releases, explosions, etc. to be categorized. Associated with each consequence category is a band of frequencies designed so that the product of consequence and frequency is sensibly constant across the table.

The term 'guide' frequency in Fig. 3.4 is employed because the method is used to obtain a rank order rather than to classify absolutely. To work in absolute frequencies would imply acceptability of the risk level and thus an 'acceptable' frequency. However, acceptance of levels of risk by the public is difficult to obtain since each person has his own idea of what he would accept. Nevertheless there is an attempt to fix guide frequencies to be comparable with toxic gas and major incident criteria agreed nationally.

Depending on the size and complexity of the plant, about 2–5 half-days will need to be set aside for the study. Half-days are preferred because it enables the plant personnel to fit in their day-to-day activities, albeit on a reduced scale. Also the study is quite intensive and concentration could diminish if full days are attempted.

Personnel involved in the study are generally the plant manager, who brings his knowledge of the plant and of the sensitivity of the surrounding public to nuisance, a second person acquainted with day-to-day running of the plant, and the analyst, who brings his general skills in the methodology. An outline diagram is required, showing the process flow and main control loops and trip systems. The analyst should be taken on a conducted tour of the plant to note its general condition and the spacing and size of plant items, in order to supplement the picture represented by the line diagram.

With the team around the drawing, the plant is divided up in terms of unit operations; thus distillation, reaction, drying, storage, etc. will each form separate plant sections for study.

For each section of plant ask, 'What serious event could occur?' The aim is to start with the most serious – perhaps total loss of containment.

For each serious event identified the analyst will ask 'What are the consequences of this?' For example:

- How much gas will escape?
- Will it form a gas cloud?
- Of what size?
- Will it reach a source of ignition?

Guide frequency – return period of event (years)	CATEGORY	FIRE AND EXPLOSION		CHEMICAL TOXIC AND NUISANCE EFFECTS	
		CONSEQUENCES	CRITERIA	CONSEQUENCES	CRITERIA
1	I	Minor damage, confined within plant section. No public damage. Minor injuries to employees. No business loss.	Replacement cost up to £1000. Few cases, no hospitalization. Sales unaffected.	Plant personnel's minor gas exposure or burns, etc. Public: detectable.	Some medical attention, no hospitalization. No more than mildly irritating.
10	II	Damage confined within plant section. Public damage very minor. Some employee injuries. No business loss.	Replacement cost £1000 – £10 000, e.g. broken window, mild local reaction. Minor compensation. Some hospitalization, no fatalities. Sales unaffected.	Some injuries to employees. Irritation to public. Occasional casualty.	Occasional hospitalization, no fatalities. Minor local outcry. Minor compensation. Minor medical treatment.
100	III	Damage to plant section and minor local works damage. Occasional employee fatality. Minor business loss.	Replacement cost £10 000 – £100 000. Considerable local outcry. Up to 1 fatality in 10 occasions. Loss of profit up to £1m.	Injuries and occasional fatalities to employees. Irritation to public. Some casualties to public.	Hospitalization necessary. One fatality in 10 occasions. National press reaction.
1000	IV	Severe plant damage and appreciable local works damage. Irritation to public. Fatality to employee or member of public. Serious business loss.	Replacement cost £100 000-£1m. Severe local reaction. Up to 1 fatality per occasion. Loss of profit £1m.	Some casualties and single fatality to plant personnel or public. Public reaction considerable.	Up to 1 fatality per occasion. Considerable national reaction. Severe local reaction.
10 000 100 000	V	Total plant destruction and severe works damage. Widespread public property damage. Multiple fatalities to public and employees. Severe irritation to public. Total loss of business.	Replacement cost £1m+. Up to 10 fatalities per occasion. Demands for permanent closure. Loss of profit £20m over 10 years due to outside pressure.	Multiple fatalities to public and employees. Public reaction profound.	Up to 10 fatalities per occasion. Severe national reaction. Termination of business under outside pressure.

Fig. 3.4 Classification of hazardous incidents

- Will it give toxic concentrations at or beyond the boundary of the works?
- How many people are likely to be exposed to the fire/toxic gas?

The team uses this list and its local knowledge to evaluate the approximate consequences of the event under consideration.

The next stage is to establish a likely cause and so deduce how often it would occur. Broadly, if the frequency assessed is greater than the guide frequency then that incident is deemed to require detailed study with the highest priority – priority A. If the assessed frequency is within the guide range the priority for study is less high – priority B. If the assessed frequency is lower than the guide frequency then the priority for study is low – priority C or D. Some events defy frequency determination despite the best endeavours because the mechanism is not clearly understood. In this case a priority for detailed study will be proposed which takes into account the severity of the consequences in conjunction with the uncertainty of the way in which the event could take place. A guide to subsequent action is shown in Fig. 3.5.

Category of potential incident	Estimated frequency relative to guide frequency			
	Smaller –	Same =	Greater +	Uncertain U
1	Such incidents do not fall within the scope of this method and many will remain unidentified. Such as are identified are recorded thus: No study	No study	Safety audit form of study	
2	No study.	Normally priority C study but if upper end of frequency potential raised to B at team's discretion.	Equally damaging hazard as those below priority A study but if lower end of frequency potential could be lowered to B at team's discretion.	Priority B study. Frequency estimates should not be difficult at this category; may be a lack of fundamental knowledge which requires research.
3	Priority C study.	Priority B study.	Major hazard. Priority A study.	Priority A/B at team's discretion. Such potential should be better understood.
4 and 5	Priority B/C team's discretion.	Normally priority B study but can be raised to A at team's discretion.	Major hazard. Priority A study.	Priority A study. Such potential should be better understood.

Fig. 3.5 Guide to subsequent action

3.6 Preliminary hazard analysis

Preliminary hazard analysis (PHA) is a technique widely used in the USA for industrial safety studies. The objective is to identify in broad terms the potential accidents associated with an installation. Essentially it is a qualitative technique which involves a disciplined analysis of the event sequences which could transform a potential hazard into an accident. The possible consequences of the accident are then considered together with the corrective measures which could contribute to its prevention. PHA studies performed in the offshore industry have tended to adopt similar hazard rankings to those employed by the aerospace industry where hazards are initially characterized by their effects, namely:

Class I hazards – catastrophic effects – likely to cause one or more deaths or total plant loss.

Class II hazards – critical effects – likely to cause severe injury, major property or system damage and total loss of output.

Class III hazards – marginal effects – likely to cause minor injury, property or system damage with some loss of availability.

Class IV hazards – negligible effects – unlikely to cause injury, property or system damage

The next step in a PHA is to decide on the accident prevention measures that can be taken to eliminate all Class I and possibly Class II and III hazards.

One sheet, taken from a preliminary hazard analysis of a gas compression system on a North Sea platform is shown in Fig. 3.6. It can be seen that the PHA is partly narrative, listing both the events and the corrective actions that might be taken to prevent an accident. The 'validation' column is used as a subsequent check on whether the recommended solution has been incorporated. In this study all of the equipment listed in Fig. 3.6 became the subject of more detailed analysis but, of course, many Class IV and Class III (negligible or marginal) situations can arise in other systems on the platform.

SUB-SYSTEM OR FUNCTION	ITEM NO.	HAZARDOUS ELEMENT	EVENT CAUSING HAZARD	HAZARDOUS CONDITION	EVENT CAUSING POTENTIAL ACCIDENT	POTENTIAL ACCIDENT	EFFECT	HAZARD CLASS	PREVENTATIVE MEASURES			VALIDATION
									HARDWARE	PROCEDURES	PERSONNEL	
Gas metering	M12	Gas pressure	Leak rupture Damage to equipment Instrument failure	Gas released to module	Spark Flame Static electricity	Fire Explosion	Injury to personnel Damage to equipment	I or II	ESD system Fire suppression system			Detailed safety study initiated on these systems
Flare KO drum	PV4	Gas pressure Oil	Relief system inoperative due to ESD 2, 3 or 4 fault Instrument or relief valve fault	Leak rupture High noise level	Excess pressure damage to equipment Excess liquid carryover to drum or boom flare-back	Fire Explosion debris and missiles	Injury to personnel Damage to equipment Damage to structure	II or III				
Produced water separator	PV8	Salt water at high temperature	Leak rupture	Hot salt water released in module	Personnel in vicinity of leak Panic reaction by personnel	Steam/water burns Water damage to equipment Electrical faults	Injury to personnel Damage to equipment	II				

Fig. 3.6 Example of preliminary hazard analysis – gas compression system

3.7 Reliability and maintainability screening

Reliability and maintainability (R and M) screening based on flowcharts and process and instrumentation diagrams (P and IDs) is often carried out on process plant installations to identify the critical sub-systems for which more detailed quantitative analysis may be required. This screening generally covers the major sub-systems in the safety and production systems such as the chemical clean-up plant, automatic protective system, electric power generation, etc. The screening is carried out for each item by assigning a rating (1 = good, to 4 = bad) under different column headings which aim to express the main attributes of safety and/or availability. Since the number of components making up the plant is large it is conducted at the level of a logical grouping of parts – frequently corresponding to equipment items grouped by function on a P and ID sheet. The criteria used to screen each item are as follows:

1. Reliability: based on mean time between failures (MTBF). Rating 1 for MTBF > 10^6 h, rating 4 for < 10^3 h.
2. Maintainability: based on mean time to recovery (MTTR), including delays for spares, repairs at manufacturer's works, etc. Rating 1 for < 1 h, 4 for > 1 day.
3. Safety effect: based on effect on personnel/equipment safety if item fails to operate on demand – redundancy is ignored at this stage. Rating 1 = negligible hazard, 4 = potential catastrophic accident.
4. Hazard class: effect of failure of the item on personnel or system (worst case is always assumed). Rating 1 = safe, 4 = catastrophic.
5. Shutdown level: rating 1 = local shutdown, 2 = process shutdown, 3 = critical shutdown, 4 = evacuate the plant.
6. Effect on production: rating 1 = no effect, 4 = production shutdown, i.e. total shutdown of production facilities.
7. Redundancy: rating 1 = 100 per cent redundancy (either auto or manual), 2 and 3 = some degradation of performance, 4 = no redundancy.
8. Complexity: rating 1 = simple system, 4 = very complex system with a variety of controls, parts, interfaces with other systems, etc.
9. Environment: sensitivity of equipment R and M to operating environment, wind, rain, sun, salt spray, etc. Rating 1 = no effect, 4 = catastrophic effect.
10. Contamination: sensitivity of systems R and M to process contamination. Rating 1 = no effect, 4 = catastrophic effect.

EQUIPMENT OR SUB-SYSTEM	1 RELIABILITY	2 MAINTAINABILITY	3 EFFECT ON SAFETY	4 HAZARD CLASS	5 SHUTDOWN LEVEL	6 EFFECT ON PRODUCTION	7 REDUNDANCY	8 COMPLEXITY	9 ENVIRONMENT	10 CONTAMINATION	HAZARD SCORE (1–6 INCLUSIVE)	TOTAL SCORE
2.2 GAS COMPRESSION												
1. HP gas header	1	4	4	3	3	4	2	1	1	1	19	24
2. LP gas header	1	4	4	3	3	4	2	1	1	1	19	24
3. LP suction scrubber, vessel, piping, etc.	1	3	4	3	3	4	2	2	1	2	18	25
4. LPSS control and instrumentation	2	2	2	1	1	1	4	2	3	2	9	20
5. First stage compression	2	3	1	2	2	2	2	2	2	3	12	21
6. Compressor control and instrumentation	2	2	1	2	2	2	4	3	3	1	11	22
7. Aftercooler (HP comp.)	1	4	4	3	3	4	2	1	1	1	19	24
8. HPSS control and instrumentation	2	2	2	1	1	1	4	2	3	2	9	20
9. Second stage compression	2	5	1	2	2	2	2	2	2	3	14	23
10. Second stage comp. control and instrumentation	2	2	1	2	2	2	4	3	3	1	11	22
11. Gas cooling, piping, etc.	1	3	4	3	3	4	2	1	1	2	18	24
12. HP treated water cooling	1	2	1	1	2	1	2	2	2	2	8	16
13. KO drum	1	3	4	3	3	4	4	1	1	1	18	25
14. Gas heat exchanger	1	3	4	3	3	4	4	1	1	1	18	25
15. Gas chiller	1	3	4	3	3	4	4	1	1	1	18	25

Fig. 3.7 Reliability/maintainability screening – gas compression system

The sub-systems and ratings are entered on a standard R and M screening form. Hazard scores – the sum of columns 1 to 6 – and total impact scores are entered in the last two columns. Figure 3.7 shows an example of R and M screening applied to the production facilities of an offshore platform. The sub-systems scoring a total of 25 or more and others with hazard scores greater than 15 were subsequently selected for quantitative reliability analysis.

3.8 Summary

Qualitative methods are employed in the early stages of reliability and risk assessments. They provide a systematic approach to the identification and evaluation of hazards and a guide to their likely frequency of occurrence. The most comprehensive method is undoubtedly the HAZOP study, which has received wide acceptance by the process industries and regulatory authorities. Other techniques can be applied and may be particularly useful for existing plant. The choice will generally depend on the perceived hazard potential and available resources.

3.9 References

(1) Norwegian Petroleum Directorate Guidelines for Safety Evaluation of Platform Conceptual Design (1981).

(2) **Lees, F. P.** (1980) *Loss Prevention in the Process Industries*. Butterworths.

(3) *A Guide to Hazard and Operability Studies* (1977) Chemical Industries Association, London.

(4) **Guy, G. B.** (1981) Rapid ranking techniques. Safety and Reliability Society Inaugural Lecture, Southport.

Chapter 4

Failure Mode and Effects Analysis

4.1 Introduction

Failure mode and effects analysis (FMEA) is a step-by-step procedure for the systematic evaluation of the severity of potential failure modes in a system. The objective is to identify the items where modification to the design or the operating, inspection, or maintenance strategies may be required to reduce the severity of the effect of specific failure modes. It can be performed to meet a variety of different objectives, for example, to identify weak areas in the design, the safety-critical components, or critical maintenance and test procedures.

Criticality analysis (FMECA) is an extension of FMEA which aims to rank each potential failure mode according to the combined influence of its severity classification and probability of failure based on the best available data. The procedures for conducting the different types of FMEA are well described in the US Department of Defense MIL-STD-1629A (**1**) for military and aerospace systems. Andrews and Moss (**2**) discuss the method for application to process equipment including repairable systems. The basic techniques are described in the first part of this chapter and illustrated with examples.

Advances in the techniques which are of interest particularly with regard to repairable systems have also taken place since 1980 including a University of South Carolina research project XFMEA for the American Society of Automobile Engineers reported in AIR (1993) (**3**). These advances, and other methods featured in an Annual

Reliability and Maintainability Symposium tutorial by Bowles and Bonnell (**4**), are discussed in the second part of the chapter.

4.2 Procedure for performing an FMEA/FMECA

There are two basic approaches, known as functional and hardware FMEAs, described in MIL-STD-1629A.

1. The functional FMEA is normally used for complex systems in a top-down approach where the system is successively decomposed down to sub-system and equipment level depending on the information available and the objective of the analysis. Sub-assemblies/equipment are treated as 'black boxes' providing some required function in the system. The reliability engineer considers the effect of loss of inputs (e.g. essential supplies) and sub-assembly/equipment failures on the functional performance of the system. Its main application has been to identify and classify equipment according to the severity of their failure modes on system operation. It is frequently employed as a precursor to a more detailed FMECA or fault tree analysis.
2. The alternative is a hardware (bottom-up) approach where each component of the system is considered in isolation for each of its physical failure modes to establish the likely effect on system operation.

In both cases the analysis can be extended to include failure rates or probabilities as well as severities to become an FMECA – a failure mode and effects criticality analysis.

The functional FMEA has been adopted by reliability-centred maintenance practitioners, for example Nowlan and Heap (**5**) and Moubrey (**6**) but in a more restricted way than described in 1629A. Their objective is to identify the hidden failures which could inhibit functioning of the system.

The functional FMECA has also been applied for the analysis of industrial systems. For risk-based inspection and risk-based maintenance applications range estimates of failure frequencies are used with severity classifications based on outage time or cost to optimize inspection intervals or maintenance.

There are many variations on the basic themes. The selection of a particular approach depends on the intended use of the FMEA/FMECA and the data available at the time of the analysis. For industrial systems the procedure involves the following steps:

1. Define the system to be analysed and its required reliability performance.
2. Construct functional and reliability block diagrams (if necessary) to illustrate how the different sub-systems and components are interconnected.
3. Note the assumptions that will be made in the analysis and the definition of system and sub-system failure modes.
4. List the components, identify their failure modes and, where appropriate, their modal failure rates (alternatively failure rate ranges can be used).
5. Complete a set of FMEA worksheets analysing the effect of each sub-assembly or component failure mode on system performance.
6. Enter severity rankings and failure rates (or ranges) as appropriate on to the worksheets and evaluate the criticality of each failure mode on system reliability performance.
7. Review the worksheets to identify the reliability-critical components and make recommendations for design improvements or highlight areas requiring further analysis.

4.2.1 System definition

The system definition is a necessary step to ensure the reliability engineer understands the system and its function in the different operational modes before proceeding to an analysis of its component failure modes and effects. A description of operation in the different operational modes must be developed, clearly identifying the function of each equipment and its interfaces with other equipment and sub-assemblies in the system.

The functional description should contain a definition of system operation in each of its operational modes. The expected operational time in service and equipment utilization between major overhauls should be included with an assessment of the effect of operational and environmental stresses which may be imposed on the system.

4.2.2 Block diagrams

During system definition functional block diagrams will generally be constructed to show diagrammatically how the system breaks down into its various sub-systems, equipment, and components. These may need to be constructed for different indenture levels in a system hierarchy since a sub-system may need to be decomposed to its equipment, sub-assembly, or component level to ensure that the effect of each potential failure on system performance can be determined from the failure mode information available. For complex system FMECAs where there is equipment redundancy the functional block diagrams may need to be supported by reliability block diagrams or failure logic diagrams.

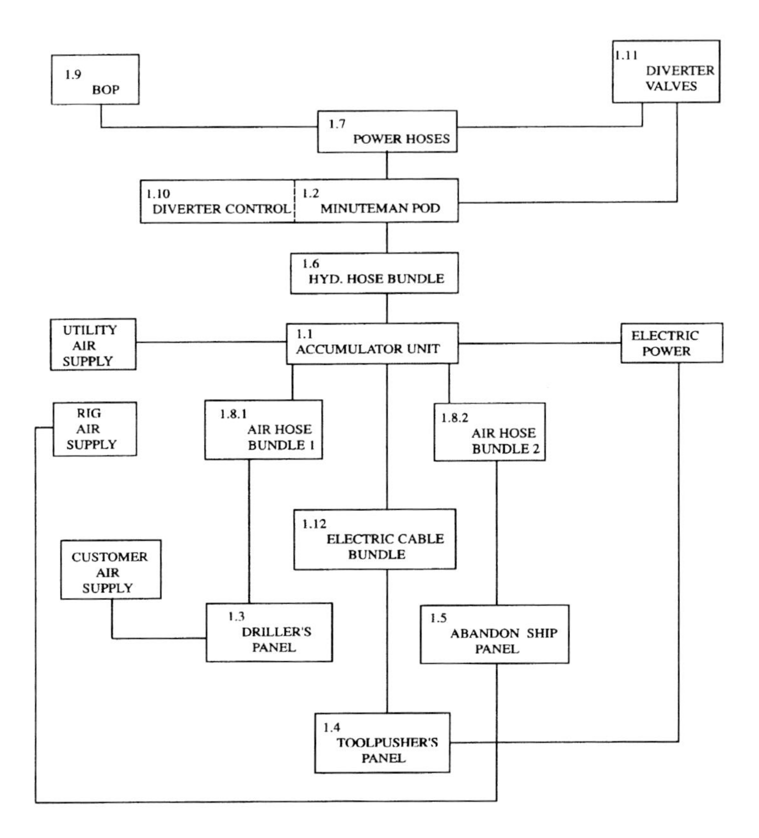

Fig. 4.1 Hydraulic control unit – functional block diagram

A functional block diagram of the hydraulic control unit of an offshore safety system is shown in Fig. 4.1. Each element in the block diagram could, if necessary, be broken down to the next hierarchical level. For the Accumulator Unit this would involve representing the different components of the Accumulator Unit, such as air and electrically actuated hydraulic pumps, pneumatically activated valves, etc., as a series of interconnected 'black boxes' within the Accumulator Unit boundary.

At the next lower hierarchical level each component may be further decomposed into its component parts. For the air-operated pumps this would require consideration of all component parts such as the body, piston, flapper valves, seals, etc. and the ways in which they could fail. The objective is to break down the system to a level at which the reliability engineer is confident that the failure information is available to allow prediction of the item's failure characteristics.

The hierarchical breakdown can be represented as shown in Fig. 4.2. Each box in the hierarchical diagram is identified by a code which indicates its function in the system. Here the Piston (1.1.1.2) is a component of the Air Pump (1.1.1) which in turn is a part of the Accumulator Unit (1.1). Thus, if failure data were only available at level 4, the reliability characteristics of the Air Pump (1.1.1) would be determined from consideration of the failure modes of the Air Pump components (1.1.1.1, 1.1.1.2, etc.) to establish the effects (in terms of the Air Pump failure modes) at the next hierarchical level. This continues at each successive Indenture Level to obtain a measure of the severity of the combined failure modes at the top level. If the analysis is an FMECA the component failure rates will also be combined to establish failure rates for the important failure modes.

4.2.3 Assumptions

It is necessary to make proper preparations before embarking on a failure mode and effects analysis. An important part of this preparation is the consideration and recording of the assumptions that will be made in the study. These assumptions will include definition of the boundaries of the system and each sub-system, equipment, etc., down to the expected level of analysis of the system. The operating states, failure rates, and failure modes at each level of analysis and the assumptions on any inputs or outputs across the boundaries which may affect system reliability performance should also be noted.

4.2.4 Reliability data

The level at which an analysis is to be carried out must be decided before the study commences. This will involve reviewing the failure information and deciding the level at which adequate data are available. When this has been agreed, a list of items with details of failure modes and, where appropriate, failure rates should be compiled for use in the analysis.

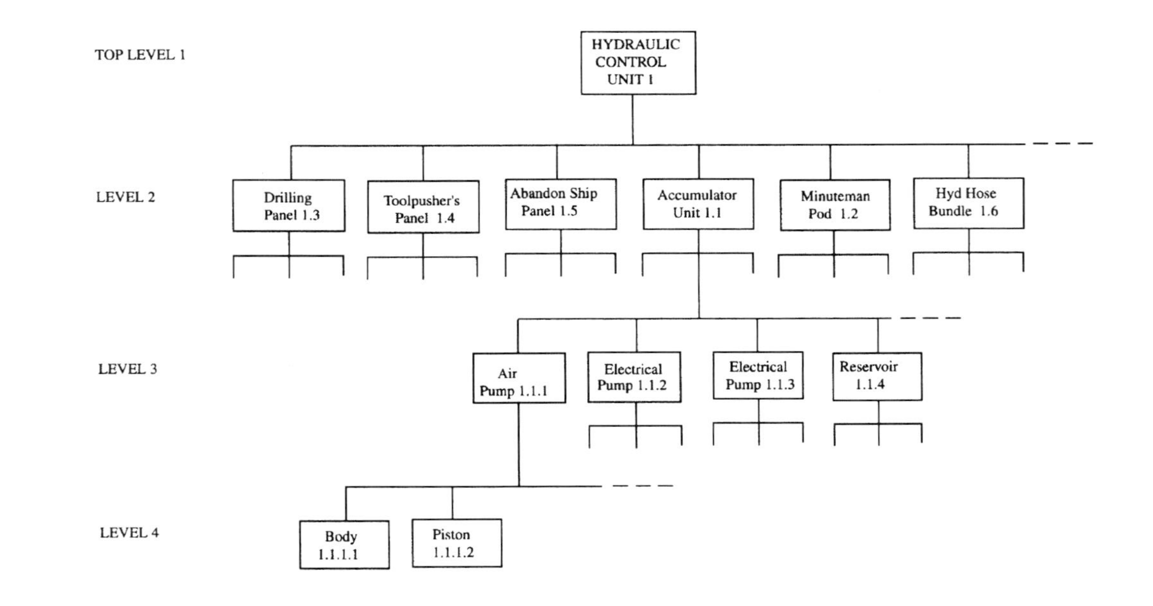

Fig. 4.2 System hierarchical breakdown

It is important to decide on the proportion of the overall failure rate to be allocated to each competing failure mode. This frequently requires the application of engineering judgement. These decisions should be made and recorded in the failure database before the study commences to ensure that a comprehensive audit trail is established at the start of the analysis. Preliminary decisions such as these will ensure that the analysis proceeds smoothly since it obviates the need to make arbitrary judgements during the course of the FMEA/FMECA when the analyst is concentrating on an engineering evaluation of each postulated failure mode.

4.2.5 FMEA worksheets

FMEA worksheets can take a variety of forms depending on the analysis requirements. They are tabular in format to foster a systematic approach to their completion and the review which is carried out at the end of each analysis.

An example of a typical FMEA worksheet format is shown in Fig. 4.3. On this form, Indenture Level is the level in the hierarchical breakdown of the system at which the FMEA has been carried out. The analysis progresses from component through equipment and subsystem to system indenture level. It is usual to produce individual sets of FMEA worksheets for each Indenture Level.

One page of the FMEA of the offshore safety system Hydraulic Control Unit carried out at Indenture Level 2 is shown in Fig. 4.4. Consider the records for the first failure mode of the Accumulator Unit. It can be seen that this failure mode has three outcomes each one has been considered separately with respect to its local effect (the effect on the Accumulator Unit) and its system effect (the effect on the Hydraulic Control Unit). Methods for detecting the occurrence of this failure mode and the compensating provisions for each outcome are also recorded and ranked according to severity.

When criticality analysis is applied in the FMEA the procedure is called failure mode and effects criticality analysis (FMECA). The FMEA worksheet is then extended beyond the severity column to include estimates of the loss frequency. A criticality number related to each failure mode and details of the data source from which these data were obtained may also be included on the worksheet.

FMEA WORKSHEET

SYSTEM: .... DATE: ....

INDENTURE LEVEL: .... SHEET: .... OF: ....

REF. DRAWING: .... ORIGINATOR: ....

OPERATING STATE: .... APPROVED: ....

IDENTIFICATION	FUNCTION	FAILURE MODE	FAILURE EFFECT		FAILURE DETECTION METHOD	COMPENSATING PROVISIONS	SEVERITY	REMARKS
			LOCAL EFFECT	SYSTEM EFFECT				

Fig. 4.3 FMEA worksheet

FMEA WORKSHEET

SYSTEM......Blowout Preventer .... DATE............

INDENTURE LEVEL ...2 – Hydraulic Control Unit ... SHEET ....1 ....OF ... ...4 ....

REF. DRAWING ......XYZ123 .... ORIGINATOR ....T R Moss ....

OPERATING STATE......Shallow Drilling............ APPROVED ....S T Naish ....

IDENTIFICATION	FUNCTION	FAILURE MODE	FAILURE EFFECT		FAILURE DETECTION METHOD	COMPENSATING PROVISIONS	SEVERITY	REMARKS
			LOCAL EFFECT	SYSTEM EFFECT				
1.1 Accumulator unit	Provides hydraulic power supply and converts electrical and pneumatic signals into hydraulic power output	1.1/1 Loss of utility air	(a) Pneumatically actuated valves fail to operate	(a) Control from toolpushers panel inhibited	(a) Diverter valve status lamp does not change over	(a) Normal operation is from driller's panel	3	System operation degraded
			(b) Loss of hydraulic pressure control	(b) Hydraulic output signals inhibited	(b) Hydraulic pressure alarm	(b) Pressure control can be switched to customer supply	3	System operation degraded
			(c) Air–driven hydraulic pump fails	(c) Control from driller's and toolpusher's panels inhibited	(c) Hydraulic pressure alarm	(c) Electric-driven pumps available	3	System operation degraded
		1.1/2						

Fig. 4.4 Example for one component of an offshore hydraulic control unit

4.3 Criticality analysis

Criticality is defined in MIL-STD-1629A as 'a relative measure of the consequences of a failure mode and its frequency of occurrence'. 1629A criticality analysis is then the procedure for ranking potential failure modes according to the combined influence of severity (the degree of damage that could ultimately occur) and the probability of occurrence or expected loss frequency. Severity ranks are defined in 1629A at four levels:

Level IV – Minor –	a failure not serious enough to cause injury, property damage or system damage but which will, however, result in unscheduled maintenance or repair.
Level III – Marginal –	a failure that may cause minor injury, minor property damage, or minor system damage which results in delay or loss of availability or mission degradation.
Level II – Critical –	a failure which may cause severe injury, property damage, or major system damage which will result in mission loss.
Level I – Catastrophic –	a failure which may cause death or weapon system loss (i.e. aircraft, missile, ship, etc.).

These definitions were modified by Andrews and Moss (**3**) for use in process plant FMECAs as follows:

Level 1 – Minor –	no significant effect.
Level 2 – Major –	some reduction in operational effectiveness.
Level 3 – Critical –	significant reduction of functional performance with an immediate change in system operating state.
Level 4 – Catastrophic –	total loss of system involving significant property damage, death of operating personnel and/or environmental damage.

These are the definitions most frequently used for industrial systems. Generally the assumption is that 'catastrophic' failures require a detailed safety assessment as defined, for example, by the COMAH regulations (**7**). However, severity and frequency ranges may need to be defined to meet a specific application.

Range estimates of failure probability can be used to rank probabilities of occurrence or, alternatively, item failure rates may be employed. Frequency ranges for process equipment are, typically:

1. Very low < 0.01 failures/year
2. Low 0.01 to 0.1 failures/year
3. Medium 0.1 to 1.0 failures/year
4. High > 1 failure/year

For component parts the units are often given in failures per million hours (f/Mh).

Where failure rate data are available, the base failure rate (λ_b) will need to be adjusted for environmental and duty stresses – typical stress factors are given in Tables 4.1 and 4.2. Further factors for the proportion of failures in the specified failure mode (α) and the conditional probability that the expected failure effect will result (β) are then applied to obtain the part failure mode rate (λ_o). This metric can be multiplied by the operating time (t) to generate a failure mode criticality number. Alternatively the failure mode rate can be used as a measure of frequency of occurrence.

Table 4.1 Environmental stress factors (k_1)

General environmental conditions	k_1
Ideal static conditions	0.1
Vibration-free, controlled environment	0.5
Average industrial conditions	1.0
Chemical plant	1.5
Offshore platforms/ships	2.0
Road transport	3.0
Rail transport	4.0

Table 4.2 Duty stress factors (k_2)

Component nominal rating (%)	k_2
140	4.0
120	2.0
100	1.0
80	0.6
60	0.3
40	0.2
20	0.1

Note: For equipment designed against a specific design code (e.g. vessels) nominal rating – 100 per cent. For other equipment (e.g. valves, pumps, etc.) a nominal rating of 80 per cent or less may be appropriate.

Another approach which is useful when criticality analysis is part of the design process is to rank frequencies as a proportion of the equipment or system failure rate or probability (F). The ranges generally employed for process systems are:

1. Very low probability $< 0.01F$
2. Low probability $0.01–0.1F$
3. Medium probability $0.1–0.2F$
4. High probability $> 0.2F$

The level of probability or failure rate is combined with the severity classification in a criticality matrix to provide a means for identifying and comparing the frequency of each failure mode to all other failure modes with respect to severity. The matrix is constructed by inserting failure mode identification numbers in the appropriate matrix frequency versus severity cell. The resulting matrix provides a model for setting corrective action priorities.

The procedure for criticality analysis can be followed from the FMECA worksheet in Fig. 4.5 of the lubrication system in the diesel engine unit of a warship.

FMECA WORKSHEET

SYSTEM ....Diesel Engine Unit ....

INDENTURE LEVEL...3 – Lubrication system

REF. DRAWING ......XYZ567 ....

OPERATING STATE ...Normal Operation ....

DATE ....

SHEET ....8 ....OF ....20..................

ORIGINATOR ....S T Naish................................

APPROVED ....T R Moss ....

IDENTIFICATION	FUNCTION	FAILURE MODE	FAILURE EFFECT		FAILURE DETECTION METHOD	COMPENSATING PROVISIONS	SEVERITY
			LOCAL EFFECT	SYSTEM EFFECT			
22.2 Oil heater	Maintain lube oil temperature	22.2/1 Heater unit failure	Violent foaming of oil on startup	Low lube oil temperature – fluctuating system pressure	Oil temperature gauge	Pre-start-up checks included oil temperature readings	3
		22.2/2 External leak	Loss of lube oil from system	Bearing or seal failure on diesel	Visual level indication on oil reservoir	Bearing temperature high alarm. Automatic S/D on HH temperature	3

FAILURE MODE	LOSS FREQUENCY			FAILURE MODE RATE λ_0	DATA SOURCE	REMARKS
	λ_b	α	β			
22.2/1	73	0.4	1.0	29.2 f/Mh	HARIS Reliability Data Handbook	Heaters maintain lube oil temperature during S/D
22.2/2	73	0.6	0.3	13.1 f/Mh		

Fig. 4.5 FMECA – example for one failure mode of a diesel engine unit FMECA

In this figure the α and β factors were determined (as is usual) by engineering judgement. The criticality matrix in Fig. 4.6 shows the distribution of failure modes with respect to frequency and severity for the diesel engine unit of which the lubrication system FMECA in Fig. 4.5 was part.

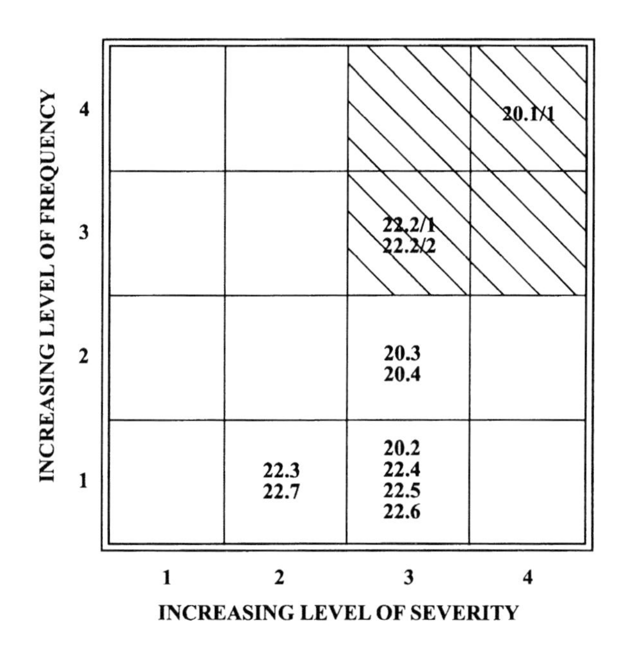

Notes:

(a) Failure modes located in the shaded area of the matrix are arbitrarily defined as unacceptable for this example, requiring further engineering investigation and correction.

(b) This matrix example was based on the relevant worksheets.

Fig. 4.6 Criticality matrix for diesel engine unit

4.4 Functional and hardware FMEA/FMECA examples

4.4.1 General

The FMEA/FMECA examples in this section were part of the case study shown in Chapter 14, concerning the Business Interruption Risk Assessment of a 300 MW combined cycle electricity generation plant. The analysis proceeds in a series of steps from a functional FMEA of the main plant items through a functional FMECA of the cooling water train and a hardware FMECA of an electric motor driven centrifugal pump set. The procedure follows the steps discussed previously in Section 4.2.

4.4.2 System definition

The plant is a 300 MW combined cycle electricity generating unit employing two 100 MW gas turbines and a 100 MW steam turbine. To aid analysis the plant was initially partitioned into five systems defined as:

System 1: gas turbine train 1 – comprising a gas turbine, electricity generator and transformer.

System 2: gas turbine train 2 – also with a gas turbine, electricity generator and transformer.

System 3: steam turbine train – comprising the steam turbine, electricity generator, and the two heat recovery steam boilers.

System 4: the feedwater train – comprising the feedwater heater boiler feed pumps, condenser and condensate pumps.

System 5: the cooling water train – which is the system analysed here in more detail.

4.4.3 Block diagrams

Functional block diagrams of the plant at system, sub-system, and equipment inventory levels are shown in Fig. 4.7. The boundaries are defined by the dotted lines around the functional blocks. System 5 supplies coolant to the condenser of the combined cycle electricity generation plant.

4.4.4 Assumptions

Essential inputs and output are shown by the arrows across the boundaries. For System 5 the inputs shown at Indenture Level 3 are returned coolant and electricity supply. The output is pressurized water to the condenser.

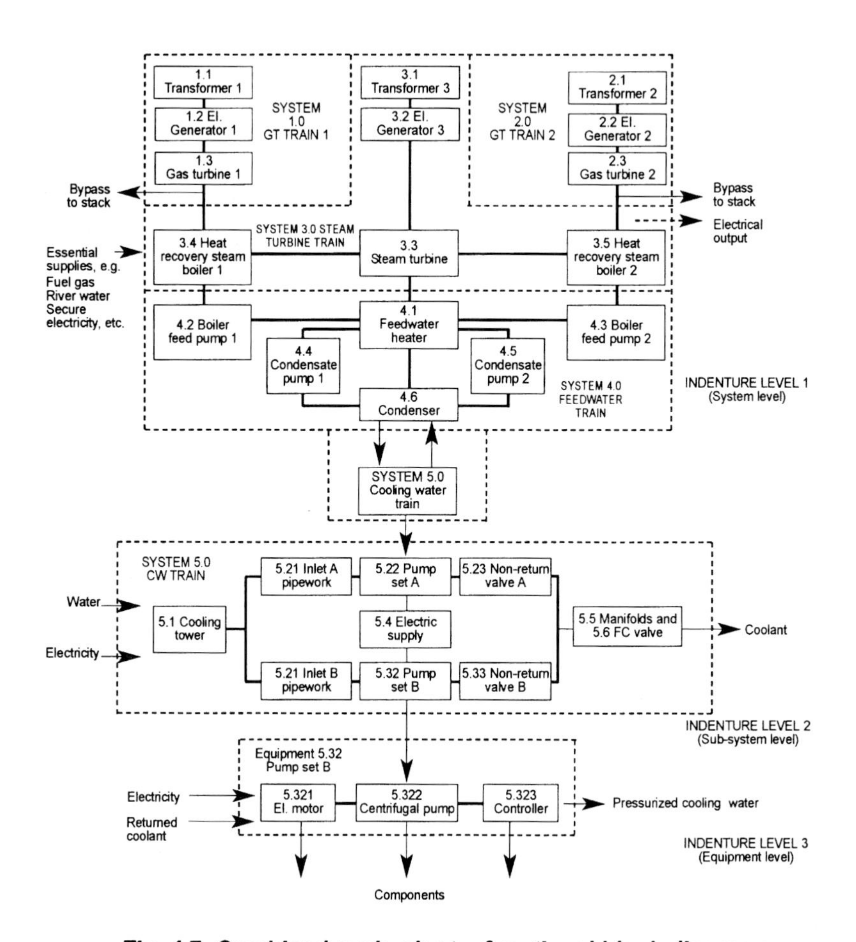

Fig. 4.7 Combined cycle plant – functional block diagrams

The operational mode for System 5 is 'Continuous Operation' with three defined operating states: 100 per cent output, 50 per cent output, and zero output. One hundred per cent output is assumed to be required to assure the design output from the combined cycle plant. Loss of one pump stream reduces the output to the condenser to 50 per cent. Total loss of output from System 5 reduces the output to zero. The failure logic for System 5 Operating State 3 (zero output) is shown in Fig. 4.8. However, it should be noted that total loss of output from System 5 does not result in total loss of output from the plant because the exhaust from both gas turbine trains can be diverted to stack. The feedwater heater can also be bypassed.

4.4.5 Reliability data

Frequency ranges are the same as proposed in Section 4.3, namely:

1. Very low < 0.01 failures/year
2. Low 0.01–0.1 failures/year
3. Medium 0.1–1.0 failures/year
4. High > 1 failure/year

Severity ranges have been modified to suit this particular analysis as follows:

Severity 1 – no effect.
Severity 2 – reduced output.
Severity 3 – no output.
Severity 4 – major equipment damage.

Failure rate range estimates used in the functional FMEA of System 5 were based on engineering judgement. Best-estimates of failure rates for the equipment in the system could be derived from the data sources discussed in Chapter 12 or from more detailed hardware FMECAs such as the example shown here for the electric motor driven centrifugal pump set. The use of failure rate ranges as against best-estimate failure rates depends on the objective and the data sources available. Where in-house sources of failure data for specific failure modes are available, these are generally to be preferred to failure rate ranges. For the hardware FMECA of the pump set the data published by Bloch and Geitner (**8**) were used.

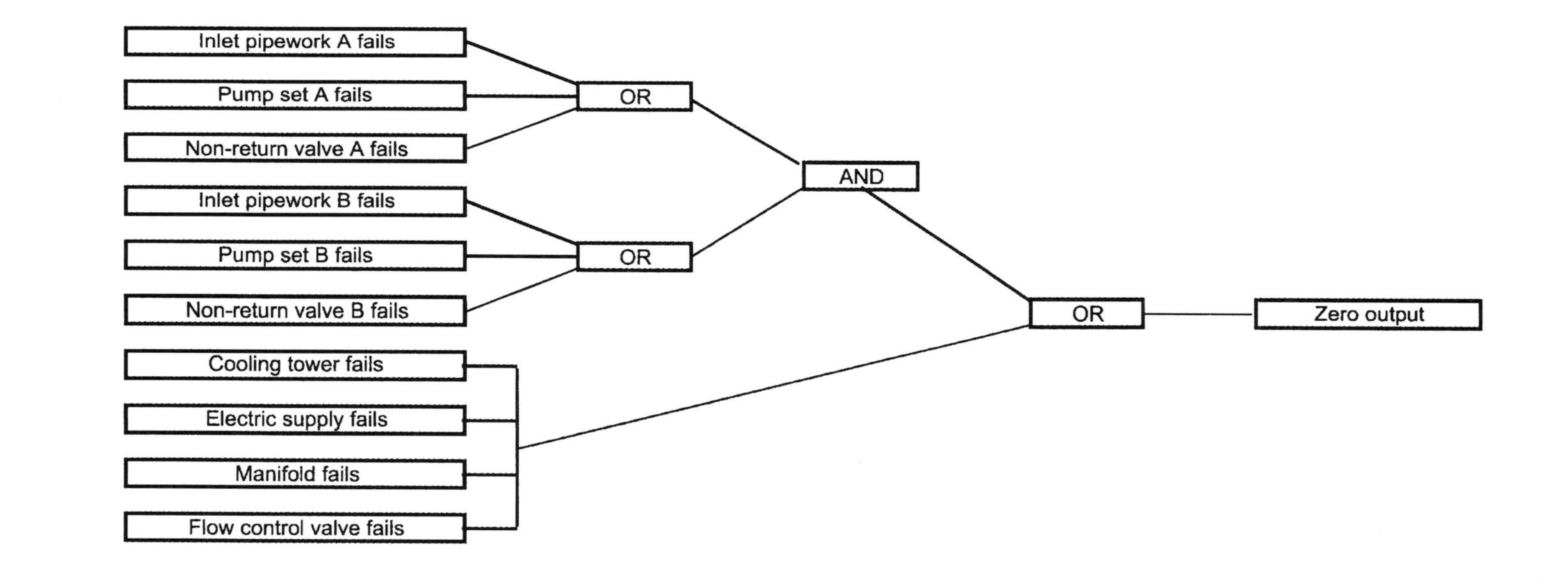

Fig. 4.8 System 5 – failure logic diagram for 'Operating State 3 – zero output'

4.4.6 Functional FMEA/FMECA worksheets

The FMEA worksheet for the functional FMEA of the combined cycle plant at Indenture Level 1 is shown in Fig. 4.9.

This worksheet considers the major items of equipment in each train and ranks each failure mode by severity. The analysis is carried out on the assumption that all equipment is functioning correctly except for the failure mode being analysed, to determine its likely effect on system performance and its severity. It can be seen that the aim of this first-level FMEA is to identify which equipment failure modes are likely to reduce the electrical output from the combined cycle plant.

The functional FMECA example at the next indenture level is concerned with System 5 – the cooling water train which provides coolant for the condenser in System 4 – the feedwater train. This system was chosen because it is relatively simple, since the objective is to illustrate the method. In practice the high severity equipment would be analysed as the first priority. The functional FMECA worksheet for System 5 at Indenture Level 2 is shown in Fig. 4.10.

SYSTEM: Combined cycle electricity generation plant
INDENTURE LEVEL: 1
REF. DRAWING: 123
OPERATING STATE: Full production

DATE:
SHEET: 1 of 1
ORIGINATOR: TR Moss
APPROVED: JD Andrews

Identity/description	Function	Failure mode	Failure effect		Severity	Remarks
			Local effect	System effect		
1.0/2.0 Gas turbine trains						
1.1/2.1 Transformers	Converts generator output to grid input	No output	No electrical output from GT train	Reduced electrical output to grid	2 or 3	Plant electrical output reduced to 50% on loss of one GT train
1.2/2.2 Electric generators	Mechanical to electrical energy conversion	No output	No electrical output from GT train	Reduced electrical output to grid	2 or 3	If both GT trains lost, plant output reduced to zero
1.3/2.3 Gas turbines	El. generator drive	Failed while running	No output from GT train	Reduced electrical output to grid	2 or 3	
3.0 Steam turbine train						
3.1 Transformer	Converts generator output to grid input	No output	No output from steam train	Reduced electrical output to grid	2	
3.2 El. generator	Mechanical to electrical conversion	No output	No output from steam train	Reduced electrical output to grid	2	Plant electrical output reduced to 66% – GTs can exhaust to stack
3.3 Steam turbine	El. generator drive	Failed while running	No output from steam train	Loss of ST train electrical output	2	
3.4/3.5 Heat recovery steam boilers	Raises steam from GT exhaust	No output	Reduced output from steam turbine train	Reduced output from ST train	2	Plant electrical output reduced to 80% – associated GT can exhaust to stack
4.0 Feedwater train						
4.1 Feedwater heater	Increases condensate temperature	No heated condensate	Reduced steam output from associated HRSBs	Small reduction in output from steam train	1	Negligible effect on plant output in the short term
4.2/4.3 Boiler feeder pumps	Feedwater supply to HRSBs	No output	Reduced steam output from associated HRSB	Reduced electrical output from steam train	2	Plant electrical output reduced to 80% – associated GT can exhaust to stack
4.4/4.5 Condensate pumps	Condensate feed from condenser via FWH	No output	Reduced steam output from associated HRSB	Small reduction in output from steam train	1	Negligible effect on plant electrical output in short term – alternative make-up of cold demin. water available
4.6 Condenser	Converts wet steam to solid water	Condensing failure	Wet steam input to BFPs	No output from steam train	2	Plant electrical output reduced to 66% – potential damage to BFPs
5.0 Cooling water train	Recircuiting raw water to condenser	No output	Cooling for condenser lost	No output from steam train	2	Plant electrical output reduced to 66% – see functional FMECA of system

Fig. 4.9 Functional FMEA example – combined cycle plant – Indenture Level 1

SYSTEM: Combined cycle electricity generation plant
INDENTURE LEVEL: 2
REF. DRAWING: XYZ123/1
OPERATING STATE: Full production

DATE:
SHEET: 1 of 1
ORIGINATOR: TR Moss
APPROVED: JD Andrews

Identity/description	Function	Failure mode	Failure effect		Severity	Frequency	Remarks
			Local effect	System effect			
5.1 Cooling tower	Water supply	Low level	Loss of inlet water supply to both pumps – possible pump damage	Loss of cooling water output	3	1	No cooling water supply and possibility of pump damage
5.21/5.31 Inlet pipes A and B	Input water conduits to pump inlet	Major leak	Loss of water supply to pump – possible pump damage	Reduced output of cooling water	2 or 3	2	Cooling water output reduced to 50% – both pumps need to fail for total loss
5.22/5.32 Pump sets A and B	Increases pressure/flow of cooling water	Fail while running	Loss of cooling water supply from pump	Reduced output of cooling water	2 or 3	4	Cooling water output reduced to 50% – both pumps need to fail for total loss
5.23/5.33 Non-return valves A and B	Prevents reverse flow	a) Fail open	No effect unless other pump failed	No effect during operation	1	2	Possible reverse flow through failed pump if mismatch of pump output pressures
		b) Fail closed	No output from pump	Reduced output of cooling water	2 or 3	3	Cooling water output reduced to 50% – both pumps need to fail for total loss
5.4 Electric supply to pump sets A and B	Power supply to pump motors A and B	Open circuit	Loss of cooling water supply from pump	Loss of cooling water output	3	2	No cooling water supply
5.5 Manifold/outlet pipework	Distributes high pressure water flow to flow control valve	Major leak	No output to flow control valve	Loss of cooling water output	3	2	No cooling water supply
5.6 Flow control valve	Controls delivery of cooling water	a) Fail open	No effect	No effect during operation	1	2	Possible increase in cooling water supply
		b) Fail closed	No output from flow control valve	Loss of cooling water output	3	3	No cooling water supply

Fig. 4.10 Functional FMECA example – cooling water system – Indenture Level 2

The standard FMEA worksheet shown in Fig. 4.3 has been modified by adding a column for failure mode frequency and deleting the columns for 'Failure Detection Method' and 'Compensating Provisions' to make it suitable for this functional FMECA. At this indenture level the failure modes become more specific and the system effect for each failure mode relates directly to the next highest indenture level – Indenture Level 1 failure modes. The failure mode identities from the System 5 FMECA are now entered into the criticality matrix shown in Fig. 4.11.

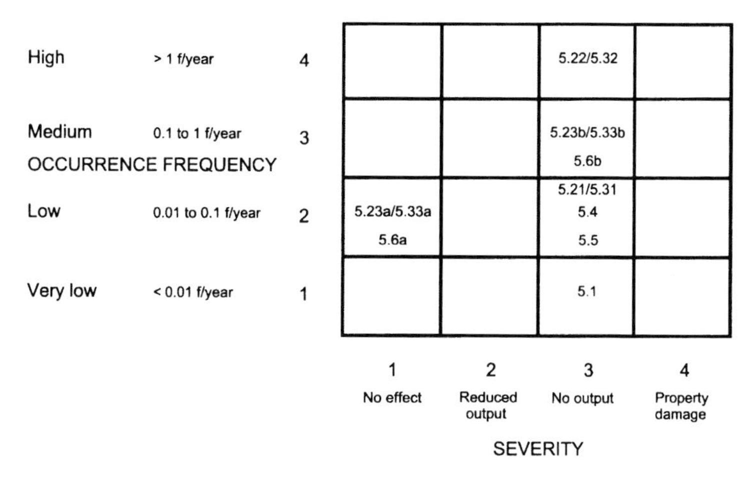

Fig. 4.11 Functional FMECA example – cooling water system – criticality matrix – Indenture Level 2

Ranking of the failure modes in the criticality matrix is based on worst-case severities so there are no entries for Severity 2 – the reduced output category. The high-frequency failure mode for the pump sets 'fail while running' is clearly the first priority arising from this FMECA; this sub-system therefore becomes the subject of the hardware FMECA at Indenture Level 3.

It can be seen that this hardware FMECA (Fig. 4.12) is significantly more specific with respect to failure modes and their likely effects. An additional column for 'Failure Detection' has also been included since early detection of a failure during operation may reduce the severity of the failure. The base failure rates (λ_b) were obtained from data published by Bloch and Geitner (**8**). These data are modified by the α and β metrics for the estimated proportion of the base failure rate and the conditional probability that the specified failure mode would cause a failure of the expected level of severity. The sum of these predicted failure rates (λ_o) provides estimates for the 'Forced-Outage' failure rates for the pump, motor, and controller which are aggregated to give an estimated failure rate for the pump set.

An alternative to the criticality matrix is to rank failure modes by their failure rates at each severity level. The rankings of failure modes by criticality and rate are given below:

1. Severity = catastrophic

5.3221	Pump casing crack/fracture	Negligible
5.3211	Motor frame crack/fracture	Negligible

2. Severity = critical

5.3213	Exciter open/short circuit	8 f/Mh
5.3216	Rotor windings open/short circuit	5 f/Mh
5.3231	Instrumentation open/short circuit	3 f/Mh
5.3217	Connections open/short circuit	2 f/Mh
5.3212	Stator open/short circuit	1 f/Mh
5.3228	Oil seals	0.16 f/Mh
5.3222	Pump shaft	0.02 f/Mh
5.3214	Motor shaft	0.02 f/Mh
5.3227	Bearing housing	Negligible
		19.2 f/Mh

3. Severity = major

$\Sigma\lambda_{major}$=16.54 f/Mh

Thus the hardware FMECA predicts an overall failure rate for this equipment of 91.45 f/Mh (0.8 f/year) and a forced outage rate (Severity = 'Catastrophic' + 'Critical' + 'Major') of 35.74 f/Mh (0.3 f/year).

The probability of a 'Catastrophic' failure of a pump set is clearly very small. The majority of 'Critical' failures result from sudden failures in the electric motor. This prediction compares well with failure rates based on large samples published in the Swedish Tbook. **(9)** and OREDA **(10)** – centrifugal pump samples of about 40 f/Mh.

These examples conclude the description of the basic methods used in FMEAs and FMECAs. The following sections review some other methods which extend the methodology beyond the 1629A approach.

INDENTURE LEVEL: 3
REF. DRAWING: XYZ123/1/A
OPERATING STATE: Continuous operation

DATE:
SHEET: 1 of 1
ORIGINATOR: TR Moss
APPROVED: JD Andrews

Description	Function	Failure mode	Failure effect		Failure detection	Severity	Loss frequency				Remarks
			Local effect	System effect			λ_b f/Mh	α %	β% Prob	λ_0 f/Mh	
5.32 PUMP SET											
5.322 PUMP											
5.3221 Casing	Containment	Crack/fracture	External leak	Forced outage	Visual	Catastrophic	0.01	100%	20%	Negligible	Small chance of fracture
5.3222 Pump shaft	Power transmission	Crack/shear	Vibration/shear	Forced outage	Noise	Critical	0.1	100%	20%	0.02	Low probability of shear
5.3223 Impeller	Energy conversion	Erosion/wear	Vibration	Degraded performance	Noise/instrumentation	Major	0.1	100%	20%	0.02	Some performance degradation should be tolerable
5.3224 Wear rings (2)	Sealing	Wear	Internal leak	Degraded performance	Instrumentation	Major	2 × 0.1 = 0.2	100%	10%	0.02	Ditto
5.3225 Stuffing box	Sealing	Wear	External leak	Degraded performance	Visual	Major	25	100%	50%	12.5	Small leaks probably rectified by local action
5.3226 Bearings (2)	Power transmission	Wear	Vibration	Degraded performance	Noise	Major	2 × 4 = 8	100%	20%	1.6	Degradation rates predictable
5.3227 Bearing housing	Containment	Crack/fracture	Vibration/seize	Forced outage	Noise/temperature	Critical	0.01	10%	10%	Negligible	Small chance of fracture
5.3228 Oil seals (2)	Sealing	Wear	Vibration/seize	Forced outage	Visual/temperature	Critical	2 × 8 = 16	10%	10%	0.16	Major leaks unlikely between services
5.3229 Drive coupling	Power transmission	Wear/misalignment	Vibration	Degraded performance	Noise	Major	8	100%	20%	1.6	Some wear/misalignment can be tolerated
5.32210 Baseplate	Structural support	Subsidence	Vibration	Degraded performance	Visual/noise	Major	0.01	100%	10%	Negligible	Small chance of subsidence
					Sub-total (all failures)		57.43 f/Mh	Forced Outage rate		15.92 f/Mh	
5.321 ELECTRIC MOTOR											
5.3211 Frame	Structural support	Crack/fracture	Vibration	Forced outage	Visual/noise	Catastrophic	0.01	100%	20%	Negligible	Small chance of fracture
5.3212 Stator	Energy conversion	Open/short circuit	Power loss	Forced outage	None	Critical	1	100%	100%	1	Possible sudden failure
5.3213 Exciter	Energy conversion	Open/short circuit	Power loss	Forced outage	None	Critical	8	100%	100%	8	Possible sudden failure
5.3214 Motor shaft	Power transmission	Crack/fracture	Vibration	Forced outage	None	Critical	0.01	100%	20%	0.02	Small chance of fracture
5.3215 Bearings (2)	Power transmission	Wear	Vibration	Degraded performance	Noise	Major	5 + 3 = 8	100%	10%	0.8	Degradation rates predictable
5.3216 Rotor windings	Energy conversion	Open/short circuit	Power loss	Forced outage	None	Critical	5	100%	100%	5	Possible sudden failure
5.3217 Electrical connections	Control	Open/short circuit	Power loss	Forced outage	None	Critical	2	100%	100%	2	Possible sudden failure
					Sub-total (all failures)		24.02 f/Mh	Forced Outage rate		16.82 f/Mh	
5.323 CONTROLLER											
5.3231 Instrumentation	Control	Open/short circuit	Power loss	Forced outage	None	Critical	10 f/Mh	100%	30%	3 f/Mh	Possible sudden failure
PUMP SET					TOTAL (all failures)		91.45 f/Mh	Forced Outages		35.74 f/Mh	

Fig. 4.12 Hardware FMECA example – cooling water pump set – Indenture Level 3

4.5 Multi-criteria Pareto ranking

As can be seen from the criticality matrix in Fig. 4.6, criticality increases from the bottom left-hand corner of the matrix to the upper right-hand corner. Bowles and Bonnell (4) proposed a method for prioritizing the failure modes using a multi-criteria ranking procedure since they claim it is generally not possible to define a meaningful cost function which combines severity and probability of occurrence into a single number which can then be ranked. The multi-criteria Pareto ranking aims to do this in a rational manner.

Consider the criticality matrix in Fig. 4.6; it can be seen that the failure mode in cell 4/4 has the highest probability of failure and the highest severity. This failure mode is said to be non-dominated because there are no failure modes that rank higher on both criteria.

The multi-criteria Pareto ranking procedure operates as follows: first all non-dominated failure modes are identified and labelled as rank 1 failure modes. These item failures are then removed from the matrix and all the non-dominated failure modes of the ones that are left are identified and labelled as rank 2 failures. The procedure continues in this way, analysing the item failure modes, identifying all non-dominated item failures, ranking them, removing them from the matrix and then analysing and ranking the remaining item failures until no item failure modes are left on the matrix. If the item severity and probability of occurrence cover an interval rather than a single point the entire interval should be considered when determining whether one item dominates another.

Every rank 2 failure mode has a lower probability of occurrence than at least one rank 1 failure mode; thus rank 1 item failure modes are given a higher priority than rank 2 failure modes. Rank 2 failure modes have a higher priority than rank 3 and so on. Within each rank the item failures may be sorted further on the basis of their severity and probability of occurrence. Otherwise they may be prioritized on other (possibly subjective) factors such as the difficulty of making the necessary corrective actions, time to take the corrective action or cost (in terms of money, reputation, or goodwill) associated with the failure effect.

The ranking process clearly forms the basis for continuous improvement of the product. Once the first rank item failures have been addressed the second rank failures move up to the first rank. Design changes introduced to eliminate or mitigate the effect of the first rank failures may introduce new failure modes but hopefully these will have lower severities and probabilities of occurrence. In general all item failure modes that represent important safety concerns or potentially hazardous conditions must be resolved by appropriate design changes, regardless of their probabilities of occurrence, before the product can be safely used.

4.6 Common cause screening

Once the high priority failure modes have been identified a common cause screening matrix can be constructed for associating the high priority concerns to their causes. This matrix identifies the causes that are common to more than one of the most critical failure modes. Figure 4.13 shows an example of a common cause screening matrix for a centrifugal pump.

	Common cause								
Pump components	Fracture	Fatigue	Wear	Creep	Corrosion/oxidation	Deposits/fouling	Flow blockage	Impact damage	Alignment problems
P1 Casing	X								
P2 Pump shaft		X							X
P3 Impeller			X			X	X		X
P4 Wear rings (2)			X						
P5 Stuffing box			X						
P6 Bearings			X						X
P7 Bearing housing		X							X
P8 Oil seals (2)			X						X
P9 Drive coupling			X						X
P10 Baseplate									X

Fig. 4.13 Common cause screening matrix (pump only)

From this matrix it can be seen that wear and alignment problems are likely to dominate the reliability performance of the pump. These are areas which should be targeted for inspection and planned maintenance to reduce the probability of failure.

4.7 Matrix method

The matrix method developed initially by Babour (**11**) for electronic systems is also covered by Bowles and Bonnell (**4**). It is a more structured and systematic approach than the FMEA method described in MIL-STD-1629A and provides more efficient accountability and traceability than is obtained by other methods. Its main use is for electronic/instrumentation system FMEAs but it could have applications for tracing important (for example, dependent) failure modes through a complex mechanical system assessment. However, it is not, perhaps, the most suitable method for criticality analysis.

Figure 4.14 taken from Babour's original paper shows the basic FMEA matrix format.

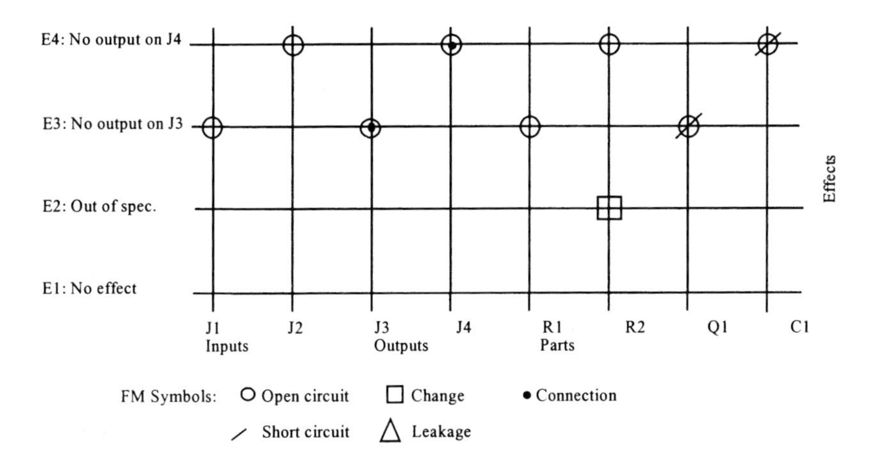

Fig. 4.14 Basic matrix FMEA block diagram

Internal components and inputs and outputs to the module are represented by vertical lines. Failure effects caused by failure of components and loss of input or output are shown as horizontal lines. Component failure modes that lead to a given effect are represented at intersections of the matrix by symbols.

For example, the circle at the intersection of R2 and E4 represents an open circuit failure mode of resistor R2 which results in loss of output from connection J4. The small dots (e.g. at the intersections of E4 and J4) show which output signals are affected by which failure effects. As can be seen from Fig. 4.14 the symbols can be overlaid when different failure modes have the same effect.

The matrix method is applied by analysing the components at each level of the hierarchical system structure in a bottom-up approach. First the inputs, outputs, and components comprising the first level modules are analysed as in the Circuit 1 FMEA of Fig. 4.15.

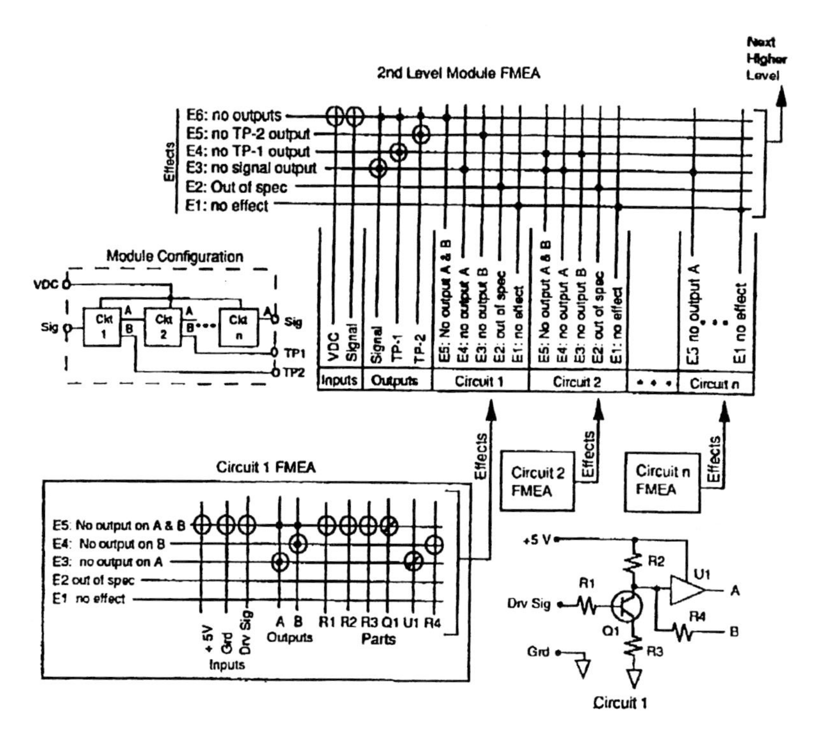

Fig. 4.15 Build-up of failure effects using the matrix method

The second level modules are then analysed. The inputs and outputs from the lower level modules which feed into the next higher level modules are listed on the higher level matrix. Instead of showing the module components, the failure effects from the first level module analysis are copied as vertical lines, and a small dot is used to show which second level effects are caused by these first level effects. This build-up is shown in Fig. 4.15. It can be seen that the open circuit in R1 of Circuit 1 becomes 'no outputs' at the second level module. The build-up of failure effects from one level to the next continues until the highest system level is reached. Combinations of lower level matrices and component analyses can be included at any level as required by the system configuration.

Although the matrix method is particularly aimed at electronic systems it has possibilities for mechanical system assessments, for example for identifying and mapping degradation mechanisms from component through to system and plant levels. The matrix FMEA may also have applications in maintainability analysis, to provide inputs for diagnostic testing and in fault tree development.

4.8 Risk priority number method of FMECA

A criticality assessment is an attempt to identify and prioritize failure mode importance as a function of the severity of the effects at system level and the probability of occurrence. To this end the automotive industry has developed an alternative to the 1629A FMECA, known as the risk priority number (RPN) method. This method is also employed by some aerospace companies. In the RPN methodology the parameters used to determine the 'criticality' of the item failure mode are its frequency of occurrence and the severity of the failure effect, as in the 1629A approach. However, another parameter is also included – the likelihood that subsequent testing, during design, will detect that the potential failure mode could actually occur. Tables 4.3, 4.4, and 4.5 show the linguistic variables proposed by Bowles and Bonnell (**4**) to describe these parameters.

Table 4.3 Frequency of occurrence criteria

Rank	Occurrence	Failure probability
1		< 1 in 10^6
2	Low	1 in 20 000
3		1 in 4000
4		1 in 1000
5	Moderate	1 in 400
6		1 in 80
7		1 in 40
8		1 in 20
9	High	1 in 8
10		> 1 in 2

Table 4.4 Severity evaluation criteria

Rank	Occurrence	Severity effect
1	Low	Unreasonable to expect any real effect on system performance
2,3		Slight deterioration of system performance
4,5,6,7	Moderate	Noticeable degradation of system performance
8,9	High	Change in operating state but does not involve safety or government regulations
10		Affects safety of personnel, the plant, and/or environment

Table 4.5 Detectability evaluation criteria

Rank	Detectability	Meaning
1	High	Almost certain to detect design weakness
2,3		Good chance of detecting design weakness
5,6,7	Moderate	May not detect design weakness
8,9	Low	Design weakness probably not detected
10		Very unlikely to detect design weakness

Occurrence is ranked according to failure probability, which represents the relative number of failures anticipated during the design life of the item. Table 4.3 describes the range of values and the linguistic terms used to rank the frequency of the failure mode occurrence. Severity is ranked according to the seriousness of the failure mode effect on the next higher level assembly, the system, or the user. The effects of a failure mode are normally described by the effects on the user of the product or as they would be seen by the user. For example, some common failure effects for automobiles are excessive noise, intermittent operation, impaired control, and rough ride. Table 4.4 shows the criteria used to rank the severity of the failure effects. Detectability is an assessment of the ability of a proposed design verification programme to identify a potential weakness before the component or assembly is released to production. Table 4.5 shows the evaluation criteria used by the automotive industry for the rankings and the corresponding linguistic terms. The RPN method is illustrated in the following example.

The RPN FMECA worksheet for the electric motor driven centrifugal pump set considered previously is shown in Fig. 4.16.

It can be seen that the first six columns are identical to the traditional 1629A worksheet. Engineering judgement estimates for occurrence frequency, severity, and detectability were taken from the appropriate tables and entered into the next three columns. These estimates are then multiplied together to obtain the risk priority number (RPN) for each failure mode. The last column of Fig. 4.16 gives a criticality ranking based on the RPN. The 'Top five' in the RPN prioritized list are all associated with the electric motor and controller, as shown in Table 4.6.

Table 4.6

Criticality rank	RPN	Equipment	Component	Failure mode
1	400	Electric motor	Exciter	Open/short circuit
2	336	Controller		Open/short circuit
3	288	Electric motor	Rotor windings	Open/short circuit
4	240	Electric motor	Stator	Open/short circuit
5	216	Electric motor	Electrical connections	Open/short circuit

The comparable list from the hardware FMECA in Fig. 4.12 is very similar, with the exception of the pump stuffing box which has the highest failure rate but is only ranked 9 in the RPN list; however, the severity rank for the stuffing box failure mode is 'Major', rather the more serious 'Critical'.

SYSTEM: Combined Cycle Electricity Generation Plant
INDENTURE LEVEL: 3 - Systems, cooling water pump set
REF. DRAWING: YXZ123/1/A
OPERATING STATE: Continuous operation

DATE:
SHEET: 1 of 1
ORIGINATOR: TR Moss
APPROVED: JD Andrews

Description	Function	Failure mode	Failure effect		Failure detection	Frequency	Severity	Detectability	RPN	Remarks	Criticality rank
			Local effect	System effect							
PUMP SET											
PUMP											
P1 Casing	Containment	Crack/fracture	External leak	Forced outage	Visual	1	3	3	9	Small chance of failure	18
P2 Pump shaft	Power transmission	Crack/shear	Vibration/shear	Forced outage	Noise	1	3	3	9	Low probability of shear	17
P3 Impeller	Energy conversion	Erosion/wear	Vibration	Degraded performance	Noise/instrumentation	1	4	3	12	Some performance degradation should be tolerable	16
P4 Wear rings (2)	Sealing	Wear	Internal leak	Degraded performance	Instrumentation	2	3	7	42	Ditto	12
P5 Stuffing box	Sealing	Wear	External leak	Degraded performance	Visual	10	3	3	90	Small leaks probably rectified by local action	9
P6 Bearings (2)	Power transmission	Wear	Vibration	Degraded performance	Noise	5	5	4	100	Degradation rates predictable	7
P7 Bearing housing	Containment	Crack/fracture	Vibration/seize	Forced outage	Noise/temperature	1	3	5	15	Small chance of fracture	15
P8 Oil seals (2)	Sealing	Wear	Vibration/seize	Forced outage	Visual/temperature	7	3	4	84	Major leaks unlikely between services	10
P9 Drive coupling	Power transmission	Wear/misalignment								Be tolerated	6
P10 Baseplate	Structural support	Subsidence	Vibration	Degraded performance	Visual/noise	1	3	9	27	Small chance of subsidence	13
ELECTRIC MOTOR											
M11 Frame	Structural support	Crack/fracture	Vibration	Forced outage	Visual/noise	1	3	9	27	Small chance of fracture	14
M12 Stator	Energy conversion	Open/short circuit	Power loss	Forced outage	None	3	8	10	240	Possible sudden fracture	4
M13 Exciter	Energy conversion	Open/short circuit	Power loss	Forced outage	None	5	8	10	400	Possible sudden fracture	1
M14 Motor shaft	Power transmission	Crack/fracture	Vibration	Forced outage	None	1	6	9	54	Small chance of fracture	11
M15 Bearings (2)	Power transmission	Wear	Vibration	Degraded performance	Noise	5	5	4	100	Degradation rates predictable	8
M16 Rotor windings	Energy conversion	Open/short circuit	Power loss	Forced outage	None	4	8	9	288	Possible sudden failure	3
M17 Electrical connections	Control	Open/short circuit	Power loss	Forced outage	None	3	8	9	216	Possible sudden failure	5
CONTROLLER											
C18 Instrumentation	Control	Open/short circuit	Power loss	Forced outage	None	6	8	7	336	Possible sudden failure	2

Fig. 4.16 RPN FMECA example – cooling water system pump set – Indenture Level 3

4.9 Fuzzy logic prioritization of failures

As in an RPN FMECA a fuzzy logic criticality assessment is based on the severity, frequency of occurrence, and detectability of item failure modes. These parameters can be represented as members of a fuzzy set, combined by matching them against rules in a rule base, and then defuzzified to assess the risk of failure. Although this approach is significantly more involved than other criticality analysis techniques it is considered to have several advantages compared with qualitative or strictly numerical methods.

The same scales for the linguistic variables employed in an RPN FMECA are employed in the fuzzy logic method but in each case the evaluation criteria are represented by distributions which overlap (here simply labelled High, Moderate, and Low). Figure 4.17 shows the distributions employed in the fuzzy logic evaluation; these are modified versions of the distributions proposed by Bowles and Bonnell (**4**).

The fuzzification process converts occurrence frequency, severity, and detectability inputs into outputs which can be matched to rules in a rule base. For the failure mode 'open/short circuit' of controller instrumentation C18 (in the cooling water pump set), the membership of each attribute set can be determined from the distributions in Fig. 4.17. Using the same figure as shown on Fig. 4.16 for occurrence frequency (6) it can be seen that this failure mode has a membership value of 0.3 in the 'High' occurrence frequency distribution and a membership value of 0.7 in the 'Moderate' distribution. Severity with a value of 8 has a membership value of 1.0 in the 'High' severity distribution and detectability with a value of 7 has a membership value of 0.6 in 'Low' and 0.4 in 'Moderate' detectability distributions.

The rule base describes the risk to the system for each combination of the input variables. They are formulated in linguistic terms as IF-THEN rules which are implemented by fuzzy conditional statements. For example:

> IF frequency is Low, severity is Moderate, and detectability is Low-to-Moderate
> THEN risk is Moderate

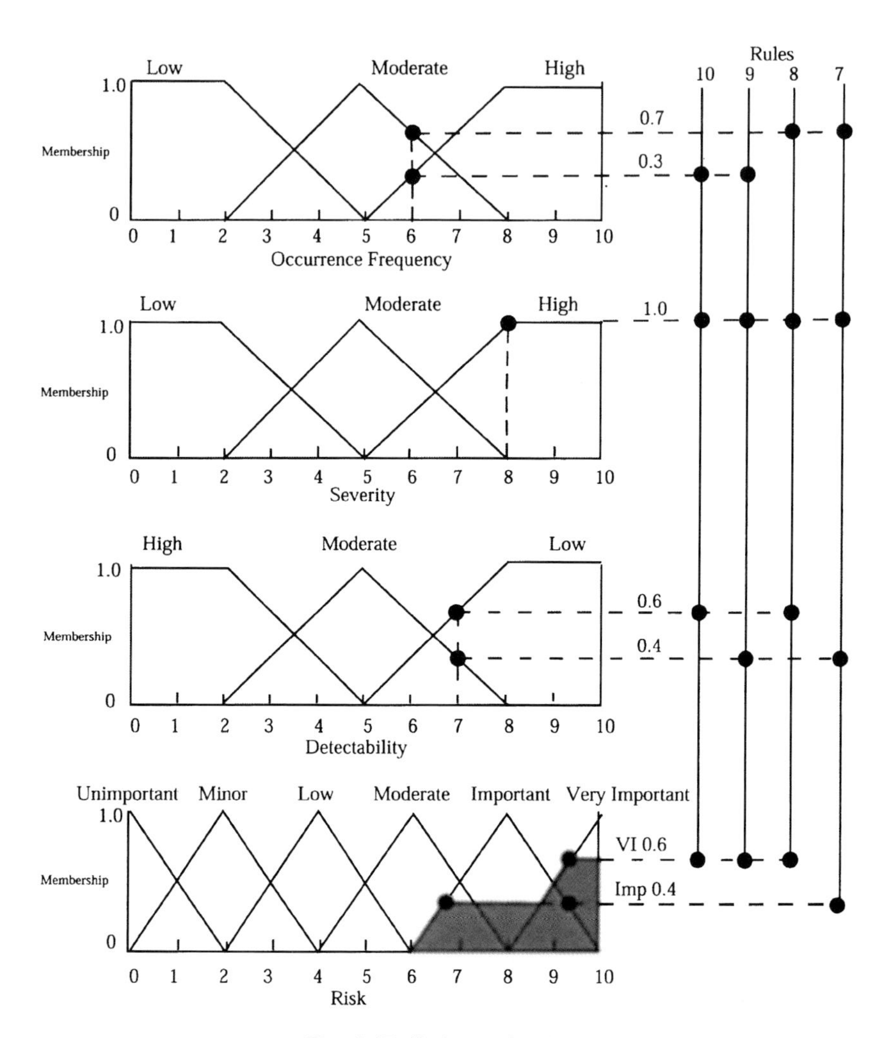

Fig. 4.17 Rule evaluation

The consistency of the rule base used for assessing criticality can be defined by examining a plot of the risk surface over all possible combinations of the input variables. Inconsistencies are revealed by abrupt changes in risk for small changes in frequency, severity, or detectability. The modification to the rules proposed by Bowles and Bonnell (4) for the assumed relationships between the different combinations of frequency, severity, and detectability with respect to risk importance are shown in Table 4.7.

Table 4.7 Sample of rule base for the evaluation of risk

Rule no.	Occurrence	Severity	Detectability	Risk
7	Moderate	High	Moderate	Important
8	Moderate	High	Low	Very important
9	High	High	Moderate	Very important
10	High	High	Low	Very important

The outputs for assessing risk importance can be defined using fuzzy sets in the same way as for the fuzzy inputs. A set of risk importance distributions is also shown in Fig. 4.17. The fuzzy inference process uses a 'min–max' approach to calculate the rule conclusions based on the input values. The outcome of this process is called the set of fuzzy conclusions.

The 'truth' value of a rule is determined from the combination of the information associated with each rule. This consists of determining the smallest (minimum) rule antecedent, which is taken to be the truth value of the rule. If any fuzzy output is the consequence of more than one rule then that output is set at the highest (maximum) truth value of all the rules that include this as a consequence. Hence, the result of the rule evaluation is a set of fuzzy conclusions that reflect the effect of all the rules with truth values greater than a defined threshold. For example, for failure mode C18 open/short circuit of the controller instrumentation, the rules that apply are 7, 8, 9, and 10. Rules 8, 9, and 10 have the outcome 'Risk = Very Important'. From Fig. 4.17 it can be seen that: for rule 8 (i.e. frequency = Moderate, severity = High, detectability = Low) the fuzzy conclusion is min(0.7, 1.0, 0.6) = 0.6; for rule 9 it is min(0.3, 1.0, 0.4) = 0.3; and for rule 10 it is min(0.3, 1.0, 0.6) = 0.3. The membership value of 'Risk = Very Important' is the maximum of these three fuzzy conclusions, i.e. max(0.3, 0.3, 0.6) = 0.6. Rule 7,

'Risk = Important', has the one fuzzy outcome min(0.7, 1.0, 0.4) = 0.4, hence max = (0.4).

The defuzzification process creates a single crisp ranking from the fuzzy conclusion set to express the inherent risk. Several defuzzification algorithms have been developed of which the weighted mean of maximums (WMoM) is probably the most commonly used. This method gives a best-estimate of the average, weighted by the degree of truth at which the membership functions reach their maximum value:

$$\text{WMoM} = \frac{\sum w_i x_i}{\sum w_i}$$

where

w_i = degree of truth of the membership function

x_i = risk rank at maximum value of the membership function

For the controller failure mode C18, therefore:

$$\begin{aligned}\text{WMoM} &= \frac{(0.6 \times 10) + (0.4 \times 8)}{(0.6 + 0.4)} \\ &= 9.2\end{aligned}$$

This represents the inherent risk of the failure mode on a scale 1–10. Criticality ranking of failure modes will be in decreasing order of risk importance. Alternatively the following relationships may be considered to apply:

Risk importance > 7 = high criticality
3–7 = moderate criticality
< 3 = low criticality

Here, criticality is a measure of the importance of the failure mode to pump set forced outage.

Clearly, fuzzy prioritization of failure modes is quite tedious to apply, nevertheless it could easily be computerized and would then provide a useful tool for relating RPNs (which are basically comparative) to defined criticality levels. It could also provide a check on completed FMECAs prepared during the design phase. However, this exercise shows that the basic MIL-STD-1629A FMECA methodology is essentially robust giving comparable results to those obtained by detailed analysis of the generic data in major reliability data handbooks. This, therefore, appears to be the best method currently available for the detailed reliability analysis of mechanical equipment and systems. The fact that there is close agreement between the FMECA prediction and the results of analysis of generic data for a class of rotating machinery where large sample data are available is considered important. It should give a degree of confidence when employing the method for other equipment (such as gas turbines) where generic and field data are significantly more limited.

Most mechanical systems can be broken down to sub-system or equipment level where components can be assumed to operate in series. This is particularly suitable for the qualitative or quantitative analysis offered by FMEA/FMECA. The need to identify all the components in the system, their failure modes and causes, and the effect on system operation means that uncertainties due to design and operational features are largely minimized. Although time-dependent effects are not specifically considered, some recent extensions to the basic methodology, such as common cause or dependency analysis, should help to identify likely degradation mechanisms which may be considered separately for critical components or sub-assemblies.

4.10 Generic parts count

Since FMEA can be a time-consuming task a number of variations on the basic theme have been developed to reduce the time required to carry out the analysis. These techniques are often employed during initial studies of the safety or availability of process plant.

The most widely used of these methods of reliability prediction is known as the generic parts count. It is based on the principle that the availability of any equipment depends on the number of its component parts and the failure rates of these parts. An important proviso is that each part is initially assumed to have generic failure characteristics which can be determined from experience of similar parts operating in approximately similar conditions.

Failure rates for the parts are obtained from published sources, reliability data banks, or engineering judgement. The proportion of the part failure rate with the potential to cause system failure is then estimated and employed with other parameters such as the operating duty, environment, maintenance policy, and quality factors to weight the part failure rate. These weighted part failure rates are then multiplied by the number of parts and aggregated for all the parts in the system to estimate the system failure rate. As with FMEA the presentation is tabular.

Generic parts count reliability predictions can be carried out in a much shorter time than FMEA. It is particularly effective for engineers with a wide experience of the type of equipment being assessed and for companies who have good service records.

An example of a generic parts count prediction for a gas compression system on an offshore platform is shown in Fig. 4.18.

COMPRESSOR SYSTEM – PARTS COUNT				
Component type	No. (N)	Forced-outage rate (λf/year)	$N\lambda$	Totals (f/year)
Compressor set:				
Compressor stages	2	0.5	1.0	
Gear box	1	0.27	0.27	
Couplings, drive	3	0.09	0.27	
Electric motor (var. speed)	1	0.98	0.98	
Motor controls	1	0.44	0.44	
	Compressor set total			**2.96**
Lube/seal oil system	1			**2.30**
System components:				
Suction scrubbers	4	0.13	0.52	
Recycle/suction coolers	4	0.30	1.20	
Control valves	9	0.19	1.71	
Blowdown valves	4	0.11	0.44	
Non-return valves	3	0.06	0.18	
Pressure-relief valves	3	0.10	0.30	
Surge controllers (SC)	3	0.44	1.32	
Temperature controllers (TC)	4	0.15	0.60	
Pressure controllers (PC)	2	0.24	0.48	
Level controllers (LC)	4	0.24	0.96	
DP transmitters (DP)	3	0.07	0.21	
Flow transmitters (FT)	3	0.26	0.78	
	System components total			**8.70**
Pipework, connections, etc. (assume 10% of process components' failure rate)				**0.87**
	Compressor system total			**14.83**

Fig. 4.18 Generic parts count failure rate estimate – compressor system

4.11 Summary

MIL-STD-1629A was the first definitive text on FMEA and criticality analysis. In the view of the authors it remains the single most effective method for the identification of important failure modes and the estimation of critical failure rates.

Alternative methods have been proposed and have advantages for specific applications. Of these the risk priority number (RPN) method, with its extension using fuzzy logic, is extensively employed on mechanical systems by the automotive and aerospace industries.

In the RPN method the use of qualitative estimates, rather than failure frequencies, and the addition of an extra parameter (detectability) to rank critical failure modes is particularly noteworthy. However, both 1629A and RPN generally rank the same failure modes as critical.

The limitation of FMEA/FMECA is its emphasis on failure. The importance of long repair times is not reflected in the definition of severity although it can clearly have a major impact on the availability of some mechanical systems.

Generic parts count is an alternative method that is sometimes employed by experienced reliability engineers to predict the failure rate of equipment or systems during initial studies; however, it does not have the force of an FMEA to identify the critical failure modes or the impact of these failure modes on system reliability.

4.12 References

(1) MIL-STD-1629A (1980); and MIL-STD-1629 Notice 2 (1984).

(2) **Andrews, J. D.** and **Moss, T. R.** (1993) *Reliability and Risk Assessment*. Longman.

(3) Air (1993) The FMECA process in the concurrent engineering environment. Society of Automotive Engineers Information Report AIR4845.

(4) **Bowles, J. B.** and **Bonnell, R. D.** (1996) Failure mode and effects analysis (What it is and how to use it) – American Reliability and Maintainability Symposium, Tutorial Notes.

(5) **Nowlan, F. S.** and **Heap, H. F.** (1980) Reliability Centred Maintenance – MSG1 and MSG2 United Airlines.

(6) **Moubrey, J.** (1991) *Reliability Centred Maintenance*. Butterworth Heinemann.

(7) Health and Safety Executive (1999) COMAH – *A Guide to the Control of Major Accident Hazards Regulations*.

(8) **Bloch, H. P.** and **Geitner, F. K.** (1990) *An Introduction to Machinery Reliability Assessment*. Van Nostrand.

(9) Vattenfall (1994), *TBook, Reliability Data of Components in Nordic Nuclear Power Plants*, Third edition 1992, Fourth edition 1994 (in Swedish). Vattenfall Energisystem AB, Sweden.

(10) OREDA 97 (1997) *Offshore Reliability Data Handbook*, Third edition. DnV, Hovik, Norway.

(11) **Babour, G.** (1977) Failure mode and effects analysis by the matrix method. In Proceedings of the Reliability and Maintainability Symposium, pp. 114–119.

Chapter 5

Quantification of Component Failure Probabilities

5.1 Introduction

The main concern of this book is to describe the methods which can be used to predict the reliability performance of a system in terms of the reliability performances of the components of which it is constructed. Fault tree analysis, network analysis, and Monte Carlo simulation are some of the methods which will later be described and are used to predict system reliability. All of these methods require that the probability of component failure can be assessed. This chapter looks at the means available for measuring component failure probability and defines the relevant parameters used to measure component performance. The distinction between components, sub-systems, and systems is only a matter of scale. It is therefore well worth spending time to fully understand the meaning of these parameters since analogous quantities will occur many times in later chapters referring to groups of components and systems.

No matter how well systems or components are designed and manufactured they will eventually fail. Nothing will last forever. It is the task of the design engineer to ensure that, although the system will fail, its frequency of failure or probability of being in the failed state at any time is acceptable. Acceptability may be judged according to either economic or safety criteria depending upon the type of system involved. As described in the introductory chapter the two most useful probabilistic terms to describe either component or system performance are **availability** and **reliability**. These will be defined and related to a typical component life history.

Consider a repairable component whose life history is represented in Fig. 5.1. The component starts its life in the working state. It remains in this state for some period of time until it undergoes failure and therefore makes a transition from the working state to the failed state.

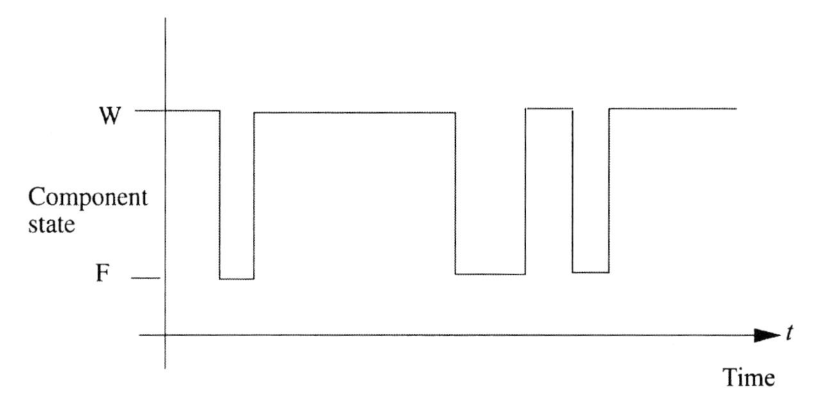

Fig. 5.1 Typical life history of a repairable component

It will reside in the failed state until a repair is completed and the component functions again. This alternating sequence of failures and repairs will continue throughout the lifetime of the component. For repairable components such as this, a relevant measure of its performance is its availability.

5.1.1 Availability

Expanding on the definition given in Chapter 1, the term availability can be used in three distinct senses.

1. Availability – the probability that a system/component works on demand.
2. Availability at time t – the probability that the system/component is working at a specific time t.
3. Availability – the fraction of the total time that a system/component can perform its required function.

The first of these definitions is appropriate for safety protection systems or for standby systems which are required to function on demand. A safety protection system monitors for the occurrence of an undesirable event and then acts to prevent a hazardous situation developing. If this protective system were in the failed state when it was required to function it would be unavailable. Since the demand can occur at any instant in time the longer the system is in the working state the better its chance of functioning, i.e. the higher its availability.

Standby systems such as back-up generators are also required to work on demand. For systems or components in this category the demand occurs when the primary system or component fails.

The nature of protection systems and standby systems means that they are usually passive, that is, they do not normally carry out any task. Components which fail in such a way as to render the systems inactive fail in an unrevealed manner. Repair times are then dependent upon the following factors. Firstly, how long it takes to discover that the component has actually failed. Since this will only be detected upon inspection the test interval will be a critical consideration. Once detected the repair process can then be initiated. How long this takes will depend on the availability of maintenance crews and the nature of the failure. If components cannot be repaired and need to be replaced, or need replacement parts, then the repair time will be dependent on whether the parts are kept in stores or need to be ordered from the supplier. Having carried out the repair there will also be some down-time while the system is tested prior to it being returned on-line.

Decreasing any of the factors described above will decrease the time that the system is down and hence increase its availability.

Definition 2 is appropriate for continuously operating systems or components whose failure is revealed. When component failure occurs it is revealed and hence the repair process can be initiated immediately.

The third definition is relevant for estimating the productivity of manufacturing processes. In this situation the fraction of the time that the system is functioning can be used to estimate the total output and therefore the expected revenue in any time period. The probability that a component/system does not work at any time t is termed **unavailability**, where:

Unavailability = 1 – availability

5.1.2 Reliability

For some systems failure cannot be tolerated. It may be that the failure would be a catastrophic event such as an explosion which destroys the plant. In this instance the non-occurrence of the failure event over the plant's expected lifetime would be an important measure of performance. Thus the system's reliability needs to be considered.

The probability that a component/system fails to work continuously over a stated time interval, under specified conditions, is known as its **unreliability**:

Unreliability = 1 – reliability

In the event that components or systems are not repairable then if it works at time t it must have worked continuously over the interval $[t_0, t]$ where t_0 is the start of the working period. Hence for non-repairable components/systems the unavailability equals the unreliability.

In the most simple models used to derive expressions for the probability of component failure it is assumed that the component can only exist in one of two states: working and failed. A component is expected to start its life in the working state. After a finite period of time the component fails and thus transfers into the failed state. If the component is non-repairable this is an absorbing state and it continues in this state forever. A repairable component will only remain in the failed state until a repair is completed, when it will transfer back to its working state. Two assumptions are made in this type of model: (a) that only one transition can occur in a small period of time dt and (b) that the change between the two discrete states is instantaneous.

As shown in Fig. 5.2 the transition process from the normal, working condition to the failed condition is termed **failure**. The reverse process transforming the component to a working state is termed **repair**. It is also assumed here that following the repair process the component is as good as new. The component life-cycle consists of a series of transitions between these two states. First we will consider the failure process, then the repair process, and finally the whole life-cycle.

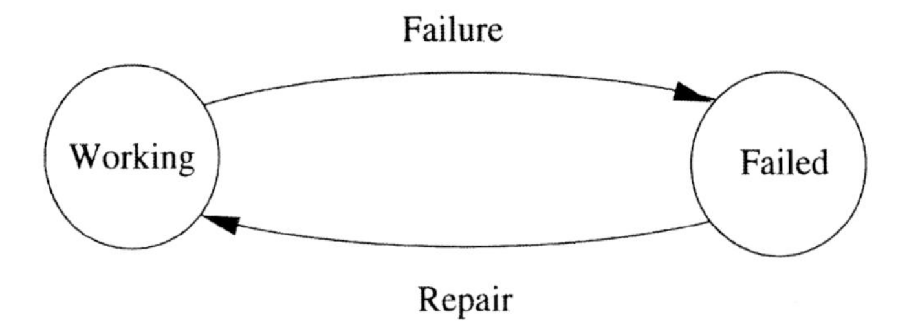

Fig. 5.2 Component failure and repair process

5.2 The failure process

The time to failure for any component cannot be predicted exactly. The failure can only be characterized by the stochastic properties of the population as a whole. Components of the same type, manufactured by the same supplier, and operated in identical conditions will fail after different time periods have elapsed. The times to failure can be used to form a probability distribution which gives the likelihood that components have failed prior to some time t. Components operated under different conditions or supplied by different manufacturers will have different time to failure distributions. These distributions can be estimated by either component testing procedures in controlled conditions or by collecting and analysing field data from functioning components.

For the failure process let $F(t)$ denote component unreliability, that is the probability that a component fails at some time prior to t, i.e.

$$F(t) = P\,[\text{a given component fails in } [0, t)]$$

The corresponding probability density function $f(t)$ is therefore

$$f(t) = \frac{\mathrm{d}F(t)}{\mathrm{d}t}$$

so $f(t)\,\mathrm{d}t = P$ [component fails in time period [t to $t + \mathrm{d}t$)]:

$$F(t) = \int_0^t f(t')\,\mathrm{d}t'$$

Transition to the failed state can be characterized by the conditional failure rate $h(t)$. This function is sometimes referred to as the hazard rate or hazard function. This parameter is a measure of the rate at which failures occur, taking into account the size of the population with the potential to fail, i.e. those which are still functioning at time t.

$$h(t)\,\mathrm{d}t = P[\text{component fails between } t \text{ and } t + \mathrm{d}t \mid \text{it has not failed in } [0, t)]$$

From equation (2.9) for conditional probabilities we have

$$P(A|B) = \frac{P(A \cap B)}{P(B)}$$

Since $h(t)\,\mathrm{d}t$ is a conditional probability we can define events A and B as follows by comparing this with the equation above:

A = component fails between t and $t + \mathrm{d}t$

B = component has not failed in $[0, t)$

With events thus defined, $P(A \cap B) = P(A)$ since if the component fails between t and $t + \mathrm{d}t$ it is implicit that it cannot have failed prior to time t.

Therefore we get

$$\begin{aligned} h(t)\,\mathrm{d}t &= \frac{P\left[\text{component fails between } t \text{ and } t + \mathrm{d}t\right]}{P[\text{no failure in } [0, t)]} \\ &= \frac{f(t)\,\mathrm{d}t}{1 - F(t)} \end{aligned} \tag{5.1}$$

integrating gives

$$\int_0^t h(t')\,\mathrm{d}t' = \int_0^t \frac{f(t')}{1 - F(t')}\,\mathrm{d}t' \tag{5.2}$$

$$\int_0^t h(t')\,dt' = -\ln\left[1 - F(t)\right] \tag{5.3}$$

$$F(t) = 1 - e^{-\int_0^t h(t')\,dt'} \tag{5.4}$$

If $h(t)$, the failure rate or hazard rate for a typical system or component, is plotted against time the relationship which commonly results, known as the 'bath-tub' curve, is shown in Fig. 5.3.

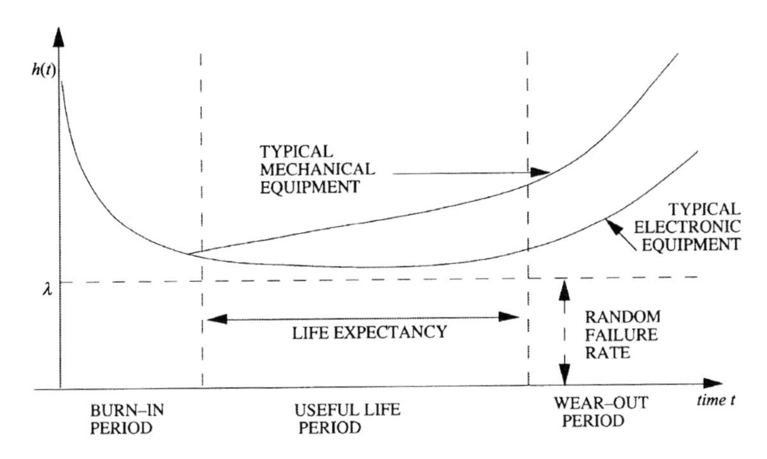

Fig. 5.3 The 'bath-tub' curve

The curve is characterized by a decreasing failure rate in the first portion, a region where the failure rate is almost constant, and finally a portion where the failure rate increases.

The curve can be considered in these three distinct parts. Part one represents the infant mortality or burn-in period. Poor welds, joints, connections, or insulation are typical causes of failures which occur in this early life period. Some quality processes carried out by manufacturers subject a product to a burn-in period to eliminate these early failures. The second portion of the curve, which exhibits a constant failure rate, corresponds to the useful life of a component. Failures in this period of time are random and are often stress related.

Finally the wear-out stage sets in and the component failure rate increases. Failures in this stage can result from such things as corrosion, oxidation, friction, wear, and fatigue. It can be seen in Fig. 5.3 that a typical electronic component follows the constant failure rate model during its useful life period more closely than mechanical equipment, where there is always an element of wear.

Neglecting the burn-in and wear-out failures and considering only the useful life of the component, $h(t) = \lambda$, i.e. the failure rate is constant.

Substituting into equation (5.4) and carrying out the integration gives

$$F(t) = 1 - \mathrm{e}^{-\lambda t} \tag{5.5}$$

Hence the reliability $R(t) = \mathrm{e}^{-\lambda t}$, the probability that the component works continuously over (0, t], is given by the exponential function. Due to the constant failure rate this function is often referred to as the 'random failure distribution' in the sense that it is independent of previous successful operating time.

5.2.1 Mean time to failure

Considering only the failure process with constant failure rate λ we obtained the expression given in equation (5.5) for $F(t)$. This has the probability density function

$$f(t) = \lambda\,\mathrm{e}^{-\lambda t}$$

The mean of the distribution can now be found which yields the mean time to failure (MTTF) which will be represented by μ

$$\mu = \int_0^\infty t f(t)\,\mathrm{d}t \tag{5.6}$$

$$= \int_0^\infty t\lambda\,\mathrm{e}^{-\lambda t}\,\mathrm{d}t \tag{5.7}$$

this can be integrated by parts to give

$$\mu = \frac{1}{\lambda} \tag{5.8}$$

This is an important result. If we have constant failure rate then the mean time to failure is simply its reciprocal.

5.2.2 *Failure data example*

In order to illustrate the component failure process parameters $F(t)$, $f(t)$, and $h(t)$ their determination is demonstrated using some hypothetical component failure data. The data relate to the number of components failing in each 50-h interval from an initial batch of 480 components. Table 5.1, column 1, indicates the time t in hours and the second column indicates the number of components failing in time period t to $t + dt$ ($dt = 50$ h).

Table 5.1 Component failure data and failure parameters

(1) Time (h) t	(2) Number of failures	(3) Cumulative failures N_f	(4) Cumulative failure distribution $F(t)$	(5) No. of survivors N_s	(6) Survival function $R(t)$	(7) Failure density function $f(t)$	(8) Conditional failure rate $h(t)$
0	70	0	0	480	1.0	2.92×10^{-3}	2.92×10^{-3}
50	42	70	0.146	410	0.854	1.74×10^{-3}	2.04×10^{-3}
100	38	112	0.233	368	0.767	1.60×10^{-3}	2.09×10^{-3}
150	34	150	0.313	330	0.687	1.40×10^{-3}	2.04×10^{-3}
200	30	184	0.383	296	0.617	1.26×10^{-3}	2.04×10^{-3}
250	26	214	0.446	266	0.554	1.08×10^{-3}	1.95×10^{-3}
300	25	240	0.5	240	0.5	1.04×10^{-3}	2.08×10^{-3}
350	19	265	0.552	215	0.448	0.80×10^{-3}	1.79×10^{-3}
400	17	284	0.592	196	0.408	0.70×10^{-3}	1.72×10^{-3}
450	15	301	0.627	179	0.373	0.62×10^{-3}	1.66×10^{-3}
500	15	316	0.658	164	0.342	0.64×10^{-3}	1.87×10^{-3}
550	20	331	0.690	149	0.310	0.84×10^{-3}	2.71×10^{-3}
600	30	351	0.732	129	0.268	1.24×10^{-3}	4.63×10^{-3}
650	37	381	0.794	99	0.206	1.54×10^{-3}	7.48×10^{-3}
700	31	418	0.871	62	0.129	1.28×10^{-3}	9.92×10^{-3}
750	21	449	0.935	31	0.065	0.88×10^{-3}	13.54×10^{-3}
800	7	470	0.979	10	0.021	0.30×10^{-3}	0.88×10^{-3}
850	3	477	0.994	3	0.006	0.12×10^{-3}	20.00×10^{-3}
900		480	1.000	0	0.0		

Column 3 contains N_f, the cumulative number of failed components at time t. So at the start of the test, $t = 0$, all components are functioning so $N_f = 0$. During the next 50 h, 70 components fail so at $t = 50$ h the cumulative number of failures is 70; 42 failures result in the next 50-h period giving $N_f = 70 + 42 = 112$ at $t = 100$ h. The number of surviving components, N_s, are listed in column 5. These are calculated by taking the cumulative number of failures (column 3) from the total population of 480. Therefore, for example, at $t = 400$ h, $N_s = 480 - 284 = 196$.

To turn the cumulative failures from column 3 into a probability distribution requires that at $t = 0$, $F(0) = 0$, and that when all components have failed $F(t) = 1$ so $F(900) = 1$. This is achieved by dividing the cumulative number of failures by the total sample size. The result of this is shown in column 4 and plotted in the graph shown in Fig. 5.4.

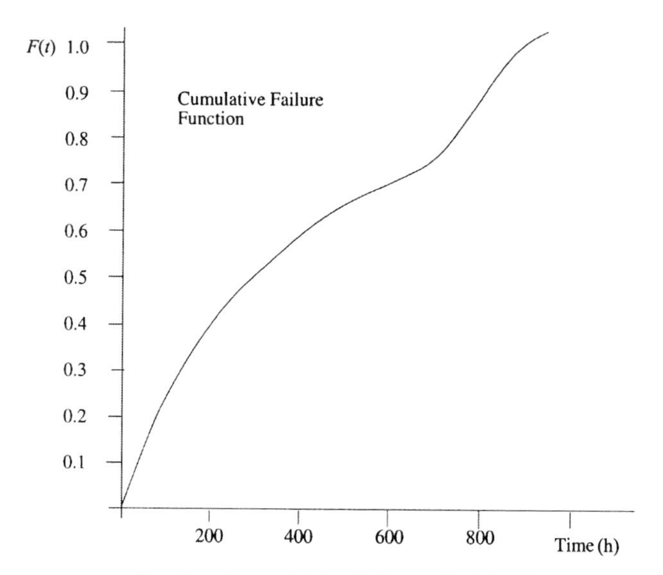

Fig. 5.4 Cumulative failure distribution

Similarly a survival function $R(t)$ can be calculated from $R(t) + F(t) = 1$. These figures are given in column 6 of the table and are plotted in Fig. 5.5.

Column 7 contains the failure density function $f(t)$ of the cumulative failure distribution $F(t)$. This can be calculated using the forward difference formula:

$$f(t) = \frac{F(t + \mathrm{d}t) - F(t)}{\mathrm{d}t}$$

So

$$f(0) = \frac{F(50) - F(0)}{50} = \frac{0.146 - 0}{50} = 2.92 \times 10^{-3}$$

and

$$f(200) = \frac{F(250) - F(200)}{50} = \frac{0.446 - 0.383}{50} = 1.26 \times 10^{-3}$$

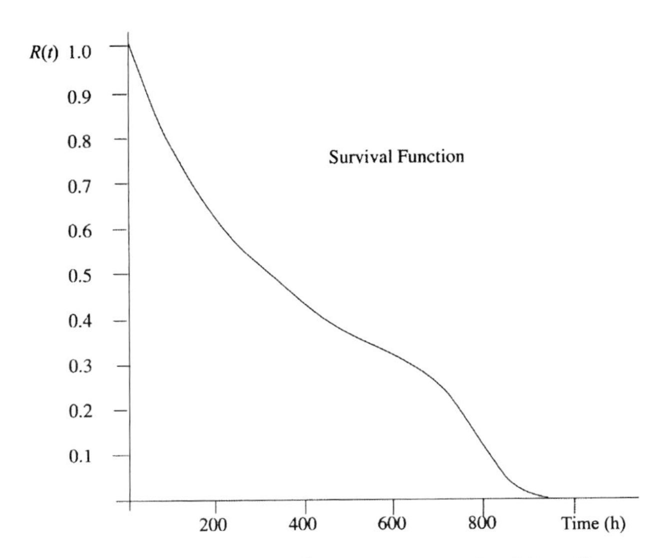

Fig. 5.5 Reliability function or survival function

Figure 5.6 shows the density function. The area under the density function between time equals zero and time equals t will give the cumulative failure distribution or unreliability. Since the total area under $f(t)$ is unity and $F(t) + R(t) = 1$, the area under the curve from time t onwards gives the component reliability.

$$F(t) = \int_0^t f(t')\,\mathrm{d}t' \quad \text{and} \quad R(t) = \int_t^\infty f(t')\,\mathrm{d}t'$$

Finally it remains to calculate the conditional failure rate $h(t)$. This is obtained from equation (5.1), i.e.

$$h(t) = \frac{f(t)}{1 - F(t)}$$

From the table values of $f(t)$ and $F(t)$, $h(t)$ can be calculated for each time point. So when $t = 0$: $f(t) = 2.92 \times 10^{-3}$ and $F(t) = 0.0$

giving

$$h(0) = f(0) = 2.92 \times 10^{-3}$$

When $t = 0$ is the only time that $h(t)$ and $f(t)$ are equal:

$$h(200) = \frac{f(200)}{1 - F(200)} = \frac{1.26 \times 10^{-3}}{1 - 0.383} = 2.04 \times 10^{-3}$$

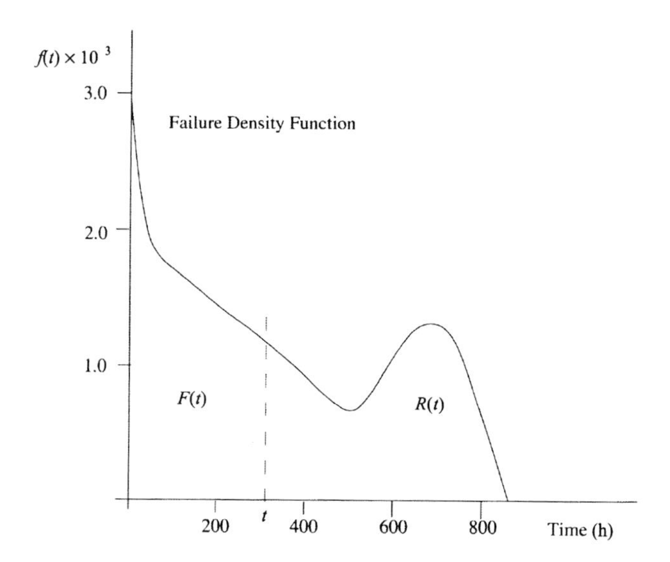

Fig. 5.6 Failure density function

The full distribution tabulated in column 8 is illustrated in Fig. 5.7. It is possible to identify the burn-in, useful life, and wear-out periods for the distribution.

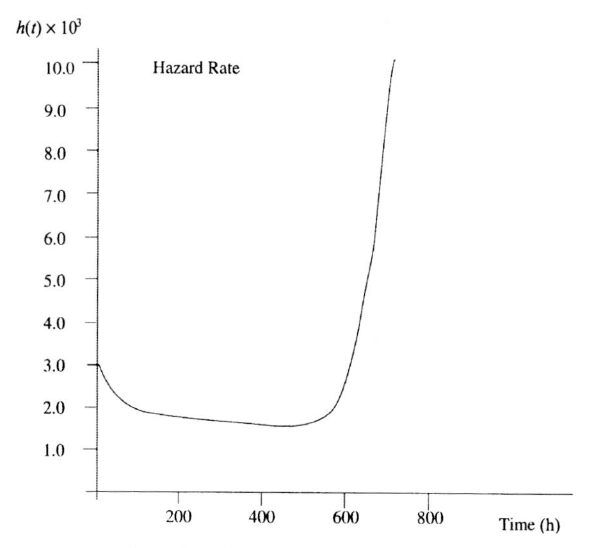

Fig. 5.7 Conditional failure rate

5.3 The repair process

Parameters which represent the repair process can be determined by following the same procedure as for the failure process. Let

$$G(t) = P \text{ [a given failed component is repaired in [0, t)]}$$

Time is now measured from the instant when the component failed; $g(t)$ is the corresponding density function and

$$m(t) = \frac{g(t)}{1 - G(t)} \tag{5.9}$$

where $m(t)$ is the conditional repair intensity.

Therefore by integrating:

$$G(t) = 1 - \mathrm{e}^{-\int_0^t m(t')\mathrm{d}t'} \tag{5.10}$$

If the repair rate can be assumed constant, $m(t) = \nu$, then

$$G(t) = 1 - \mathrm{e}^{-\nu t}$$

If the repair rate is constant then we also know that the mean time to repair (τ) is the reciprocal of the repair rate:

$$\tau = \frac{1}{\nu} \tag{5.11}$$

5.4 The whole failure/repair process

The previous two sections have considered the failure and repair processes individually. To calculate component unavailability the entire life-cycle of a component needs to be modelled and these two processes considered simultaneously.

5.4.1 Component performance parameters

Before going on to develop relationships which express the probability or frequency of component failure, the parameters which affect the component performance need to be defined. To aid in the understanding of these definitions they will be illustrated with reference to the ten component histories shown in Fig. 5.8.

Availability A(t)

The probability that a component is working at time t. For example

$$A(4) = \frac{6}{10} = 0.6$$

$$A(8) = \frac{4}{10} = 0.4$$

Unavailability Q(t)

The probability that a component is failed at time t $[Q(t) = 1 - A(t)]$:

$$Q(4) = 1 - 0.6 = 0.4$$

$$Q(8) = 1 - 0.4 = 0.6$$

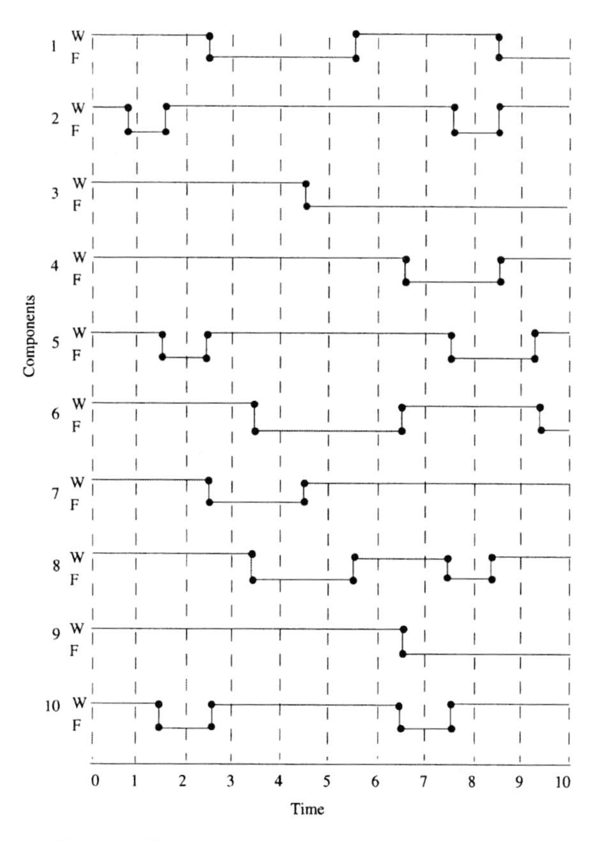

Fig. 5.8 Ten typical component life histories

Unconditional failure intensity w(t)

The probability that a component fails per unit time at t given that it was working at $t = 0$; i.e. the probability that the component fails in $[t, t + dt)$ given that it worked at $t = 0$:

$$w(4) = \frac{1}{10} = 0.1 \quad \text{[component 3 fails in the interval [4, 5)]}$$

$$w(8) = \frac{1}{10} = 0.1 \quad \text{[component 1 fails in the interval [8, 9)]}$$

Expected number of failures $W(t_0, t_1)$

The expected number of failures during $[t_0, t_1]$ given that the component worked at $t = 0$:

$$W(t,\ t+\mathrm{d}t) = \text{expected number of failures during } [t,\ t+\mathrm{d}t)$$
$$= \sum_{i=1}^{\infty} i.P\ [i \text{ failures during } [t,\ t+\mathrm{d}t)]$$

Since at most one failure occurs during the small time interval dt

$$W(t,\ t+\mathrm{d}t) = P\ [\text{one failure during } [t, t+\mathrm{d}t)] \qquad (5.12)$$
$$= w(t)\ \mathrm{d}t$$

Therefore

$$W(t_0,\ t_1) = \int_{t_0}^{t_1} w(t)\ \mathrm{d}t \qquad (5.13)$$

If a component is non-repairable then $W(0, t) = F(t)$.

From the ten component histories in Fig. 5.8

$$W(4,5) = \int_4^5 w(t)\ \mathrm{d}t = [0.1t]_4^5 = 0.1$$
$$W(0,10) = w(0)\ \times\ 1 + w(1)\ \times\ 1 + \cdots + w(9)\ \times\ 1$$
$$= 0.1 + 0.2 + 0.2 + 0.2 + 0.1 + 0 + 0.3 + 0.3 + 0.1 + 0.1$$
$$= 1.6$$

Unconditional repair intensity $v(t)$

The probability that a failed component is repaired per unit time at t given that it worked at $t = 0$:

$v(4) = \frac{1}{10} = 0.1$ [component 7 is repaired during time interval [4, 5)]

$v(8) = \frac{3}{10} = 0.3$ [components 2, 4, and 8 are repaired during time interval [8,9)]

Expected number of repairs $V(t_0, t_1)$

The expected number of times a failed component is repaired in the time period t_0 to t_1.

The argument which relates the expected number of repairs and the repair intensity is the same as that developed for the expected number of failures, giving:

$$V(t,\ t+\mathrm{d}t) = v(t)\ \mathrm{d}t \tag{5.14}$$

Integrating

$$V(t_0,\ t_1) = \int_{t_0}^{t_1} v(t)\ \mathrm{d}t \tag{5.15}$$

$$V(4,\ 5) = \int_4^5 v(t)\ \mathrm{d}t = [0.1t]_4^5 = 0.1$$

$$\begin{aligned} V(0,10) &= \int_0^{10} v(t)\ \mathrm{d}t \\ &= v(0)\times 1 + v(1)\times 1 + \cdots + v(9)\times 1 \\ &= 0 + 0.1 + 0.2 + 0 + 0.1 + 0.2 + 0.1 + 0.1 + 0.3 + 0.1 \\ &= 1.2 \end{aligned}$$

Conditional failure rate $\lambda(t)$

The probability that a component fails per unit time at t given that it was working at time t and working at time 0.

The difference between this and the unconditional failure intensity w is that λ is the failure rate based only on those components which are working at time t. w is based on the whole population.

$\lambda(4) = \dfrac{1}{6} = 0.167$ [component 3 fails in time interval [4,5) but only components 2, 3, 4, 5, 9, and 10 were working at $t = 4$]

$\lambda(8) = \dfrac{1}{4} = 0.5$ [component 1 fails in time interval [8,9) but only components 1, 6, 7, and 10 were working at $t = 8$]

so $\lambda(t)\ \mathrm{d}t$ is the probability that a component fails during $[t, t + \mathrm{d}t)$ given that the component worked at time t and was working at $t = 0$.

Note that $h(t) \neq \lambda(t)$. $h(t)\ \mathrm{d}t$ is the probability that a component fails during $[t, t + \mathrm{d}t)$ given that the component has been working **continuously** from $t = 0$ to time t.

The relationship between the conditional failure rate $\lambda(t)$ and the unconditional failure intensity $w(t)$ can be derived by considering conditional probabilities:

$$\lambda(t)\ \mathrm{d}t = P[\text{component fails in } [t, t+\mathrm{d}t) \text{ given that it was working at } t = 0]$$

$$= \frac{P[\text{component fails in } [t,\ t+\mathrm{d}t)]}{P[\text{component is working at } t]}$$

$$= \frac{w(t)\ \mathrm{d}t}{A(t)}$$

$$= \frac{w(t)\ \mathrm{d}t}{1-Q(t)} \qquad (5.16)$$

Therefore

$$w(t) = \lambda(t)\ [1 - Q(t)] \qquad (5.17)$$

Conditional repair rate $\rho(t)$

The probability that a component is repaired per unit time at t given that it is failed at time t and was working at time 0:

$\rho(4) = \frac{1}{4} = 0.25$ [component 7 is repaired in time interval [4, 5) and only components 1, 6, 7, and 8 were failed at $t = 4$]

$\rho(8) = \frac{3}{6} = 0.5$ [components 2, 4, and 8 are repaired in time interval [8, 9) and only components 2, 3, 4, 5, 8, and 9 were failed at $t = 8$]

5.5 Calculating unconditional failure and repair intensities

For a component which is continually subjected to the failure and repair process

$$w(t)\,\mathrm{d}t = P[\text{failure in } [t, t+\mathrm{d}t) \mid \text{working at } t=0] \tag{5.18}$$

The situation where components fail in $[t, t + \mathrm{d}t)$ can arise in two ways as illustrated by options 1 and 2 in Fig. 5.9. The first situation is where the component works continuously from 0 to t and fails for the first time in $[t, t + \mathrm{d}t)$. The probability of this is given by $f(t)\,\mathrm{d}t$ where $f(t)$ is the failure density function of $F(t)$.

The second way in which a component fails in the time interval $[t, t + \mathrm{d}t)$ is when this failure is not the first and the component has undergone one or more repairs prior to this failure. If the component was last repaired in interval $[u, u + \mathrm{d}u)$ and has therefore worked continuously from that repair to time t then the probability of this is

$$v(u)\,\mathrm{d}u \times f(t-u)\,\mathrm{d}t \tag{5.19}$$

where

$v(u)\,\mathrm{d}u$ = probability of repair in time $[u, u + \mathrm{d}u)$ given it was working at $t = 0$.

$f(t - u)\,\mathrm{d}t$ = probability of failure in $[t, t + \mathrm{d}t)$ having worked continuously since being repaired in $[u, u + \mathrm{d}u)$ given it was working at $t = 0$

The repair time u can occur at any point between 0 and t and so summing all possibilities gives

$$w(t)\,\mathrm{d}t = f(t)\,\mathrm{d}t + \int_0^t f(t-u)\,v(u)\,\mathrm{d}u\,\mathrm{d}t \tag{5.20}$$

Repair can only occur in the interval $[t, t + \mathrm{d}t)$ providing failure has occurred at some interval $[u, u + \mathrm{d}u)$ prior to t. This situation is illustrated in component history 3 of Fig. 5.9. The probability of this is

$$g(t-u)\,\mathrm{d}t \times w(u)\,\mathrm{d}u \tag{5.21}$$

where

$w(u)\,\mathrm{d}u$ = probability of failing in $[u, u + \mathrm{d}u)$ given it was working at $t = 0$
$g(t-u)\,\mathrm{d}t$ = probability of repair in $[t, t + \mathrm{d}t)$ given it has been in the failed state since last failure in $[u, u + \mathrm{d}u)$ and that it was working at $t = 0$
$g(t)$ = repair density function)

Since u can again vary between 0 and t:

$$v(t)\,\mathrm{d}t = \int_0^t g(t-u)\,w(u)\,\mathrm{d}u\,\mathrm{d}t \qquad (5.22)$$

Cancelling dt from equations (5.20) and (5.22) we get the simultaneous integral equations defining the unconditional failure and repair intensities:

$$w(t) = f(t) + \int_0^t f(t-u)v(u)\,\mathrm{d}u$$

$$\qquad (5.23)$$

$$v(t) = \int_0^t g(t-u)w(u)\,\mathrm{d}u$$

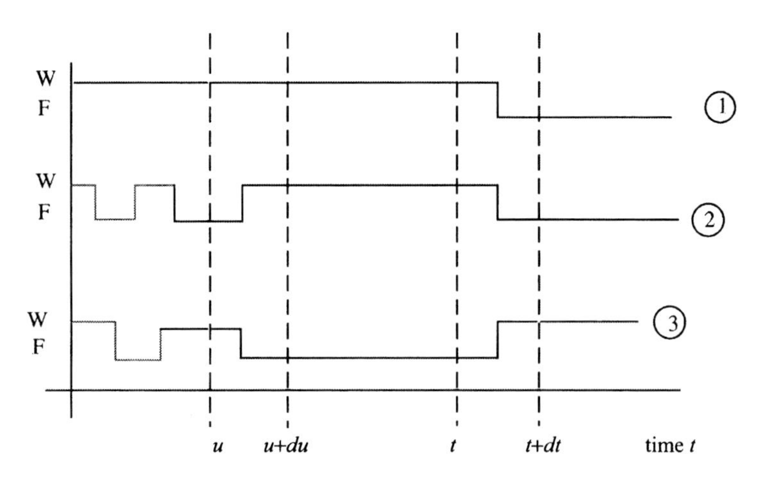

Fig. 5.9 Component histories

5.5.1 Expected number of failures and repairs

Once the unconditional failure and repair intensities have been calculated by solving equations (5.23), the expected number of failures and repairs can be determined over any time period from the following equations:

$$W(0,\ t) = \int_0^t w(u)\ \mathrm{d}u \tag{5.24}$$

and

$$V(0,\ t) = \int_0^t v(u)\ \mathrm{d}u \tag{5.25}$$

5.5.2 Unavailability

Component unavailability can also be derived from $w(t)$ and $v(t)$ as described below. Let $x(t)$ be an indicator variable which defines the component state, i.e.

$$x(t) = \begin{cases} 1 & \text{component in a failed state} \\ 0 & \text{component in a working state} \end{cases}$$

and let $x_F(t)$ and $x_R(t)$ indicate the number of failures and repairs carried out between time equals 0 and t respectively. Therefore

$$x(t) = x_F(t) - x_R(t) \tag{5.26}$$

Taking the expectation of equation (5.26) gives

$$E[x(t)] = E[x_F(t)] - E[x_R(t)] \tag{5.27}$$

The expectation of the indicator variable $x(t)$, the left-hand side of this equation, is the component unavailability since

$$E[x(t)] = \sum_{x_i} x_i P\left(x(t) = x_i\right)$$

$$= 1 \times P\left[x(t) = 1\right] + 0 \times P\left[x(t) = 0\right]$$

$$= P[x(t) = 1]$$

$$= Q(t)$$

The right-hand side of equation (5.27) is $W(0,\ t) - V(0,\ t)$. Therefore $Q(t) = W(0,\ t) - V(0,\ t)$ or

$$Q(t) = \int_0^t [w(u) - v(u)]\ \mathrm{d}u \tag{5.28}$$

Example 1
Find the unavailability of a non-repairable component which has an exponential failure distribution, i.e.

$$f(t) = \lambda \mathrm{e}^{-\lambda t}$$

Since it is non-repairable $g(t)$, and therefore $v(t)$, in equations (5.23) are zero. Hence

$$w(t) = f(t) = \lambda \mathrm{e}^{-\lambda t}$$

The unavailability can then be obtained from equation (5.28):

$$Q(t) = \int_0^t [w(u) - v(u)]\ \mathrm{d}u$$

$$= \int_0^t \lambda \mathrm{e}^{-\lambda u}\ \mathrm{d}u$$

$$= 1 - \mathrm{e}^{-\lambda t}$$

Example 2
Find the unavailability of a component which has exponential failure and repair intensities, i.e.

$$f(t) = \lambda \mathrm{e}^{-\lambda t}$$

$$g(t) = \nu \mathrm{e}^{-\nu t}$$

The solutions of equations (5.23) for $w(t)$ and $v(t)$ now become more difficult but can be carried out by using Laplace transforms (see Appendix B).

Taking Laplace transforms of equations (5.23) gives

$$\mathcal{L}[w(t)] = \mathcal{L}[f(t)] + \mathcal{L}[f(t)] \, . \, \mathcal{L}[v(t)]$$

$$\mathcal{L}[v(t)] = \mathcal{L}[g(t)] \, . \, \mathcal{L}[w(t)] \tag{5.29}$$

Substituting the Laplace transforms of the failure density function $f(t)$ and repair density function $g(t)$ into these equations gives

$$\mathcal{L}[w(t)] = \frac{\lambda}{s+\lambda} + \frac{\lambda}{s+\lambda} \; \mathcal{L}[v(t)]$$

$$\mathcal{L}[v(t)] = \frac{\nu}{s+\nu} \; \mathcal{L}[w(t)]$$

Since

$$\mathcal{L}[f(t)] = \mathcal{L}[\lambda e^{-\lambda t}] = \frac{\lambda}{s+\lambda}$$

and

$$\mathcal{L}[g(t)] = \mathcal{L}[\nu e^{-\nu t}] = \frac{\nu}{s+\nu}$$

solving these two simultaneous algebraic equations for $\mathcal{L}[w(t)]$ and $\mathcal{L}[v(t)]$ gives

$$\mathcal{L}[w(t)] = \frac{\lambda\nu}{\lambda+\nu}\left(\frac{1}{s}\right) + \frac{\lambda^2}{\lambda+\nu}\left(\frac{1}{s+\lambda+\nu}\right)$$

$$\mathcal{L}[v(t)] = \frac{\lambda\nu}{\lambda+\nu}\left(\frac{1}{s}\right) + \frac{\lambda\nu}{\lambda+\nu}\left(\frac{1}{s+\lambda+\nu}\right)$$

Taking inverse Laplace transforms gives

$$w(t) = \frac{\lambda\nu}{\lambda+\nu}\mathcal{L}^{-1}\left(\frac{1}{s}\right) + \frac{\lambda^2}{\lambda+\nu}\mathcal{L}^{-1}\left(\frac{1}{s+\lambda+\nu}\right)$$

$$v(t) = \frac{\lambda\nu}{\lambda+\nu}\mathcal{L}^{-1}\left(\frac{1}{s}\right) + \frac{\lambda\nu}{\lambda+\nu}\mathcal{L}^{-1}\left(\frac{1}{s+\lambda+\nu}\right)$$

Simplifying further

$$w(t) = \frac{\lambda\nu}{\lambda+\nu} + \frac{\lambda^2}{\lambda+\nu}\mathrm{e}^{-(\lambda+\nu)t}$$

$$v(t) = \frac{\lambda\nu}{\lambda+\nu} - \frac{\lambda\nu}{\lambda+\nu}\mathrm{e}^{-(\lambda+\nu)t}$$

Using equation (5.28) the unavailability is then

$$\begin{aligned} Q(t) &= \int_0^t [w(u) - v(u)]\ \mathrm{d}u \\ &= \int_0^t \left(\frac{\lambda^2}{\lambda+\nu} + \frac{\lambda\nu}{\lambda+\nu}\right)\mathrm{e}^{-(\lambda+\nu)u}\ \mathrm{d}u \\ &= \lambda\left[\frac{-\mathrm{e}^{-(\lambda+\nu)u}}{\lambda+\nu}\right]_0^t \\ &= \frac{\lambda}{\lambda+\nu}\left[1-\mathrm{e}^{-(\lambda+\nu)t}\right] \end{aligned} \tag{5.30}$$

When component failure and repair distributions are more complex and Laplace transforms cannot be used, the pair of equations (5.23) which give the unconditional transition rates must be solved using numerical methods. A method which can be used to obtain the numerical solution is given in Section 5.7 below.

In certain situations accurate approximations can be made which enable the component unavailability to be calculated by a simpler formula than that given in equation (5.30). If it is only required to calculate the steady-state component unavailability when the component has been operable for a reasonable period of time (i.e. $t \to \infty$)

then the bracketed expression in equation (5.30) becomes 1 and the unavailability can be estimated by

$$Q = \frac{\lambda}{\lambda + \nu} \tag{5.31}$$

With constant failure rate and constant repair rate then the mean time to failure

$$\mu = \frac{1}{\lambda} \tag{5.32}$$

and the mean time to repair

$$\tau = \frac{1}{\nu} \tag{5.33}$$

Substituting equations (5.32) and (5.33) into equation (5.31) gives

$$Q = \frac{\tau}{\tau + \mu} = \frac{\text{MTTR}}{\text{MTTR} + \text{MTTF}} \tag{5.34}$$

For most components it is reasonable to assume that the mean time to failure will be significantly larger than the mean time to repair. If this assumption is appropriate then

$$\mu \gg \tau \quad \text{and} \quad \tau + \mu \approx \mu \tag{5.35}$$

Using equation (5.32) gives a very simple formula for the steady-state component unavailability

$$Q = \lambda \tau \tag{5.36}$$

This approximation is pessimistic.

5.6 Maintenance policies

The way in which components or systems are maintained has a large influence over the unavailability. In this section three basic types of maintenance repair policies are considered.

1. No repair.
2. Unscheduled maintenance – the repair process is initiated as soon as a failure is revealed.
3. Scheduled maintenance – maintenance or inspection takes place at fixed, scheduled time intervals. When failure is discovered by this process the repair is then initiated.

Maintenance type 1, no repair, is typical of remotely controlled systems such as satellites or 'one-shot' machines such as spacecraft. In this case repair cannot be carried out and a system failure loses the device for good. If this type of system is working at a particular time t then since it is unrepairable it must have been working continuously until this time. Therefore its availability is equal to its reliability. For constant failure rate devices

$$Q(t) = F(t) = 1 - e^{-\lambda t} \tag{5.37}$$

As shown in Fig. 5.10, for large values of time the unavailability is close to one, i.e. as $t \to \infty$, $Q(t) \to 1$.

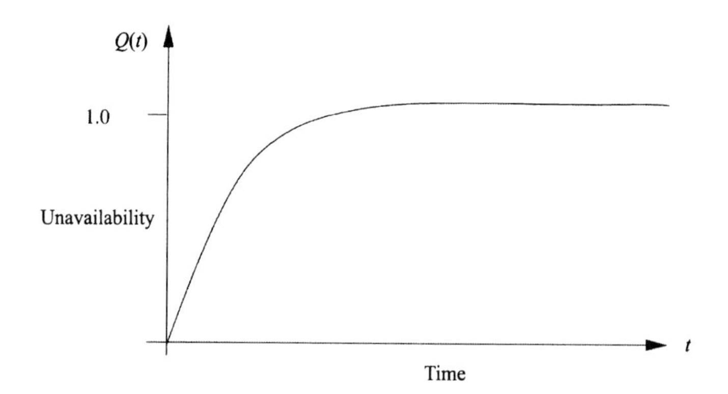

Fig. 5.10 Unavailability as a function of time for non-repairable components

Repairs which are carried out when a failure is revealed are those covered in category 2. This type of maintenance policy is common when some kind of continuous monitoring is being used or when the failure will reveal itself by causing a noticeable system failure such as shutdown. In this situation the time to repair will not include any detection time since the failure will be immediately known and the time it takes to rectify the fault will depend on the repair time alone. For constant failure and repair rates the unavailability will be given by equation (5.30), as shown in Fig. 5.11, i.e.

$$Q(t)=\frac{\lambda}{\lambda+\nu}\left(1-e^{-(\lambda+\nu)t}\right) \tag{5.38}$$

as $t \to \infty, Q(t) \to \dfrac{\lambda}{\lambda+\nu} \leq \lambda\tau$

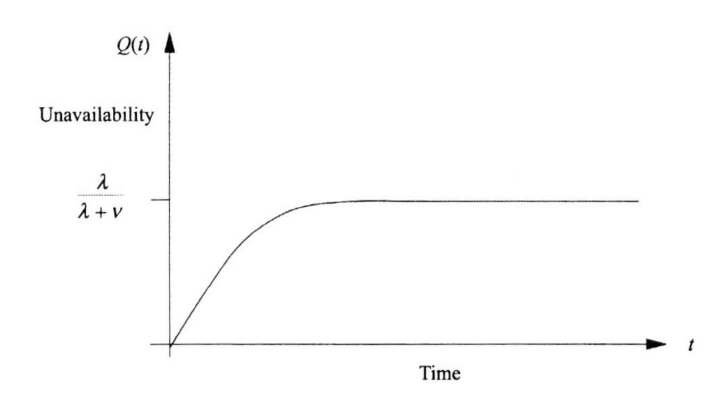

Fig. 5.11 Unavailability as a function of time for revealed failures

Category 3, scheduled maintenance, is common for standby or safety protection systems which are not continuously operational. For this type of system a failure will not be noticed until the systems are maintained or a demand occurs for them to function. The time that such a system resides in the down state will depend on both the time it takes to identify that a failure has occurred and the time then needed to carry out the repair. If a system is inspected every θ time units its unavailability is shown as a function of time, in Fig. 5.12.

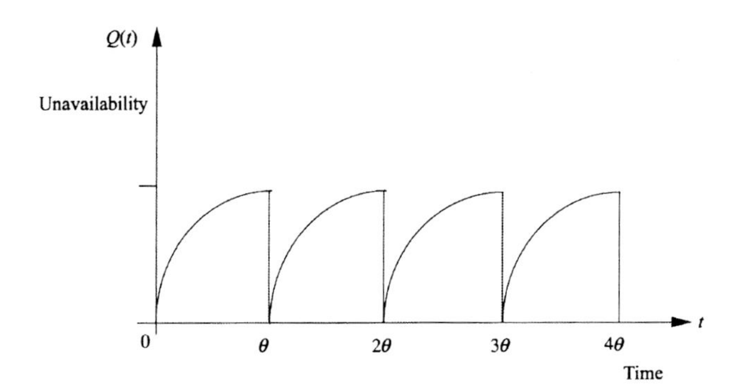

Fig. 5.12 Unavailability as a function of time for scheduled maintenance

In this case

$$Q(t) = 1 - e^{-\lambda[t-(n-1)\theta]}$$
$$(n-1)\theta \le t \le n\theta \qquad n = 1,\ 2,\ \ldots \tag{5.39}$$

The average unavailability is then

$$\begin{aligned} Q_{AV} &= \frac{\lambda\theta}{2} + \lambda\tau \\ &= \lambda\left(\frac{\theta}{2} + \tau\right) \end{aligned} \tag{5.40}$$

This results since any failure is equally likely to occur at any time and hence the average time for failure will be half-way between inspections. This failure will have existed for $\theta/2$ prior to the repair being initiated. The mean time to repair will then be

$$\frac{\theta}{2} + \tau$$

It is usual for the inspection period to be much greater than the mean repair time, i.e.

$$\theta \gg \tau$$

in which case

$$\frac{\theta}{2}+\tau \approx \frac{\theta}{2}$$

and the average unavailability can be approximated by

$$\theta_{AV}=\frac{\lambda\theta}{2} \tag{5.41}$$

Equations (5.40) and (5.41) are approximations which, for some combinations of λ and θ, will give probability values greater than 1.0! In these circumstances, the approximation is obviously invalid.

If we return to the component failure probability graph for scheduled inspection, Fig. 5.12, we know that between inspections the component is effectively non-repairable. To get an average unavailability then consider the first inspection period on the graph, i.e. between time = 0 and θ. In this period

$$\begin{aligned} Q_{AV} &= \frac{1}{\theta}\int_0^{\theta} 1-e^{-\lambda t}\,dt \\ &= \frac{1}{\theta}\left[t+\frac{e^{-\lambda t}}{\lambda}\right]_0^{\theta} \end{aligned} \tag{5.42}$$

$$\begin{aligned} Q_{AV} &= \frac{1}{\theta}\left[\theta+\frac{e^{-\lambda\theta}}{\lambda}-\frac{1}{\lambda}\right] \\ &= 1-\frac{1}{\lambda\theta}\left(1-e^{-\lambda\theta}\right) \end{aligned} \tag{5.43}$$

This is a more accurate equation to use than equation (5.41).

5.7 Failure and repair distribution with non-constant hazard rates

The component failure probability models discussed so far all make the assumption that the component failure process has a constant hazard rate [equations (5.37), (5.38), (5.41), and (5.43)]. In the event that times to failure or repair are governed by some distribution other than the negative exponential distribution (perhaps normal, log normal, or Weibull) we can still derive a value for the failure probability but, in most cases, this needs to be achieved numerically rather than analytically.

We start from equations (5.23), which give the unconditional failure and repair intensities $w(t)$ and $v(t)$ in terms of the failure time and repair time density functions $f(t)$ and $g(t)$:

$$w(t) = f(t) + \int_0^t f(t-u)v(u)\,\mathrm{d}u$$
$$v(t) = \int_0^t g(t-u)w(u)\,\mathrm{d}u$$

5.7.1 Method 1

We can see that at time $t = 0$:

$$v(0) = 0 \quad \text{and} \quad w(0) = f(0)$$

and at time $t = h$:

$$\begin{aligned} w(h) &= f(h) + \int_0^h f(h-u)v(u)\,\mathrm{d}u \\ v(h) &= \int_0^h g(h-u)w(u)\,\mathrm{d}u \end{aligned} \qquad (5.44)$$

If h is small enough we can approximate these integrals using a simple two-step trapezoidal rule, giving

$$\begin{aligned} w(h) &= f(h) + \frac{h}{2}\left[f(h)v(0) + f(0)v(h)\right] \\ v(h) &= \frac{h}{2}\left[g(0)w(h) + g(h)w(0)\right] \end{aligned} \qquad (5.45)$$

Solving simultaneously we get

$$w(h) = \frac{f(h) + \frac{h}{2}\left[f(0)g(h)w(0)\right]}{1 - \frac{h^2}{4} f(0)g(0)}$$

and

$$v(h) = \frac{\frac{h}{2}[g(0)f(h) + g(h)w(0)]}{1 - \frac{h^2}{4}f(0)g(0)} \tag{5.46}$$

Now if we increment t by another h:

$$\begin{aligned} w(2h) &= f(2h) + \int_0^{2h} f(t-u)v(u)\ \mathrm{d}u \\ &= f(2h) + \int_0^{h} f(t-u)v(u)\ \mathrm{d}u + \int_h^{2h} f(t-u)v(u)\ \mathrm{d}u \\ v(2h) &= \int_0^{2h} g(t-u)w(u)\ \mathrm{d}u \\ &= \int_0^{h} g(t-u)w(u)\ \mathrm{d}u + \int_h^{2h} g(t-u)w(u)\ \mathrm{d}u \end{aligned} \tag{5.47}$$

which gives

$$\begin{aligned} w(2h) &= f(2h) + \frac{h}{2}[f(0)v(h)] \\ &\quad + \frac{h}{2}[f(2h)v(h) + f(h)v(2h)] \\ v(2h) &= \frac{h}{2}[g(0)w(h) + g(h)w(0)] \\ &\quad + \frac{h}{2}[g(h)w(2h) + g(2h)w(h)] \end{aligned} \tag{5.48}$$

Solving simultaneously we get

$$w(2h)=\frac{\begin{aligned}&f(2h)+\frac{h}{2}\Big[f(0)v(h)+f(2h)v(h)+\frac{h}{2}f(h)g(h)w(h)\\&\qquad+\frac{h}{2}f(h)g(h)w(0)+\frac{h}{2}f(h)g(2h)w(h)\Big]\end{aligned}}{1-\frac{h^2}{4}f(h)g(h)}$$

$$v(2h)=\frac{\begin{aligned}&f(2h)+\frac{h}{2}\Big[g(0)w(h)+g(h)w(0)+g(h)f(2h)\\&+\frac{h}{2}f(0)v(h)g(h)+\frac{h}{2}g(h)f(2h)v(h)+g(2h)w(h)\Big]\end{aligned}}{1-\frac{h^2}{4}f(h)g(h)} \tag{5.49}$$

Following the same system as for $t = 2h$ we get, for $t = 3h$

$$w(3h)=\frac{\begin{aligned}&f(3h)+\frac{h}{2}\Big[f(0)v(h)+f(2h)v(h)+f(h)v(2h)+f(3h)v(2h)\\&+\frac{h}{2}f(2h)g(0)w(h)+\frac{h}{2}f(2h)g(h)w(0)+\frac{h}{2}f(2h)g(h)w(2h)\\&+\frac{h}{2}f(2h)g(2h)w(h)+\frac{h}{2}f(2h)g(3h)w(2h)\Big]\end{aligned}}{1-\frac{h^2}{4}g(2h)f(2h)}$$

$$v(3h)=\frac{\begin{array}{l}\frac{h}{2}\Big[g(0)w(h)+g(h)w(0)+g(h)w(2h)+g(2h)w(h)\\ +g(3h)w(2h)+g(2h)f(3h)+\frac{h}{2}g(2h)f(0)v(h)\\ +\frac{h}{2}g(2h)f(2h)v(h)+\frac{h}{2}g(2h)f(h)v(2h)+\frac{h}{2}g(2h)f(3h)v(2h)\Big]\end{array}}{1-\frac{h^2}{4}g(2h)f(2h)} \tag{5.50}$$

From these results we can formulate the general results thus:

$$w(nh)=\frac{\begin{array}{l}f(nh)+\frac{h}{2}\Big[f(0)v(h)+\cdots+f(mh)v((m-1)h)+f((m-1)h)v(mh)\\ +\cdots+f(nh)v((n-1)h)\Big]+\frac{h^2}{4}f((n-1)h)\Big[g(0)w(h)+g(h)w(0)\\ +\cdots+g((m-1)h)w(mh)+g(mh)w((m-1)h)+\cdots+g(nh)w((n-1)h)\Big]\end{array}}{1-\frac{h^2}{4}g((n-1)h)f((n-1)h)}$$

where m goes from 2 to $(n-1)$ and

$$v(nh)=\frac{\begin{array}{l}\frac{h}{2}g((n-1)h)f(n)+\frac{h}{2}\Big[g(0)w(h)+g(h)w(0)+\cdots+\\ g((m-1)h)w(mh)\ +g(mh)w((m-1)h)+\cdots+g(nh)w((n-1)h)\Big]\\ +\frac{h^2}{4}g((n-1)h)\Big[f(0)v(h)+\cdots+f(mh)v((m-1)h)+\\ f((m-1)h)v(mh)+\cdots+f(nh)v((n-1)h)\ \Big]\end{array}}{1-\frac{h^2}{4}g((n-1)h)f((n-1)h)} \tag{5.51}$$

where m goes from 2 to $(n - 1)$.

Having evaluated equation (5.51) to get $w(t)$ and $v(t)$, these can be substituted into equation (5.28) to obtain the failure probability $Q(t)$.

5.7.2 *Method 2*

Again starting from equations (5.23) we can differentiate to get

$$
\begin{aligned}
\frac{\mathrm{d}w}{\mathrm{d}t} &= f'(t) + f(0)v(t) + \int_0^t f'(t-u)v(u)\,\mathrm{d}u \\
\frac{\mathrm{d}v}{\mathrm{d}t} &= g(0)w(t) + \int_0^t g'(t-u)w(u)\,\mathrm{d}u
\end{aligned}
\tag{5.52}
$$

To use these equations we need the differentials of $f(t)$ and $g(t)$. Now we can calculate at any point (nh) the gradients of $w(t)$ and $v(t)$ using the trapezoidal rule, thus

$$
\begin{aligned}
w'(nh) = f'(nh) + f(0)v(nh) \\
&+ \left[f'(nh)v(0) + 2f'((n-1)h)v(h)\right. \\
&\left. + \cdots + 2f'(nh)v((n-1)h) + f'(0)v(n)\right]
\end{aligned}
$$

and

$$
\begin{aligned}
v'(nh) = g(0)w(nh) + \left[g'(nh)w(0) + 2g'((n-1)h)w(1)\right. \\
\left. + 2g'(nh)w((n-1)h) + g'(0)w(nh)\right]
\end{aligned}
\tag{5.53}
$$

Providing that we know all the points $w(0)$ to $w(nh)$ and $v(0)$ to $v(nh)$, we can calculate the points $v((n+1)h)$ and $w((n+1)h)$ using

$$
\begin{aligned}
w((n+1)h) &= w(nh) + hw'(nh) \\
v((n+1)h) &= v(nh) + hv'(nh)
\end{aligned}
\tag{5.54}
$$

As before, equation (5.28) yields the component failure probability.

5.8 Weibull analysis

5.8.1 Introduction

The Weibull distribution is employed extensively in the analysis of equipment lifetime data. The features which make it attractive to reliability and maintenance engineers include its flexibility – it can deal with decreasing, constant, and increasing failure rates and can consequently model all phases of the reliability 'bath-tub' curve. It has also been shown to fit most lifetime data better than other distributions and thus special probability papers have been developed to facilitate graphical analysis.

Graphical Weibull analysis involves plotting times to failure to obtain estimates of the distribution parameters. Although analytical methods are available, graphical analysis has the advantage of producing additional subjective information about the data beyond that contained in the parameter estimates alone, for example, the identification of outliers.

5.8.2 The Weibull distribution

For the two-parameter Weibull distribution the failure probability, in time t, is given by the cumulative distribution function:

$$F(t) = 1 - e^{-(t/\eta)^{\beta}} \tag{5.55}$$

where $F(t)$ is the probability of failure prior to time t, η is the scale parameter, and β the shape parameter.

The scale parameter (η) – or characteristic life – is the time taken for 63.2 per cent of the original population to fail, since

$$1 - e^{-(t/\eta)^{\beta}} = 1 - e^{-1} = 0.632 \quad \text{when } t = \eta$$

The Weibull distribution function is a simple exponential when $\beta = 1$, a reasonable approximation to the log-normal distribution when $\beta = 2$ and a close approximation to the normal distribution when $\beta = 3{\cdot}57$. For values of β less than unity the curve is characteristic of early-life or commissioning failures. The variation in the shape of the Weibull cumulative distribution function for identical values of the characteristic life but with different shape factors is shown on linear-scaled paper in Fig. 5.13. On Weibull probability paper they appear as shown in Fig. 5.14. Weibull probability paper has a double natural log scale for the ordinate and a single log scale for the abscissae. The scales are derived from the cumulative distribution function in order to linearize the plot:

$$1-F(t)=\exp\left[-(t/\eta)^{\beta}\right]$$

$$\ln\frac{1}{1-F(t)}=(t/\eta)^{\beta} \tag{5.56}$$

$$\ln\ln\frac{1}{1-F(t)}=\beta(\ln t-\ln\eta)$$

This expression is in the form $y = mx + c$, where m is the gradient of the line and c is the intercept on the y-axis. The slope of the Weibull line produced by plotting $\ln\ln[1/1 - F(t)]$ against $\ln t$ therefore gives the value of β which, in turn, indicates the shape of the failure distribution.

If a smooth curve rather than a straight line is produced it indicates that the time base is displaced. In such cases the three-parameter version of the distribution should be employed with the failure distribution function in the form

$$F(t)=1-\exp\left[-\left(\frac{t-\gamma}{\eta}\right)^{\beta}\right] \tag{5.57}$$

where each value of t is modified by a location constant (γ). The best value of γ can be derived by trial and error; Bompas-Smith (**1**) also gives two methods for its calculation. The latter are often used when developing computer programs for fitting the three-parameter Weibull distribution. Here only the two-parameter Weibull distribution is considered.

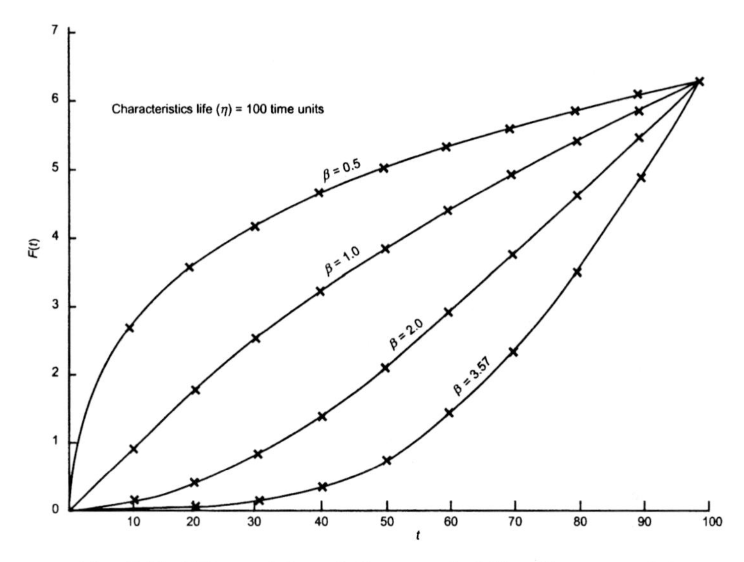

Fig. 5.13 Effect of β on failure probability (linear scale)

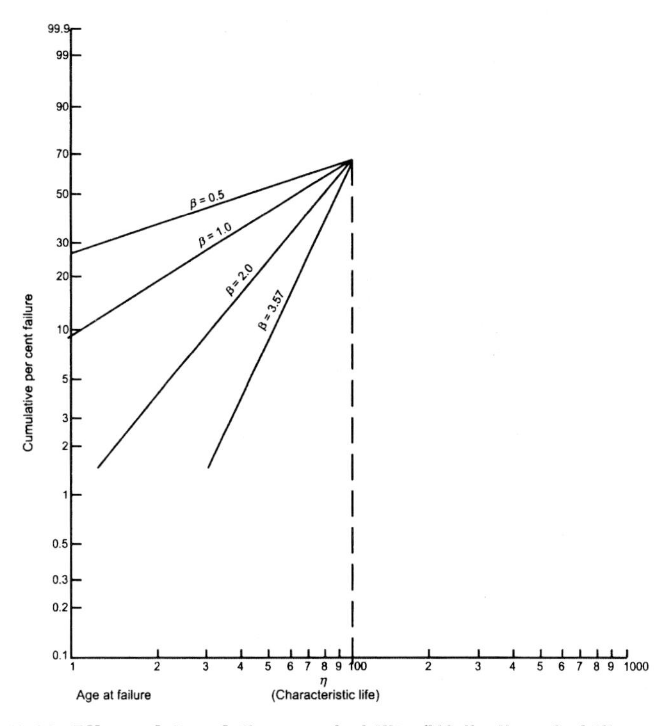

Fig. 5.14 Effect of β on failure probability (Weibull probability paper)

5.8.3 *Graphical analysis*

Graphical analysis needs to be considered under two headings:

1. Complete data – when all the items in the sample have failed
2. Censored data – when the sample data include unfailed items

In both cases the cumulative percentage failed at a particular value of t is used as an estimate of the failure distribution function $F(t)$ at that time.

From a large number of samples of the same size there will obviously be some variation in the values observed for the first and subsequent failures. Mean or median ranking takes this variability into account. The simplest approach is to use mean ranks. For a complete set of n failures with no censoring, the mean rank of the ith failure is $i/(n + 1)$. However, mean ranks have a number of drawbacks, particularly if confidence limits are to be calculated about the estimated line. A better estimator is the median rank, which is the 50 per cent point of the assumed beta distribution of the ith event out of n. Median ranks may be calculated by Benard's equation:

$$M_i = \frac{(i - 0.3)}{(n + 0.4)} \tag{5.58}$$

The formula is accurate to 1 per cent for $n > 5$ and 0.1 per cent for $n > 50$. Median ranks for sample sizes up to 20 are given in Table 5.2.

Table 5.2 Median ranks (numbers 1–20)

	Median ranks sample size									
Rank order	1	2	3	4	5	6	7	8	9	10
1	50.0	29.2	20.6	15.9	12.9	10.9	9.4	8.3	7.4	6.6
2		70.7	50.0	38.5	31.3	26.4	22.8	20.1	17.9	16.2
3			79.3	61.4	50.0	42.1	36.4	32.0	28.6	25.8
4				84.0	68.6	57.8	50.0	44.0	39.3	35.5
5					87.0	73.5	63.5	55.9	50.0	45.1
6						89.0	77.1	67.9	60.6	54.8
7							90.5	79.8	71.3	64.4
8								91.7	82.0	74.1
9									92.5	83.7
10										93.3

Table 5.2 Continued

Rank order	Median ranks sample size									
	11	12	13	14	15	16	17	18	19	20
1	6.1	5.6	5.1	4.8	4.5	4.2	3.9	3.7	3.5	3.4
2	14.7	13.5	12.5	11.7	10.9	10.2	9.6	9.1	8.6	8.2
3	23.5	21.6	20.0	18.6	17.4	16.3	15.4	14.5	13.8	13.1
4	32.3	29.7	27.5	25.6	23.9	22.4	21.1	20.0	18.9	18.0
5	41.1	37.8	35.0	32.5	30.4	28.5	26.9	25.4	24.1	22.9
6	50.0	45.9	42.5	39.5	36.9	34.7	32.7	30.9	29.3	27.8
7	58.8	54.0	50.0	46.5	43.4	40.8	38.4	36.3	34.4	32.7
8	67.6	62.1	57.4	53.4	50.0	46.9	44.2	41.8	39.6	37.7
9	76.4	70.2	64.9	60.4	56.6	53.0	50.0	47.2	44.8	42.6
10	85.2	78.3	72.4	67.4	63.0	59.1	55.7	52.7	50.0	47.5
11	93.8	86.4	79.9	74.3	69.5	65.2	61.5	58.1	55.1	52.4
12		94.3	87.4	81.3	76.0	71.4	67.2	63.6	60.3	57.3
13			94.8	88.2	82.5	77.5	73.0	69.0	65.5	62.2
14				95.1	89.0	83.6	78.8	74.5	70.6	67.2
15					95.4	89.7	84.5	79.9	75.8	72.1
16						95.7	90.3	85.4	81.0	77.0
17							96.0	90.8	86.1	81.9
18								96.2	91.3	86.8
19									96.4	91.7
20										96.5

For comparison the proportion, mean, and median rank points to two significant figures for a sample size $n = 5$ are given in Table 5.3.

Table 5.3

	Number of failures				
	1	2	3	4	5
Proportion	0.20	0.40	0.60	0.80	1.00
Mean rank	0.17	0.33	0.50	0.67	0.83
Mean rank (calculated)	0.13	0.31	0.50	0.68	0.87
Median rank (tables)	0.13	0.31	0.50	0.68	0.87

5.8.4 Censored samples

In the general case the sample may have to be progressively censored – that is, when components are removed from test or operation for reasons other than chance failure (for example, failures induced by human error). The experience contained in the censored component should, however, be taken into account otherwise valuable information that the item did not fail in the specified period of time will be thrown away. Usually the assumption is made that the censored components have an equal probability of failing in the interval between failures in which they were removed. Components that remain unfailed at the end are treated similarly.

The mean order number for the failed and censored components is calculated from the formula

$$MO_i = MO_{i-1} + \frac{N+1-MO_{i-1}}{1+S_i} \tag{5.59}$$

where MO_i is the mean order number of the ith failure out of the total sample of N components and S_i is the number of survivors at the time of the ith failure. This formula incorporates the assumption of equal probability. For censored samples MO_i is substituted for i in the median rank approximation formula (5.58) given previously. Extending the example in the previous section a little further: if the second failure is assumed not to be due to chance causes (say, a failure which can reasonably be assumed not to come from the same distribution) then the mean order numbers for the remainder are as shown below:

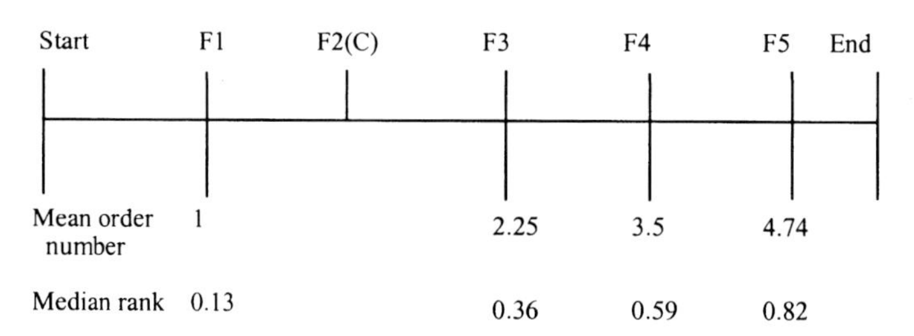

Further useful information on censoring is given by Bompas-Smith (**1**) and Epstein (**2**).

5.8.5 Probability plotting

Special probability papers have been developed to facilitate Weibull analysis. An example of Weibull probability paper by Chartwell (Graph Data Ref. 6572) is shown in Fig. 5.15. Failure times (age at failure) are plotted against the median rank of the cumulative percentage of the population failed and the best straight line drawn through the points. A perpendicular is erected from this line to the estimation point in the top left-hand corner of the paper. The intersection of this line with the two scales marked β and P_μ provide estimates of the shape factor and the percentage point of the mean (μ) of the derived distribution.

The characteristic life (η) estimator is shown as a dotted line across the paper. The intersection of this dotted line with the plotted line gives the value of the characteristic life.

For the distribution shown in Fig. 5.15 the Weibull parameters are therefore

Shape factor, $\beta = 3.3$
Characteristic life, $\eta = 30\,000$ h
Mean life $\mu = 27\,000$ h

Since the shape factor is 3.3 then the distribution of times to failure is approximately normal for the electric motor bearings in this case. A clear wear-out pattern has been identified and scheduled replacement will be the best maintenance policy to adopt.

To limit the failure in service probability to 1 in 10 the B10 life (10 per cent probability of failure) should be adopted as the maximum scheduled maintenance interval. Here the B10 life is 15 500 h so that replacement of bearings should be carried out at 20-month intervals. If 18 months is a more convenient interval for scheduled maintenance then the probability of an in-service failure reduces to 0.06.

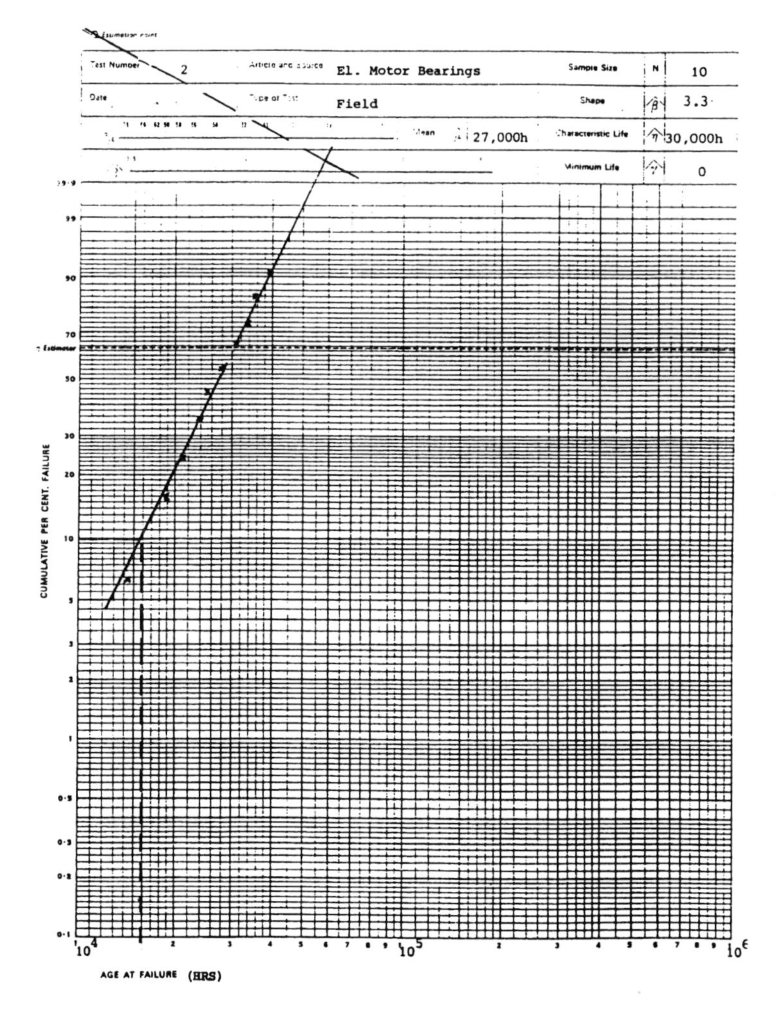

Fig. 5.15 Bearing examples

An example of Weibull plotting with censoring is shown in Fig. 5.16. In this example two items out of a total population of ten were replaced and two items were still operating when the test was truncated at 5000 h. The list of failure and censored times, estimates of mean order numbers, and median ranks are shown below in Table 5.4.

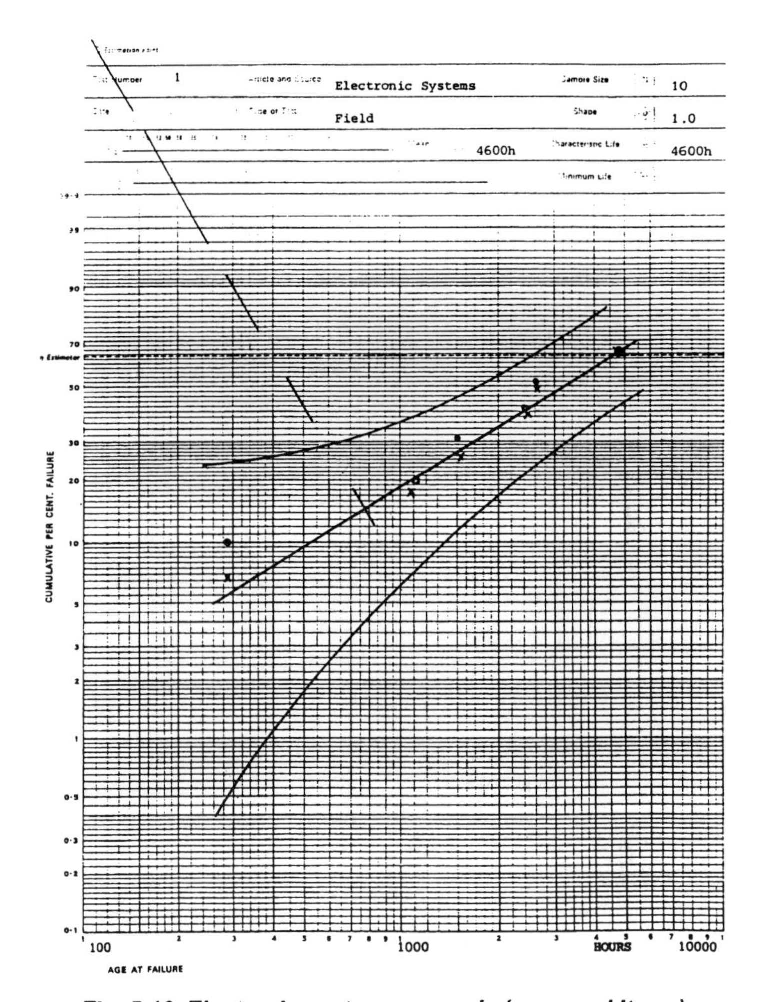

Fig. 5.16 Electronic systems example (censored items)

Table 5.4 Electronic systems

Item number	Failure time (t_i)*	MO_i	S_i	Median rank (%)	
1	290	1	10	6.7	
2	676(c)	–	–	–	Replaced before failure
3	1090	2.1	8	17.3	
4	1514	3.2	7	27.9	
5	2422	4.3	6	38.5	
6	2601	5.4	5	49.0	
7	4157(c)	–	–	–	Replaced before failure
8	4976	6.8	3	62.5	
9	5000(c)	–	–	–	
10	5000(c)	–	–	–	Test suspended at 5000 h

* Censored items are marked (c).

For this distribution the Weibull parameters are:

Shape factor, $\beta = 1.0$
Characteristic life, $\eta = 4600$ h
Mean life, $\eta = 4600$ h

Since the shape factor is 1.0, the mean and characteristic life are the same. The times to failure follow a simple exponential distribution indicating a constant failure rate – scheduled maintenance is not worthwhile and therefore breakdown maintenance is the preferred policy for these electronic systems.

In this example the 90 per cent confidence interval is defined by the 5 per cent and 95 per cent bounds plotted on either side of the Weibull line. These are produced by plotting failure times against 5 per cent and 95 per cent median ranks. Tables of 95 per cent and 5 per cent median ranks are given in Tables 5.5 and 5.6 respectively.

Table 5.5 95 per cent ranks (numbers 1 – 20)

Rank order	Sample size 1	2	3	4	5	6	7	8	9	10
1	95.0	77.6	63.1	52.7	45.0	39.3	34.8	31.2	28.3	25.8
2		97.4	86.4	75.1	65.7	58.1	52.0	47.0	42.9	39.4
3			98.3	90.2	81.0	72.8	65.8	59.9	54.9	50.6
4				98.7	92.3	84.6	77.4	71.0	65.5	60.6
5					98.9	93.7	87.1	80.7	74.8	69.6
6						99.1	94.6	88.8	83.1	77.7
7							99.2	95.3	90.2	84.9
8								99.3	95.8	91.2
9									99.4	96.3
10										99.4

Rank order	Sample size 11	12	13	14	15	16	17	18	19	20
1	23.8	22.0	20.5	19.2	18.1	17.0	16.1	15.3	14.5	13.9
2	36.4	33.8	31.6	29.6	26.3	27.9	25.0	23.7	22.6	21.6
3	47.0	43.8	41.0	38.5	36.3	34.3	32.6	31.0	29.5	28.2
4	56.4	52.7	49.4	46.5	43.9	41.6	39.5	37.6	35.9	34.3
5	65.0	60.9	57.2	54.0	51.0	48.4	46.0	43.8	41.9	40.1
6	72.8	68.4	64.5	60.9	57.7	54.8	52.1	49.7	47.5	45.5
7	80.0	75.4	71.2	67.4	64.0	60.8	58.0	55.4	52.9	50.7
8	86.4	81.8	77.6	73.6	70.0	66.6	63.5	60.7	58.1	55.8
9	92.1	87.7	83.4	79.3	75.6	72.1	68.9	65.9	63.1	60.6
10	96.6	92.8	88.7	84.7	80.9	77.3	73.9	70.8	67.9	65.3
11	99.5	96.9	93.3	89.5	85.8	82.2	78.8	75.6	72.6	69.8
12		99.5	97.1	93.8	90.3	86.7	83.3	80.1	77.0	74.1
13			99.6	97.4	94.3	90.9	87.6	84.3	81.2	78.2
14				99.6	97.5	94.6	91.5	88.3	85.2	82.2
15					99.6	97.7	95.0	92.0	89.0	86.0
16						99.6	97.8	95.2	92.4	89.5
17							99.6	97.9	95.5	92.8
18								99.7	98.0	95.7
19									99.7	98.1
20										99.7

Table 5.6 5 per cent ranks (numbers 1 – 20)

Rank order	Sample size 1	2	3	4	5	6	7	8	9	10
1	5.0	2.5	1.6	1.2	1.0	0.8	0.7	0.6	0.5	0.5
2		22.3	13.5	9.7	7.8	6.2	5.3	4.6	4.1	3.6
3			36.8	24.8	18.9	15.3	12.8	11.1	9.7	8.7
4				47.2	34.2	27.1	22.5	19.2	16.8	15.0
5					54.9	41.8	34.1	28.9	25.1	22.2
6						60.6	47.9	40.0	34.4	30.3
7							65.1	52.9	45.0	39.3
8								68.7	57.0	49.3
9									71.6	60.5
10										74.1

Rank order	Sample size 11	12	13	14	15	16	17	18	19	20
1	0.4	0.4	0.3	0.3	0.3	0.3	0.3	0.2	0.2	0.2
2	3.3	3.0	2.8	2.6	2.4	2.2	2.1	2.0	1.9	1.8
3	7.8	7.1	6.6	6.1	5.6	5.3	4.9	4.7	4.4	4.2
4	13.5	12.2	11.2	10.4	9.6	9.0	8.4	7.9	7.5	7.1
5	19.9	18.1	16.5	15.2	14.1	13.2	12.3	11.6	10.9	10.4
6	27.1	24.5	22.3	20.6	19.0	17.7	16.6	15.6	14.7	13.9
7	34.9	31.5	28.7	26.4	24.3	22.6	21.1	19.8	18.7	17.7
8	43.5	39.0	35.4	32.5	29.9	27.8	26.0	24.3	22.9	21.7
9	52.9	47.2	42.7	39.0	35.9	33.3	31.0	29.1	27.3	25.8
10	63.5	56.1	50.5	45.9	42.2	39.1	36.4	34.0	32.0	30.1
11	76.1	66.1	58.9	53.4	48.9	45.1	41.9	39.2	36.8	34.6
12		77.9	68.8	61.4	56.0	51.5	47.8	44.9	41.8	39.3
13			79.4	70.3	63.6	58.3	53.9	50.2	47.0	44.1
14				80.7	72.0	65.6	60.4	56.1	52.4	49.2
15					81.8	73.6	67.3	62.3	58.0	54.4
16						82.9	74.9	68.9	60.4	59.8
17							83.8	76.2	70.4	65.6
18								84.6	77.3	71.7
19									85.4	78.3
20										86.0

5.8.6 *Hazard plotting*

An alternative and simpler method of applying multiple censoring to failure data is the method due to Nelson (**3**) known as hazard plotting. The method is widely used in reliability analysis and is particularly applicable to the Weibull distribution.

When a failure occurs an estimate of the hazard rate, $h(t)$, at the time of failure is given by the reciprocal of the number of survivors immediately before the failure. The cumulative hazard function $H(t)$ is then the sum of these hazard rates at time t. Thus for the electronic systems example given previously, the hazard rate and cumulative hazard functions are as shown in Table 5.7.

Table 5.7

No. of failures (i)	Failure time t_i (h)*	No. of survivors (s_i)	Hazard rate $h(t)$	Cumulative hazard $H(t) = \sum h(t)$
1	290	10	0.1	0.1
–	676(c)			
1	1090	8	0.125	0.225
1	1514	7	0.143	0.368
1	2422	6	0.167	0.535
1	2601	5	0.200	0.735
–	4157(c)			
1	4976	3	0.33	1.07
–	5000(c)			
–	5000(c)			

* Censored items are marked (c).

The relationship between the failure probability $F(t)$ and the cumulative hazard function $H(t)$ is

$$F(t) = 1 - \exp\{-H(t)\} \tag{5.60}$$

Hence the values of $H(t)$ and $F(t)$ are as shown in Table 5.8 with the median rank values calculated previously for comparison.

Table 5.8

Failure time (t_i)	Cumulative hazard $H(t_i)$	Failure probability $F(t_i)$	Median (M_i)
290	0.100	0.10	0.067
1090	0.225	0.20	0.172
1514	0.368	0.31	0.277
2422	0.535	0.41	0.385
2601	0.735	0.52	0.490
4976	1.070	0.66	0.625

The $F(t_i)$ values are plotted on Fig. 5.16 as circles (median ranks as crosses). It can be seen that the Weibull distributions obtained by cumulative hazard and median rank plotting would be virtually identical. Cumulative hazard plotting can thus be seen to be a useful method which can cope with multiple censored samples. Median rank plots are more accurate and contain more information (for example confidence bounds) but are more time-consuming to perform. They should be used in preference to hazard plotting when suitable computer programs are available for calculating median ranks and mean order numbers.

5.8.7 Standard deviation

The characteristic life and shape factor can be employed to estimate the standard deviation of the Weibull distribution. For this use is made of Table 5.9.

Table 5.9

Shape factor (β)	Estimation factor (B)
1.0	1.0
1.5	0.61
2.0	0.51
2.5	0.37
3.0	0.32
3.5	0.28
4.0	0.21

$$\text{Standard deviation } \sigma = B\eta \qquad (5.61)$$

Linear interpolation is employed to estimate B for intermediate values of the shape factor β. For the bearing example in Fig. 5.15 where $\beta = 3.3$ and $\eta = 30\,000$ h:

$$\sigma = 0.29 \times 30\,000 = 8700 \text{ h}$$

The standard deviation is a measure of the variability in a population and for most distributions 95 per cent of all observations lie within $\pm 2\sigma$ of the mean (that is, within the $2^1/_2$ per cent and $97^1/_2$ per cent limits). For this distribution with a mean of 27 000 h the 2σ upper bound is

$$27\,000 + 2(8700) = 44\,400 \text{ h}$$

which compares closely with the $97^1/_2$ percentile point on the Weibull distribution plotted in Fig. 5.15.

As can be seen from the foregoing example, Weibull analysis is a powerful technique for identifying reliability characteristics of components and equipment. It must be applied with care so that non-relevant failure modes are censored (that is, not counted as failures for the analysis). The shape factor β indicates the form of the lifetime distribution. When it is less than unity the failure rate is decreasing with age, when equal to unity it is constant, and when greater than unity increasing with age. The shape factor thus gives a clear indication of the type of maintenance required and, for equipment exhibiting wear-out, the optimum frequency for scheduled maintenance.

5.9 Summary

This chapter has defined the basic characteristics which can be used to measure the reliability performance of components or systems. The failure process and the repair process have been investigated individually. They have then been considered in combination to model a complete component life-cycle. Finally, expressions for the probability of component failure have been derived when the component is unrepairable or repairable for both revealed and unrevealed failure modes.

The expressions given in this chapter for component failure probabilities can now be utilized in the methods described in the succeeding chapters to obtain the probability of system failure.

5.10 References

(1) **Bompas-Smith, J. H.** (1973) *Mechanical Survival*. McGraw-Hill, London.

(2) **Epstein, B.** (1954) Truncated lifetests in the exponential case. *Ann. Math. Stat.*, **125**, 555–564.

(3) **Nelson, L. S.** (1969) Hazard plotting for incomplete data. *J. Qual. Technology*, **I**, 27–52.

5.11 Bibliography

Barlow, R. E. and **Proschan, F.** (1965) *Mathematical Theory and Reliability*. John Wiley.

Barlow, R. E. and **Proschan, F.** (1975) *Statistical Theory of Reliability and Life Testing*. Holt, Rinehart and Winston.

Frankel, E. G. (1988) *Systems Reliability and Risk Analysis*. Kluwer Academic Publishers.

Henley, E. J. and **Kumamoto, H.** (1981) *Reliability Engineering and Risk Assessment*. Prentice-Hall.

Villemeur, A. (1991) *Reliability, Availability, Maintainability and Safety Assessment*. John Wiley.

Chapter 6

Reliability Networks

6.1 Introduction

The previous chapter discussed methods by which component failure probabilities could be determined. The overall objective of a risk assessment or reliability study is to predict the performance of the complete system. Several methods will now be presented which enable the system failure probability to be evaluated in terms of the component performance characteristics. These methods comprise reliability networks, fault tree analysis, Markov models, and simulation. The first of these methods, reliability networks, is the subject of this chapter.

A **reliability network** is a representation of the reliability dependencies between components of a system. Dependencies are used in such a way as to represent the means by which the system will function. Such a network used to assess the probability of failure of a system consists of the following features:

(a) a start node
(b) an end node
(c) a set of nodes V
(d) a set of edges E
(e) an incidence function ϕ which associates with each edge an ordered pair of nodes.

The set E of edges are used to represent the components which comprise the system. The nodes are the means by which the system structure is defined; they provide the points at which components are joined together. For example, a simple reliability network is shown in Fig. 6.1. This system is constructed from three components which make up the set of edges:

$$E = \{A, B, C\}$$

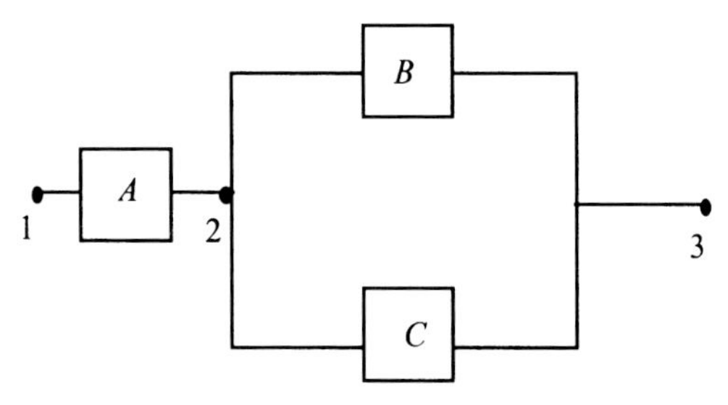

Fig. 6.1 Simple reliability network

The nodes in the network are the component link points:

$$V = \{1, 2, 3\}$$

ϕ, the incidence function, identifies an ordered pair of link points for each component:

$$\begin{aligned} \phi = A &\rightarrow (1, 2) \\ B &\rightarrow (2, 3) \\ C &\rightarrow (2, 3) \end{aligned}$$

The leftmost node, node 1, is the start node in the network and the rightmost node, node 3, is the end node.

The edges $Z \rightarrow (i, j)$ represent the component's functionality in the following way. When a component is working normally then there is a path from node i to node j through component Z. When Z fails, that path from node i to node j is broken.

Paths through the network then represent the functioning of the system. If there exists some path from the start node to the end node the system is considered to be in the working state. When component failures break edges such that there is no path through the network, and the start node and end node are isolated, forming two disjoint sections of the network, then the system is in the failed state.

The path through any component is unidirectional going from left to right. An exception to this occurs when components are placed vertically on the network diagram. Vertical components represent bidirectional components where 'flow' can be considered either way through the component.

6.2 Simple network structures

Components are placed in the networks by considering only their contribution to the system failure mode under investigation. The resulting network has components connected together in a manner which exhibits differing degrees of complexity. In this initial section the type of network will be restricted to what will be categorized as 'simple' networks. These 'simple' networks will encompass only those networks whose structure is formed by series and parallel combinations of the components. This type of system can be analysed by a reduction process which gradually redefines the network into a successively simpler structure. If networks do not conform to the series/parallel requirement then they are categorized as 'complex networks' and their analysis requires more sophisticated approaches which will be discussed later in this chapter.

6.2.1 *Series networks*

In systems where component failure cannot be tolerated, that is non-redundant systems, then the components are placed in series to represent the system reliability. The failure of any component in a non-redundant system will cause system failure. In a series network any component failure will break the single path through the network.

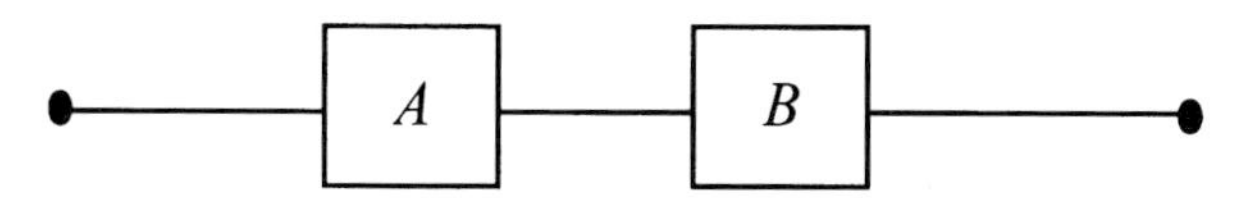

Fig. 6.2 Two-component series network

For the two-component series network shown in Fig. 6.2 the system failure and success probabilities can be expressed in terms of the component failure and success likelihoods. If R_A and R_B are probabilities of component success and Q_A and Q_B and are the probabilities of component failure ($Q_A + R_A = 1$ and $Q_B + R_B = 1$). The system will fail if either component A fails (A_F) or component B fails (B_F).

So

$$
\begin{aligned}
Q_{SYS} &= P(A_F \text{ OR } B_F) \\
&= P(A_F) + P(B_F) - P(A_F \text{ AND } B_F) \\
&= Q_A + Q_B - Q_A \cdot Q_B \\
&= 1 - (1 - Q_A)(1 - Q_B) \qquad (6.1)
\end{aligned}
$$

The system will only work if both A works (A_w) and B works (B_w). Therefore

$$
\begin{aligned}
R_{SYS} &= P(A_w \text{ AND } B_w) \\
&= R_A \cdot R_B \qquad (6.2)
\end{aligned}
$$

In a general series network containing n components the system failure and success expressions are

$$Q_{SYS} = 1 - \prod_{i=1}^{n} (1 - Q_i) \qquad (6.3)$$

$$R_{SYS} = \prod_{i=1}^{n} R_i \qquad (6.4)$$

Since each R_i is less than one, as the number of components increases the system reliability decreases.

6.2.2 Parallel networks

For situations where the system will function provided that any one of its components functions, a parallel network structure results. For the two-component parallel network shown in Fig. 6.3 there is a path through the network even if component A or component B has failed. It requires both components to fail before there are no pathways from the start point to the finish. This represents a fully redundant system.

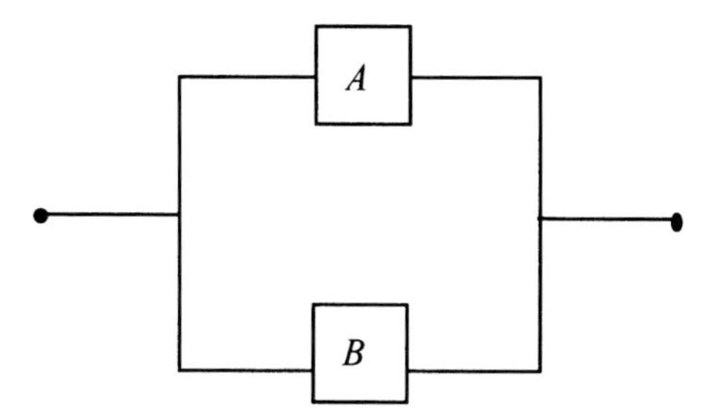

Fig. 6.3 Two-component parallel network

For system failure both components A and B must fail. Therefore

$$Q_{SYS} = P(A_F \text{ AND } B_F)$$

$$= Q_A \cdot Q_B \tag{6.5}$$

Provided that either A or B works, then the system will function, so

$$R_{SYS} = P(A_w \text{ OR } B_w)$$

$$= 1 - (1 - R_A)(1 - R_B) \tag{6.6}$$

Extending equations (6.5) and (6.6) to n components gives

$$Q_{SYS} = \prod_{i=1}^{n} Q_i \tag{6.7}$$

and

$$R_{SYS} = 1 - \prod_{i=1}^{n} (1 - R_i) \tag{6.8}$$

If the equations for the unreliability and reliability of the series system, equations (6.3) and (6.4) respectively, are compared with the equivalent equations stated above for the parallel system, it can be seen that the expressions are the same for each type of network with R and Q interchanged.

From equation (6.7), since each Q_i is less than 1, the unreliability of a parallel system decreases as the number of components increases. It is not, however, a good idea to improve system performance by adding ever-increasing levels of redundancy. The initial increase of incorporating parallel components has the greatest benefit. Further parallel additions provide a diminishing return. For example the system reliability and the degree of improvement attained by a system where components with reliability 0.9 are placed in parallel are shown in Table 6.1.

Table 6.1 System performance for increased redundancy levels

Number of parallel components	System reliability	Incremental reliability	Comparative reliability (%)
1	0.9	—	—
2	0.99	0.09	10
3	0.999	0.009	11
4	0.9999	0.0009	11.1
5	0.999 99	0.000 09	11.11
6	0.999 999	0.000 009	11.111

A single component gives a system reliability of 0.9. Adding a second in parallel to provide a first level of redundancy gives a system success probability of 0.99, an improvement of 10 per cent on the single-component system. Utilizing a third component to provide a second redundancy level gives a system reliability of 0.999. This is only an 11 per cent improvement on the single-component situation. So the addition of a third component has yielded only a 1 per cent improvement from the two-component option. Further parallel additions give the diminishing returns shown in Table 6.1.

For safety systems it is the probability of the system failing to function which is of interest. Again taking a component with reliability 0.9, the probability of failure for a single component is 0.1. Placing a second in parallel gives a failure probability 0.01 and so on, such that six parallel safety systems would give a failure probability of 1×10^{-6}.

Therefore, to improve the reliability of a product there is a limit to the degree of improvement that can be obtained by building in redundancy. However, adding redundancy can provide a significant relative improvement to the failure probability.

A second reason that increasing redundancy levels do not always give a significant improvement in system performance is common cause failures where a single common failure of all identical components can occur at the same time. This topic is covered in more detail in Chapter 8.

6.2.3 Series/parallel combinations

When a network to represent system reliability consists of only combinations of series and parallel structures its analysis can be carried out in stages. Each stage simplifies the network by combining series and parallel sections. By using equations (6.3) and (6.4) for series combinations and (6.7) and (6.8) for parallel combinations, expressions for the reliability or unreliability of the simplified network sections can be obtained at each stage. The reduction process continues until one 'super-component' remains which links the start and end nodes. The performance of this 'super-component' is then identical to that of the system.

Example

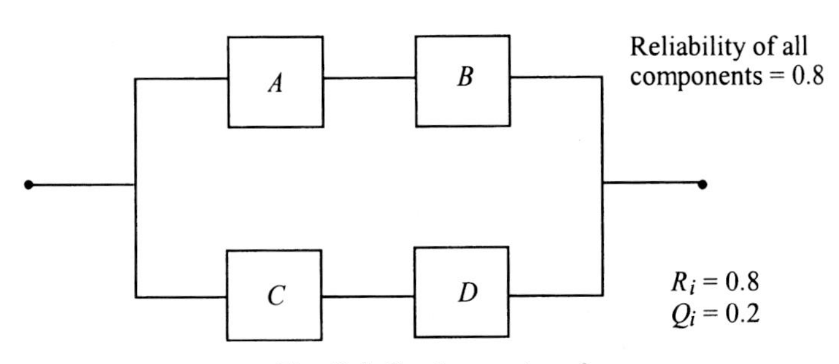

Fig. 6.4 System network

Since the simple network in Fig. 6.4 is constructed of only series and parallel combinations it satisfies the requirement for reduction in the manner described above.

Step 1. Combine series components A and B to form AB and series components C and D to form CD.

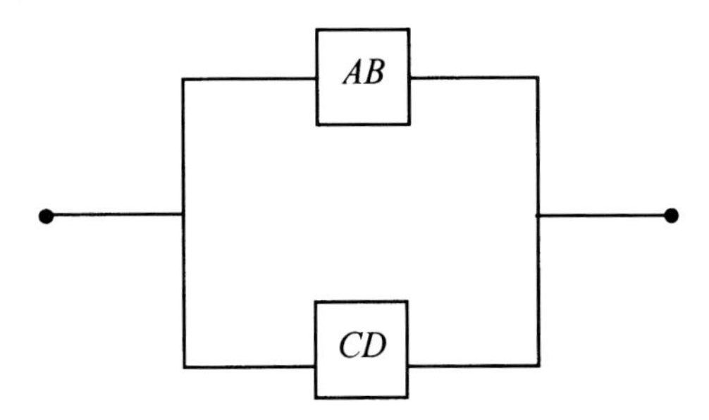

where

$$R_{AB} = R_A \,.\, R_B$$

$$= 0.64$$

$$Q_{AB} = 1 - (1 - Q_A)\,(1 - Q_B)$$

$$= 0.36$$

$$R_{CD} = R_C \,.\, R_D$$

$$= 0.64$$

$$Q_{CD} = 1 - (1 - Q_C)(1 - Q_D)$$

$$= 0.36$$

Step 2. Combine parallel sub-systems *AB* and *CD* to form a single 'super-component' SYS:

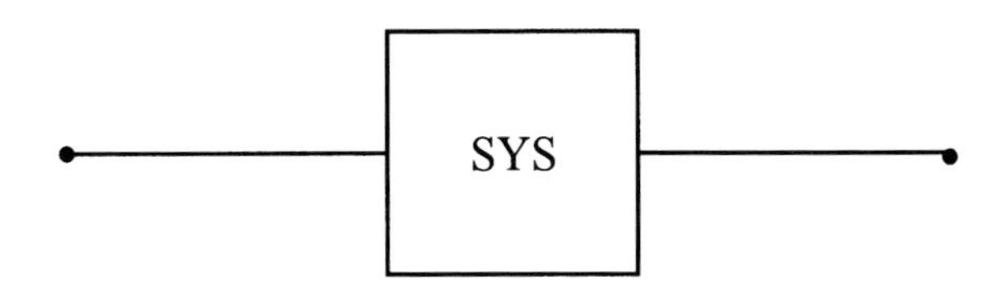

where

$$R_{\mathrm{SYS}} = 1 - (1 - R_{AB})\,(1 - R_{CD})$$

$$= 0.8704$$

and

$$Q_{\mathrm{SYS}} = Q_{AB} \cdot Q_{CD}$$

$$= 0.1296$$

6.2.4 Voting systems

It is common practice for safety shutdown systems to use voting logic to initiate the trip. Voting systems are designed such that a number of sensing devices are installed, and a function is performed when at least a set number of the sensors register the trip condition. This type of design can be used to reduce the chances of spurious system faults. A k-out-of-n voting system has n components, k of which must indicate a trip condition for the system to function. A partially redundant system of this type is shown in Fig. 6.5 for a two-out-of-three voting situation.

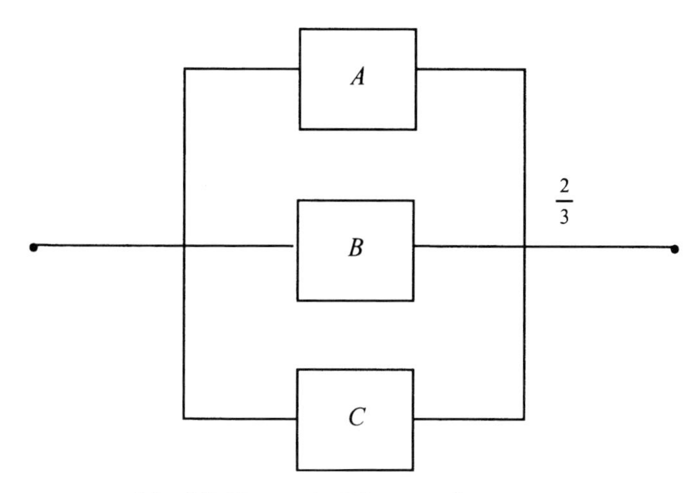

Fig. 6.5 Two-out-of-three voting system

Identical components

If the system in Fig. 6.5 is constructed using identical components then it features redundant components and

$$R_A = R_B = R_C = R \quad \text{and} \quad Q_A = Q_B = Q_C = Q$$

System success occurs provided that at least two of the components function. Therefore if either two or three components work then the system will also perform its required task. To derive the system failure probability, the binomial distribution can be applied. We have n trials (the number of components in the voting system), each trial can result in success (component works) or failure (component fails) with identical failure probabilities. For each trial we are requiring r successes (where r is two or three components to work) for the system to function.

$$P(\text{two components work}) = {}^3C_2R^2Q = 3R^2Q$$

$$P(\text{three components work}) = {}^3C_3R^3 = R^3$$

Adding the probabilities of these two mutually exclusive events in which the system will work gives

$$R_{\text{SYS}} = 3R^2Q + R^3$$

For this system, failure will result if only one component or none of the components function. Another way of viewing these situations is that two or three of the components fail. Again using the binomial distribution:

$$P(\text{2 components fail}) = 3Q^2R$$

$$P(\text{3 components fail}) = Q^3$$

$$Q_{\text{SYS}} = 3Q^2R + Q^3$$

Non-identical components

By using non-identical components in the voting system diversity is introduced. The probabilities of each component functioning or failing will not now be the same and the probability of each combination of component states which yields the system outcome required will have to be calculated separately.

Therefore, for system success we have

$$P(\text{two components work}) = R_AR_BQ_C + R_AR_CQ_B + R_BR_CQ_A$$

$$P(\text{three components work}) = R_AR_BR_C$$

So

$$R_{\text{SYS}} = R_AR_BQ_C + R_AR_CQ_B + R_BR_CQ_A + R_AR_BR_C$$

For system failure

$$P(\text{two components fail}) = Q_A Q_B R_C + Q_A Q_C R_B + Q_B Q_C R_A$$

$$P(\text{three components fail}) = Q_A Q_B Q_C$$

So

$$Q_{SYS} = Q_A Q_B R_C + Q_A Q_C Q_B + Q_B Q_C Q_A + Q_A Q_B Q_C$$

6.2.5 Standby systems

The redundancy represented by the use of parallel networks may be used to model systems which are either operating continuously but are kept in a standby mode or those which are only placed in operating mode when the normally operating component fails. When redundant systems are being modelled the reliability of the selector or switch-over component must also be considered. Three ways that the switch can be included in the analysis are detailed below.

Case 1 – perfect switching

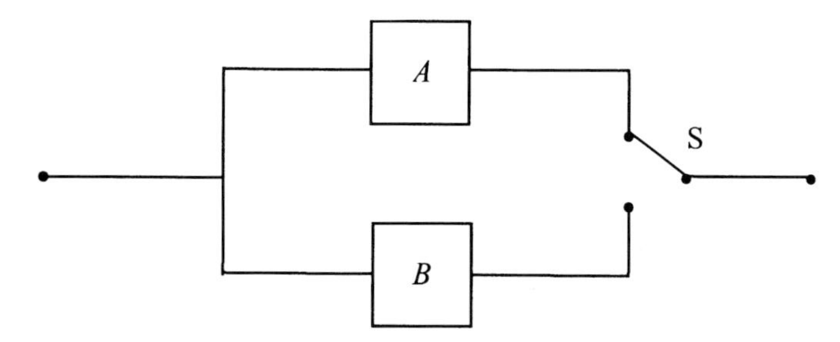

Fig. 6.6 Standby system

For perfect switching the switch S is considered to be 100 per cent reliable and working both during normal operation and during the changeover to the back-up system when failure of the primary system occurs.

If B does not fail when in standby mode then it can only fail when called upon to function, i.e. after A has failed:

$$P(\text{system failure}) = P(A \text{ fails and then } B \text{ fails})$$

$$Q_{SYS} = Q_A Q_{B|A}$$

Case 2 – imperfect switching

If the switching or changeover system is modelled as an item subject to failure then this component must also be considered in the calculations. System failure can occur given a successful changeover (which occurs with probability R_S) OR the changeover can fail (with probability Q_S) so

$$
\begin{aligned}
P(\text{system failure}) = {} & P(\text{system failure given successful changeover}) \\
& \times P(\text{successful changeover}) \\
& + P(\text{system failure given unsuccessful changeover}) \\
& \times P(\text{unsuccessful changeover})
\end{aligned}
$$

$$
\begin{aligned}
Q_{SYS} &= Q_A Q_B R_S + Q_A Q_S \\
&= Q_A Q_B (1 - Q_S) + Q_A Q_S \\
&= Q_A (Q_B + Q_S - Q_B Q_S)
\end{aligned}
$$

Case 3 – switch failures during normal operation and switching

If the switching mechanism can fail in normal operation and fail to switch to the standby on demand then since it can fail when either sub-system A or B is operating it has to be included as an additional component in series in the network.

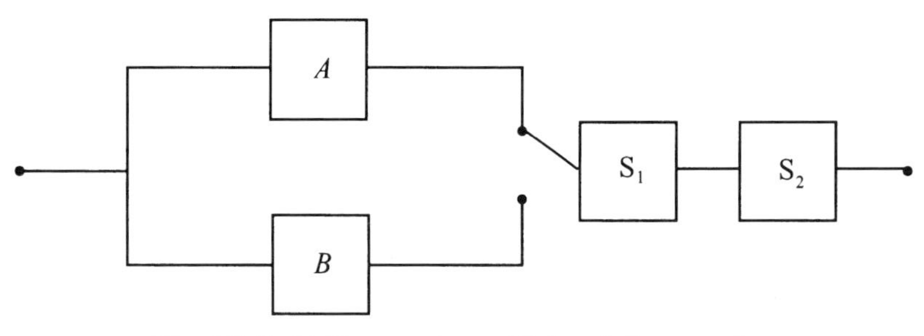

Fig. 6.7 Switching system with two failure modes

The switch is represented by two components in the network to model the distinct failure modes. S_1 models imperfect switching and S_2 models the switch failure during normal operation.

If the probability that the switching fails is Q_{S1}, and the probability that the switch fails in normal operation is Q_{S2}, then

$R_{S1} = 1 - Q_{S1}$ and $R_{S2} = 1 - Q_{S2}$

P(system failure) = P[(A fails and changeover fails) OR
(A fails, changeover works, and B fails) OR
(switch fails in normal operation)]

$$Q_{SYS} = Q_A Q_{S1} + Q_A R_{S1} Q_B + Q_{S2}$$

$$= Q_A Q_{S1} + Q_A Q_B (1 - Q_{S1}) + Q_{S2}$$

Example – system network quantification
Derive expressions for the system failure and success probabilities in terms of the component failure and success probabilities for the following network:

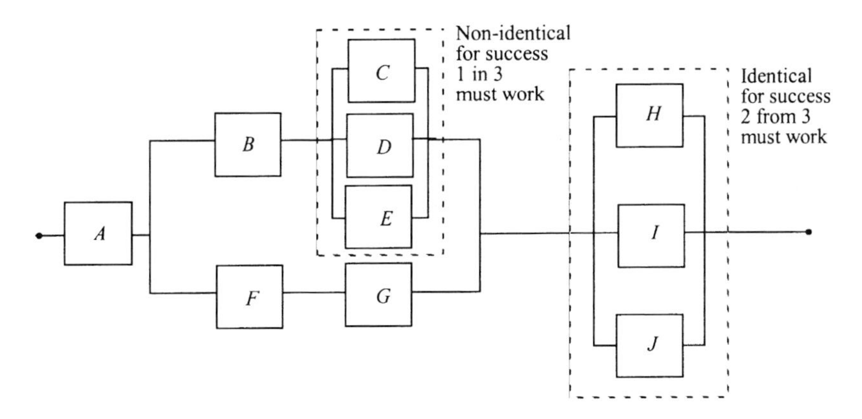

Step 1. Reduce sub-systems C, D, E, and H, I, J.

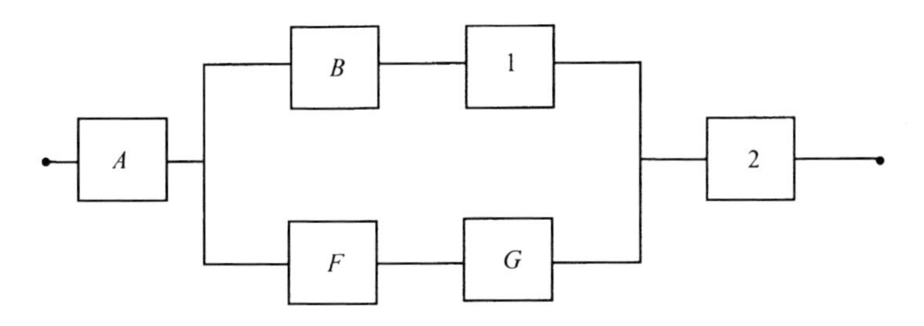

P(sub-system 1 works) = P[(any one component works) OR
(any two components work) OR
(all three components work)]

$$R_1 = R_C Q_D Q_E + R_D Q_C Q_E + R_E Q_D Q_C + R_C R_D Q_E + R_C R_E Q_D + R_D R_E Q_C + R_C R_D R_E$$

$$Q_1 = P(\text{all three components fail})$$

$$= Q_C Q_D Q_E$$

For sub-system 2, let

$$Q = Q_H = Q_I = Q_J \quad \text{and} \quad R = R_H = R_I = R_J$$

then

$$P(\text{sub-system 2 works}) = P[(\text{any two components work}) \text{ OR } (\text{all three components work})]$$

$$R_2 = 3R^2Q + R^3$$

$$Q_2 = 3Q^2R + Q^3$$

Step 2. Combine the two series sections B and 1, and also F and G:

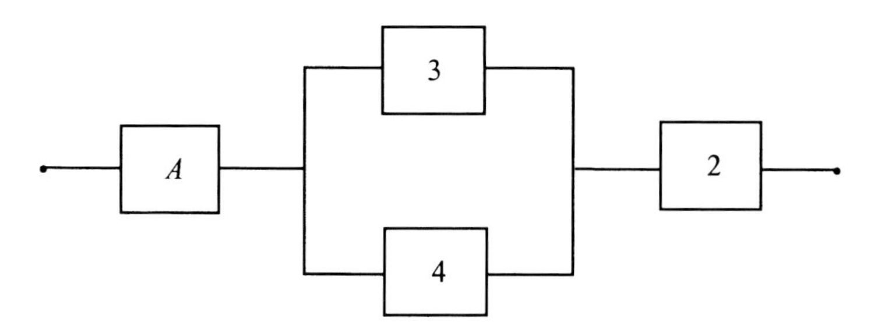

$$Q_3 = Q_B + Q_1 - Q_1 Q_B$$

$$R_3 = R_B R_1$$

$$Q_4 = Q_F + Q_G - Q_F Q_G$$

$$R_4 = R_F R_G$$

Step 3. Combine parallel section with components 3 and 4:

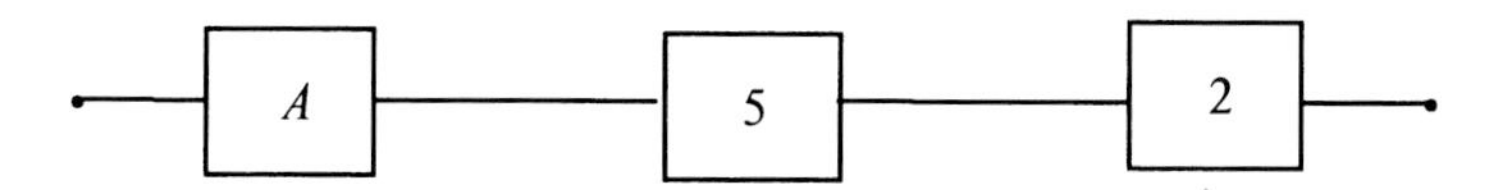

$$Q_5 = Q_3Q_4$$

$$R_5 = R_3 + R_4 - R_3R_4$$

Step 4. Finally combine the three series sections to produce system representation:

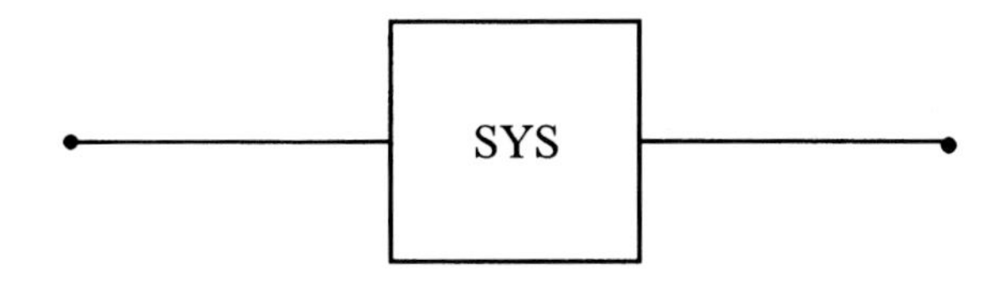

$$Q_{SYS} = Q_A + Q_5 + Q_2 - Q_A\,Q_5 - Q_A\,Q_2 - Q_2\,Q_5 + Q_A\,Q_5\,Q_2$$

$$R_{SYS} = R_AR_5R_2$$

6.3 Complex networks

When the structure of the network is of a form which cannot be sequentially reduced considering alternate series and parallel sections, it comes into the classification of a 'complex' network. Series/parallel networks are not always adequate to represent the functioning of a system. One example of such a network is the bridge network shown in Fig. 6.8. This network has no series section nor parallel section which can enable simplification. To analyse networks of this type different approaches are required. Three possible alternatives will be described in the remainder of this chapter. The first two, the conditional probability approach and the conversion from delta to star configurations, are appropriate for fairly small networks which contain only one or two sections that cannot be easily simplified. The third approach involves finding all possible paths through the network. This third option can be implemented on a computer and hence can be applied to any size and complexity of network.

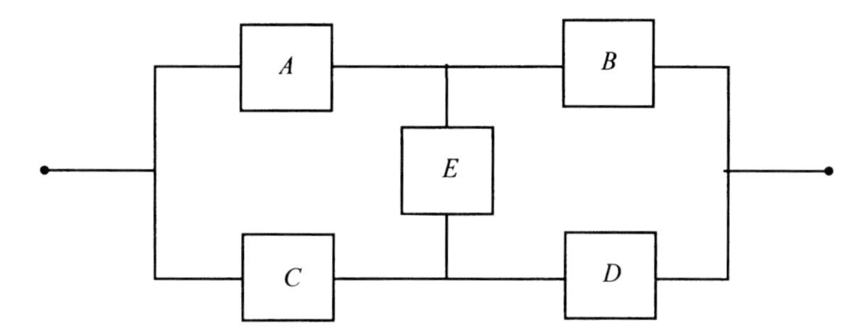

Fig. 6.8 Bridge network

6.3.1 Conditional probability approach

The difficulty with the bridge network which prevents it being simplified is the location of component *E*. If *E* were not in the network at all and there were no link between the top and bottom paths the network could be easily simplified. Also if *E* was replaced by a perfectly reliable link between the top and bottom paths the network could again be simplified. Since *E* is the problem component the conditional probability approach considers all the possible states which *E* can exist in. For each state of *E* the reliability of the resulting network is produced. These can then be combined to provide the system reliability.

For example, if *E* works the network becomes the parallel/series network which we will call system *X*, shown in Fig. 6.9. This is a simple network which can be evaluated:

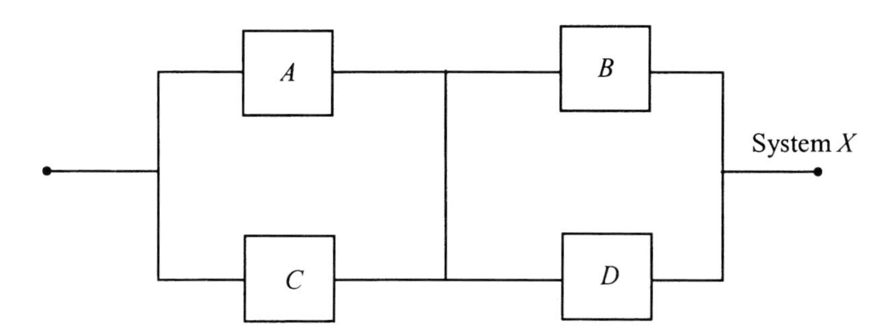

Fig. 6.9 System X

$P(\text{system } X \text{ fails}) = P[(A \text{ AND } C \text{ fail}) \text{ OR } (B \text{ AND } D \text{ fail})]$

$Q_X = Q_A Q_C + Q_B Q_D - Q_A Q_B Q_C Q_D$

$P(\text{system } X \text{ works}) = P[(A \text{ OR } C \text{ works}) \text{ AND } (B \text{ OR } D \text{ works})]$

$R_X = (R_A + R_C - R_A R_C)(R_B + R_D - R_B R_D)$

If component E fails the structure becomes that labelled system Y shown in Fig. 6.10.

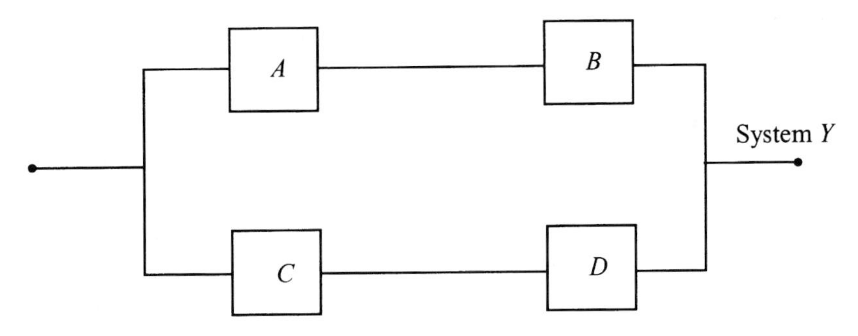

Fig. 6.10 System Y

$P(\text{system } Y \text{ fails}) = P[(A \text{ OR } B \text{ fails}) \text{ AND } (C \text{ OR } D \text{ fails})]$

$Q_Y = (Q_A + Q_B - Q_A Q_B)(Q_C + Q_D - Q_C Q_D)$

$P(\text{system } Y \text{ works}) = P[(A \text{ AND } B \text{ work}) \text{ OR } (C \text{ AND } D \text{ works})]$

$R_Y = (R_A R_B + R_C R_D - R_A R_B R_C R_D)$

Combining these results for the whole system using conditional probabilities gives

$$\begin{aligned} P(\text{system success}) = {} & [P(\text{system success given component } E \text{ works}) \\ & \times P(\text{E works})] \\ & + [P(\text{system success given component } E \text{ fails}) \\ & \times P(\text{E fails})] \end{aligned}$$

So

$$R_{SYS} = (R_A + R_C - R_A R_C)(R_B + R_D - R_B R_D) R_E \\ + (R_A R_B + R_C R_D - R_A R_B R_C R_D)(1 - R_E)$$

$$P(\text{system failure}) = [P(\text{system failure given } E \text{ works}) \times P(E \text{ works})] \\ + [P(\text{system failure given } E \text{ fails}) \times P(E \text{ fails})]$$

$$Q_{SYS} = (Q_A Q_C + Q_B Q_D - Q_A Q_B Q_C Q_D)(1 - Q_E) \\ + (Q_A + Q_B - Q_A Q_B)(Q_C + Q_D - Q_C Q_D) Q_E$$

This approach is sometimes known as the 'key element' method since it involves identifying the key element or elements which can be manipulated to produce simple network structures.

6.3.2 Star and delta configurations

A technique which is sometimes useful for simplifying reliability calculations for complex networks is that of converting three components in a delta configuration to another set of three imaginary components in a star configuration. The two structures need to be equivalent from the reliability viewpoint. Therefore, if components *A*, *B*, and *C* form the delta as shown in Fig. 6.11, the problem becomes one of finding the reliabilities which can be assigned to imaginary components *D*, *E*, and *F* in the star such that the networks are equivalent and can be interchanged.

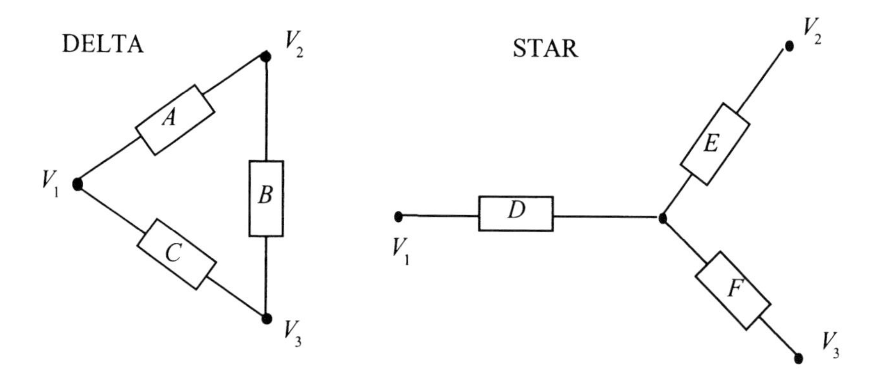

Fig. 6.11 Left-hand stars and deltas

The nodes on both star and delta are labelled V_1, V_2, and V_3. Since the paths are in the direction left to right, V_1 is the input node to this section and output is at both V_2 and V_3.

There are three mutually exclusive events to consider for 'energy flow' between nodes V_2 and V_3 in the delta configuration.

(a) 'Energy flow' through B from V_2 to V_3
(b) 'Energy flow' through B from V_3 to V_2
(c) No 'energy flow' through B.

To build the equations which will yield the reliabilities for imaginary components D, E, and F in terms of the reliability of A, B, and C these situations will be considered separately.

'Energy flow' through B from V_2 to V_3

If the flow direction through B is from V_2 to V_3 then there are two paths from input node V_1 to output node V_3 in the delta configuration. One route is through component A and then B, the other is through C alone. For the star the equivalent path from V_1 to V_3 requires both imaginary components D and F to function. The delta and star configurations resulting in this case are shown in Fig. 6.12.

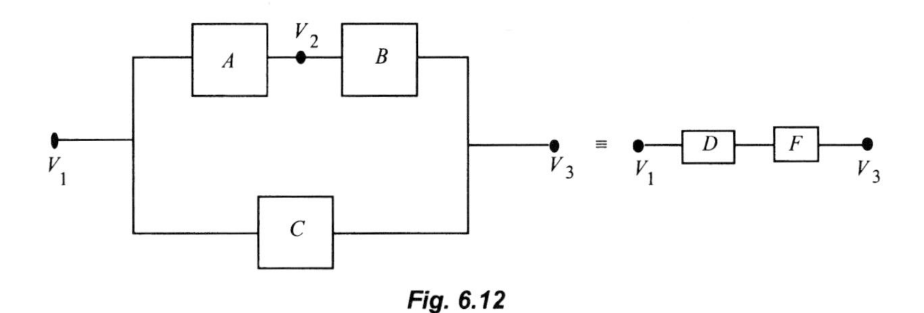

Fig. 6.12

Paths through delta: AB Path through star: DF
C

Since reliability of delta equals reliability of star

$$R_A R_B + R_C - R_A R_B R_C = R_D R_F \tag{6.9}$$

'Energy flow' through B from V_3 to V_2

The resulting star and delta for this situation are shown in Fig. 6.13. Since V_2 becomes the output node the paths through the delta are A and CB. For the star, the path from V_1 to V_2 contains components D and E.

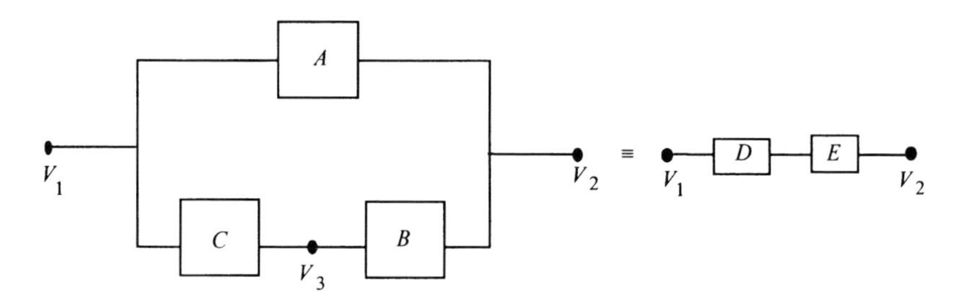

Fig. 6.13

Paths through delta: A Path through star: DE
CB

Equating reliabilities gives

$$R_A + R_B R_C - R_A R_B R_C = R_D R_E \tag{6.10}$$

No 'energy flow' through B

The final situation to consider means that component B in the delta network is irrelevant and this results in two output nodes V_2 and V_3, as shown in Fig. 6.14.

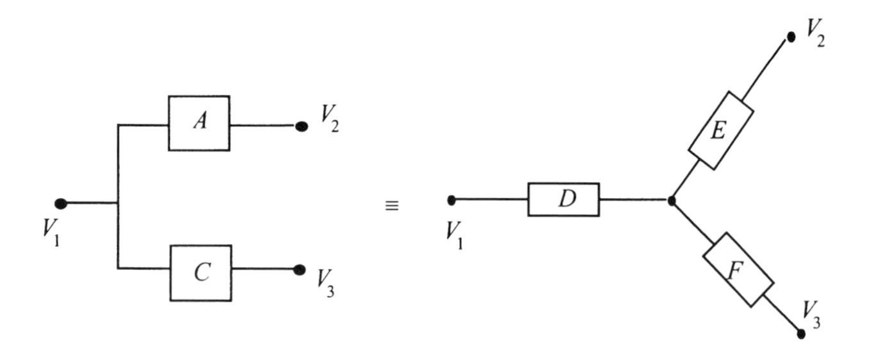

Fig. 6.14

Paths through delta: A Paths through star: DE
C DF

Therefore

$$R_A + R_C - R_A R_C = R_D(R_E + R_F - R_E R_F) \tag{6.11}$$

Equations (6.9), (6.10), and (6.11) give three equations with three unknowns, R_D, R_E, and R_F. Let

$$K_1 = R_A R_B + R_C - R_A R_B R_C$$

$$K_2 = R_A + R_B R_C - R_A R_B R_C$$

$$K_3 = R_A + R_C - R_A R_C$$

then we have

$$R_D \,.\, R_F = K_1$$

$$R_D \,.\, R_E = K_2$$

and

$$R_D(R_E + R_F - R_E R_F) = K_3 \tag{6.12}$$

Solving gives

$$R_D = \frac{K_1 K_2}{K_1 + K_2 - K_3}$$

$$R_E = \frac{K_1 + K_2 - K_3}{K_1}$$

$$R_F = \frac{K_1 + K_2 - K_3}{K_2} \tag{6.13}$$

In the situation that components A, B, and C are identical and $R_A = R_B = R_C = r$, then

$$K_1 = r + r^2 - r^3$$

$$K_2 = r + r^2 - r^3$$

$$K_3 = 2r - r^2$$

and

$$R_D = \frac{\left(r + r^2 - r^3\right)^2}{3r^2 - 2r^3} = \frac{\left(1 + r - r^2\right)^2}{3 - 2r}$$

$$R_E = R_F = \frac{3r^2 - 2r^3}{r + r^2 - r^3} = \frac{r(3 - 2r)}{1 + r - r^2}$$

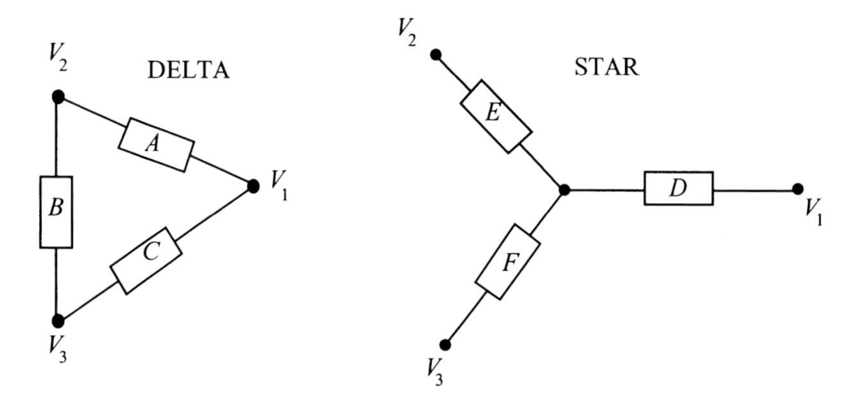

Fig. 6.15 Right-hand stars and deltas

Figure 6.15 shows right-hand delta and star structures with input nodes V_2 and V_3 and output node V_1. By again considering the three 'energy flow' alternatives for component *B* in the delta, equations which will yield the reliabilities of imaginary components *D*, *E*, and *F* can be derived. With reference to the locations of *A*, *B*, *C*, *D*, *E*, and *F* placed as shown in Fig. 6.15 the resulting equations will again be (6.9), (6.10), and (6.11) with solutions (6.13).

By formulating a star structure which is equivalent to a delta in reliability terms the two configurations can be interchanged in networks. This can be utilized to transform a complex network into a simple network suitable for direct analysis. For example consider the double-bridge network of Fig. 6.16.

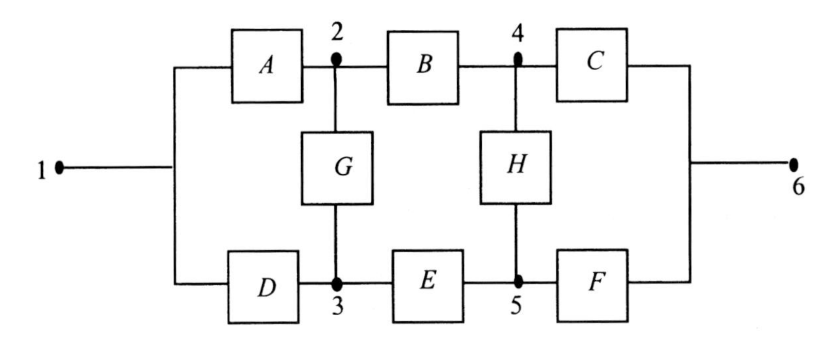

Fig. 6.16 Double-bridge network

This can be reduced by first replacing the left-hand delta (input node 1, output nodes 2 and 3, and components A, D, and G) by an equivalent star with imaginary components X, Y, and Z, to give the network shown in Fig. 6.17.

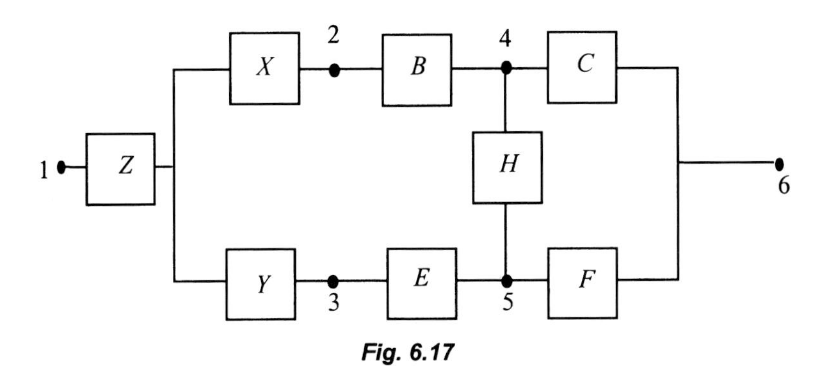

Fig. 6.17

The reliabilities of imaginary components X, Y, and Z are obtained from equation (6.13). Replacing the right-hand delta (input nodes 4 and 5, output node 6 and components H, C, and F) by an equivalent star of imaginary components R, S, and T gives the network in Fig. 6.18.

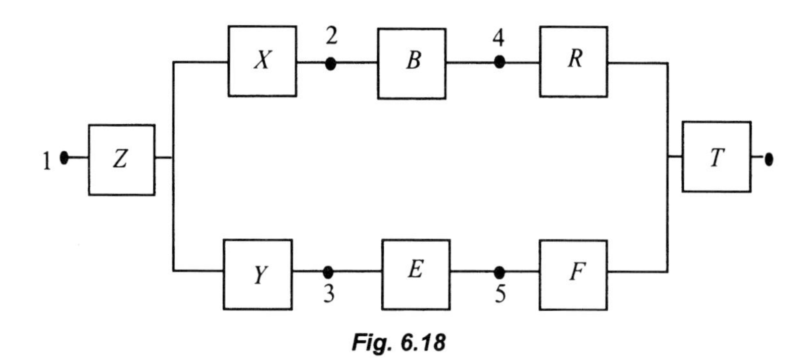

Fig. 6.18

Since the double bridge has now been converted to an equivalent series/parallel network the techniques for simple networks can now be used for its analysis.

6.4 Network failure modes

The final approach presented to obtain the system reliability from a network representation is generally applicable. It can be applied to simple or complex networks of any size. For the method to be versatile enough for all types of network to be solved it is consequently detailed and therefore tedious to apply manually. Involved procedures such as this would normally be implemented on a computer. Before explaining the algorithm there are some definitions which need to be presented.

Path set

A path set is a list of components such that if they all work then the system is also in the working state.

A minimal path set

A minimal path set is a path set such that if any item is removed the system will no longer function.

Cut set

A cut set is a list of components such that if they all fail then the system is also in the failed state.

A minimal cut set

A minimal cut set is a cut set such that if any item is removed from the list the system will no longer fail.

*Connectivity matrix [**C**]*
The connectivity matrix is an $n \times n$ matrix where n is the number of nodes in the network. Matrix elements are defined by

$$c_{ij} = k$$

where k is the number of edges from node i to node j.

Example
The definitions above can be illustrated with reference to the bridge network shown in Fig. 6.19.

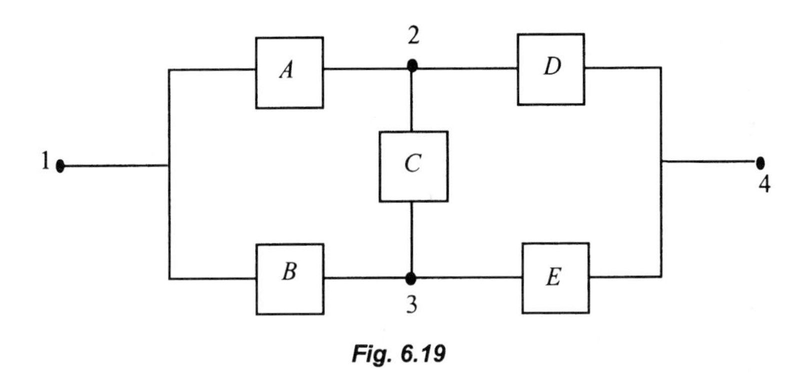

Fig. 6.19

An example of a path set of this network is $\{A, B, C, D, E\}$, i.e. all components work. A second example is $\{A, D, C\}$. In both of these cases the functioning of the components listed provides at least one path from the start node 1 to the end node 4. There are many more path sets which could be listed. They provide the conditions for the existence of at least one path through the network. Since there are many possible path sets for the network in Fig. 6.19 they will not all be listed here.

A minimal path set is more restrictive. Items contained in minimal path sets provide both necessary and sufficient conditions for each path to exist. For example in the path set $\{A, D, C\}$, component C is not necessary. Provided that components A and D are both functioning there is a path along the top branch of the network through node 2. Hence it does not matter whether C functions or not and it can be removed from the path set. This will leave $\{A, D\}$ which is also a path set but it is now a minimal path set since neither of the two items it contains can be removed and the system remain functioning. A full list of minimal path sets is:

1. $\{A, D\}$
2. $\{B, E\}$
3. $\{A, C, E\}$
4. $\{B, C, D\}$

Path sets contain lists of working components and provide a system success representation. Cut sets yield a system failure representation and contain lists of components which if they fail will leave the system in the failed state. Referring to the bridge network of Fig. 6.19 again, an example of a cut set is $\{A, B, C\}$; another example is $\{A, C, D, E\}$. Many more possibilities can be listed. Minimal cut sets represent both the necessary and sufficient conditions for the system failure. For the bridge network there are a total of four minimal cut sets:

1. $\{A, B\}$
2. $\{D, E\}$
3. $\{A, C, E\}$
4. $\{B, C, D\}$

No element in any of the minimal cut sets can be removed and the system remain in the failed state. These system failure modes once evaluated can be used to quantify the system failure probability since for the system to fail at least one minimal cut set must exist. Minimal cut set evaluation is an intermediate step in the quantitative analysis of a network.

6.4.1 Minimal path sets using the connectivity matrix

Any method which can be used to evaluate the minimal path sets of a network will first require some means of defining the system network. In the method which is described in this section the network is specified by its connectivity matrix together with a component list. The component list simply identifies which component is located on any edge between nodes. Each path through the network is traced using the connectivity matrix. Resulting paths are expressed in terms of the list of nodes through which the path travels. By using the component list to identify the components which are located between the path nodes, the minimal path sets are produced.

ISTART = start node

IEND = end node

ICUR = current node in path being traced

IPOIN(n) = pointer at next node in path

IPREV(n) = pointer at previous node in path

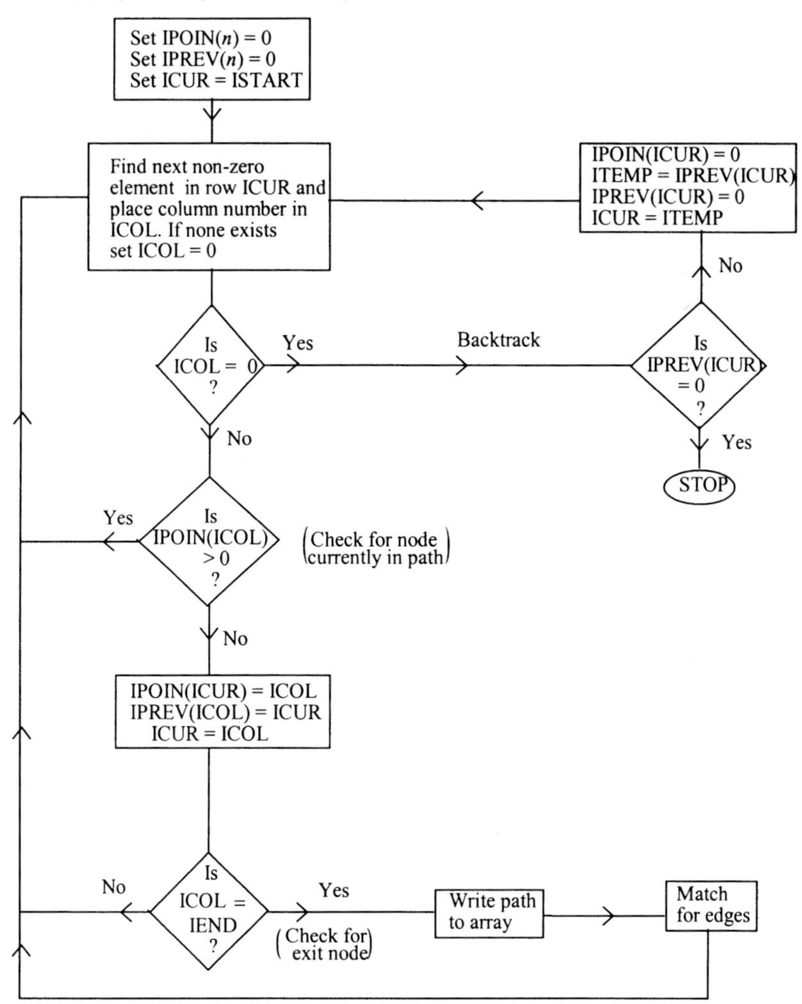

Fig. 6.20 Algorithm to determine the minimal path sets of a network

The algorithm for minimal path set evaluation is presented in Fig. 6.20. This method is ideally suited for computer implementation. Two arrays are used to store information required at intermediate stages in determining the minimal path sets. IPOIN is a variable and contains a pointer which points from any node in the current path being investigated to its successor in the path. This enables the path to be traced forwards through the network. Backward tracking is facilitated by the IPREV array which contains a pointer indicating the predecessor of any node in a path. Variable ICUR identifies the current node being considered in the path.

Initially all entries in the IPOIN and IPREV arrays are set to zero and the current node for consideration is set to the network start node number. The algorithm is explained considering the simple network shown in Fig. 6.21. The variable ICUR indicates the current node number and is initially set to 1, the start node number.

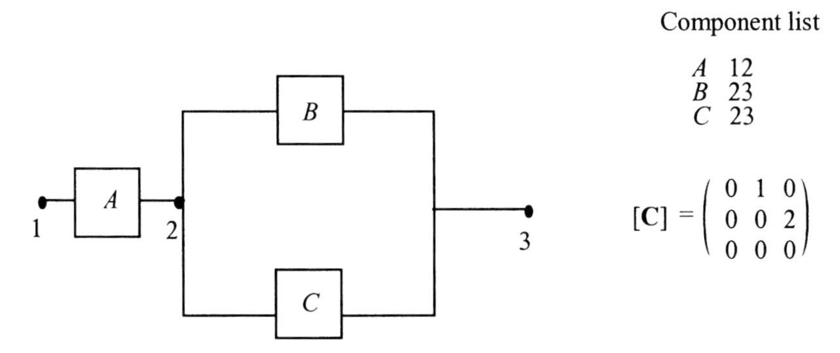

Fig. 6.21 Simple network specification

By scanning row 1 (ICUR) of the connectivity matrix the first non-zero entry indicates the first possible pathway from node 1. This first non-zero element occurs in column 2. This column number is then entered into the variable ICOL. If only zero entries are encountered a zero is entered into ICOL to indicate that all possible pathways leading from node ICUR have been considered.

Prior to progressing further in the current path a test is made to ensure that this node does not previously appear in the current path, i.e. a circular path is being developed. Any node included in the current path will have an entry in the IPOIN array which indicates its successor. This node should only be considered further provided it does not already appear in this path as indicated by IPOIN(ICUR) being zero. If this test is passed this node is incorporated into the current path list.

Once a node is accepted as a member of the current path it is entered as a successor to the previous node in the IPOIN array and its predecessor node number placed in the IPREV array. This newly accepted node now becomes the current node for consideration and the process repeated, i.e. ICUR = 2. Scanning the second row of the connectivity matrix reveals node 3 to be the first node connected to node 2. The array contents at the end of this tracking process are illustrated in Fig. 6.22. ICOL now contains the node number 3 which is the terminal node in the network.

Back track $\quad$ IPOIN $\quad$ IPREV

$$ICUR = 1 \qquad ICOL = 2 \qquad \begin{pmatrix} 2 \\ 0 \\ 0 \end{pmatrix} \qquad \begin{pmatrix} 0 \\ 1 \\ 0 \end{pmatrix}$$

Back track $\quad$ IPOIN $\quad$ IPREV

$$ICUR = 2 \qquad ICOL = 3 \qquad \begin{pmatrix} 2 \\ 3 \\ 0 \end{pmatrix} \qquad \begin{pmatrix} 0 \\ 1 \\ 2 \end{pmatrix}$$

Back track $\quad$ IPOIN $\quad$ IPREV

$$ICUR = 3 \qquad ICOL = 0 \qquad \begin{pmatrix} 2 \\ 0 \\ 0 \end{pmatrix} \qquad \begin{pmatrix} 0 \\ 1 \\ 0 \end{pmatrix}$$

Back track $\quad$ IPOIN $\quad$ IPREV

$$ICUR = 2 \qquad ICOL = 0 \qquad \begin{pmatrix} 0 \\ 0 \\ 0 \end{pmatrix} \qquad \begin{pmatrix} 0 \\ 0 \\ 0 \end{pmatrix}$$

$$ICUR = 1$$

Fig. 6.22 Variable contents at intermediate steps

The path consisting of nodes 1, 2, 3 is now written off to the list of network paths. From the component list the connection from node 1 to node 2 represents component *A*. Row 2, column 3 of the connectivity matrix contains a 2, indicating that there are two possible components connecting nodes 2 and 3. Scanning the component list shows that these are *B* and *C*. Minimal path sets {*A*, *B*} and {*A*, *C*} can now be included in the minimal path set list.

Row 3 of the connectivity matrix shows no possible connections to node 3 and ICOL is returned as zero. At this point backtracking starts. The progression of backtracking is illustrated in Fig. 6.22 which displays the IPOIN and IPREV array contents. The algorithm returns from the terminal node, node 3, to the previous node in the path, node 2, and the connectivity matrix is re-examined. Since no non-zero entries are encountered beyond column 3 all possible connections from node 2 have also been considered. Employing the backtracking phase again returns the current node number to the start point. Since all options for exit paths from node 1 have been dealt with the algorithm terminates and the current minimal path set list is now complete.

6.4.2 Transform minimal path sets to minimal cut sets

A system will fail, that is no path can be traced from the start node to the end node of its reliability network, providing at least one element in each minimal path set has failed.

This is used as a means of determining the minimal cut sets of a network from the minimal path sets. The method for doing this is described as follows.

1. Form minimal path set matrix **[P]** where rows represent each minimal path set and columns represent components:

$$P_{ij} = \begin{cases} 1 & \text{if element } j \text{ is a member of minimal path set } i \\ 0 & \text{otherwise} \end{cases}$$

2. Sum elements of each column:
   (a) If sum of any column is equal to the number of minimal path sets (N_p) then since the event appears in each minimal path set it is a single-order minimal cut set.
   (b) Add all single-order minimal cut sets to the cut set list and delete columns representing these events from the matrix.
3. Logically OR the columns together in all combinations up to N_p at a time. If the logical OR of the column combination produces a

vector whose elements sum to N_p then this is a cut set and should be included in the list.

4. Remove all non-minimal cut sets from the list using the absorption law, $A + A \, . \, B = A$.

Example

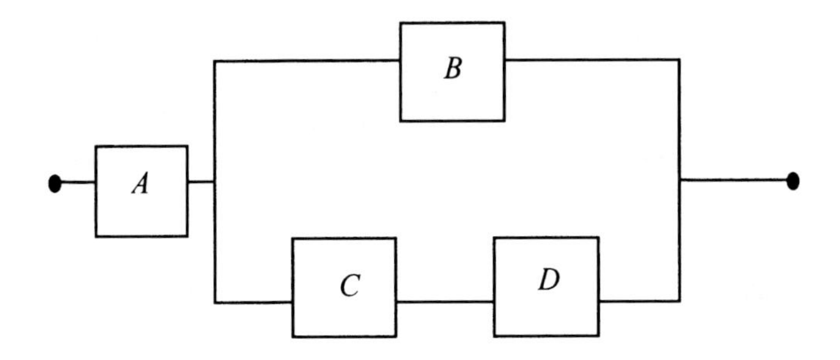

Minimal path sets:

1 $\{A, B\}$

2 $\{A, C, D\}$

Step 1

$$[\mathbf{P}] = \begin{matrix} & A & B & C & D \\ 1 & 1 & 1 & 0 & 0 \\ 2 & 1 & 0 & 1 & 1 \\ \hline & 2 & 1 & 1 & 1 \end{matrix}$$

Step 2
Since

$$\sum_{\substack{j=A \\ i=1}}^{N_P} P_{ij} = N_p$$

A is a single-order minimal cut set and is included in the cut set list.
Reduce matrix

$$\begin{matrix} B & C & D \end{matrix}$$
$$\begin{pmatrix} 1 & 0 & 0 \\ 0 & 1 & 1 \end{pmatrix}$$

Step 3

B OR **C** = 1 $B \, . \, C$ is a cut set
B OR **D** = 1 $B \, . \, D$ is a cut set
C OR **D** $\neq$ 1

Combinations of more than $N_p = 2$ are not needed.

The cut set list is:

$\{A\}$
$\{B, C\}$
$\{B, D\}$

Step 4

The cut set list is already in its simplest form and further minimization is not possible.

6.5 Network quantification

The probability of system failure can be calculated by quantifying its reliability network. In order to do this a qualitative analysis to produce either the minimal cut sets, or the minimal path sets of the network, must first be carried out. Component failure/success probabilities can then be used to evaluate the minimal cut/path set probabilities. These can in turn be combined to produce the system failure/success probability.

As an example let us return to the bridge network of Fig. 6.19 where the probability that any component functions normally is 0.9 and the component failure probability is 0.1.

6.5.1 *Minimal cut set calculations*

System minimal cut sets:

1. $\{A, B\}$
2. $\{D, E\}$
3. $\{A, C, E\}$
4. $\{B, C, D\}$

$$P\left[\text{system failure}\right] = P\left[\left(A_{\mathrm{F}} \text{ AND } B_{\mathrm{F}}\right) \text{ OR } \left(D_{\mathrm{F}} \text{ AND } E_{\mathrm{F}}\right) \text{ OR } \left(A_{\mathrm{F}} \text{ AND } C_{\mathrm{F}} \text{ AND } E_{\mathrm{F}}\right) \text{ OR } \left(B_{\mathrm{F}} \text{ AND } C_{\mathrm{F}} \text{ AND } D_{\mathrm{F}}\right)\right]$$

$$\begin{aligned}
&= [q_A q_B + q_D q_E + q_A q_C q_E + q_B q_C q_D] \\
&\quad - [q_A q_B q_D q_E + q_A q_B q_C q_E + q_A q_B q_C q_D \\
&\qquad + q_A q_C q_D q_E + q_B q_C q_D q_E + q_A q_B q_C q_D q_E] \\
&\quad + [q_A q_B q_C q_D q_E + q_A q_B q_C q_D q_E \\
&\qquad + q_A q_B q_C q_D q_E + q_A q_B q_C q_D q_E] \\
&\quad - [q_A q_B q_C q_D q_E]
\end{aligned}$$

Since $q_i = 0.1$

$$Q_{\mathrm{SYS}} = [0.022] - [0.000\,51] + [0.000\,04] - [0.000\,01]$$

$$= 0.021\,52$$

6.5.2 Minimal path set calculations

System minimal path sets:

1. $\{A, D\}$
2. $\{B, E\}$
3. $\{A, C, E\}$
4. $\{B, C, D\}$

$$P\left[\text{system success}\right] = P\left[\left(A_{\mathrm{W}} \text{ AND } D_{\mathrm{W}}\right) \text{ OR } \left(B_{\mathrm{W}} \text{ AND } E_{\mathrm{W}}\right) \text{ OR } \left(A_{\mathrm{W}} \text{ AND } C_{\mathrm{W}} \text{ AND } E_{\mathrm{W}}\right) \text{ OR } \left(B_{\mathrm{W}} \text{ AND } C_{\mathrm{W}} \text{ AND } D_{\mathrm{W}}\right)\right]$$

$$\begin{aligned}
&= [r_A r_D + r_B r_E + r_A r_C r_E + r_B r_C r_D] \\
&\quad - [r_A r_B r_D r_E + r_A r_C r_D r_E + r_A r_B r_C r_D \\
&\qquad + r_A r_B r_C r_E + r_B r_C r_D r_E + r_A r_B r_C r_D r_E] \\
&\quad + [r_A r_B r_C r_D r_E + r_A r_B r_C r_D r_E \\
&\qquad + r_A r_B r_C r_D r_E + r_A r_B r_C r_D r_E] \\
&\quad - [r_A r_B r_C r_D r_E]
\end{aligned}$$

Since $r_i = 0.9$

$$R_{SYS} = [3.078] - [3.870\ 99] + [2.361\ 96] - [0.590\ 491]$$

$$= 0.978\ 48$$

Therefore

$$Q_{SYS} = 1 - 0.978\ 48$$

$$= 0.021\ 52$$

These calculations show that the system failure probability can be obtained from either minimal path sets or minimal cut sets. In this case to get these exact probabilities required virtually the same amount of work whether path sets or cut sets were used. However, this is a very simple 'text book'-type problem. The analysis of engineering systems can yield very large numbers of minimal path sets and minimal cut sets. When this situation applies it is not practical to calculate the probability of the combinations of minimal cut sets/path sets occurring two at a time, three at a time, and so on. This means that all terms in the probability expansion cannot be calculated.

When using minimal cut sets the terms in the probability expansion become numerically less significant. Taking only the first two terms gives an approximation to the system failure probability of 0.021 49. This convergence rate is not obtained for the minimal path set calculations. For this reason quantification is usually performed using minimal cut sets in preference to minimal path sets. A detailed discussion of approximate methods of quantifying the system performance probabilities using minimal cut sets is given in Chapter 7, fault tree analysis.

6.6 Summary

The use of reliability networks as a means of calculating the system failure probability has been presented in this chapter. Several different methods have been covered which enable both the qualitative evaluation of the network in terms of its minimal path sets and cut sets, and its subsequent quantification. The most efficient methods of analysing a network are dependent on its structure. Those networks which can be categorized as 'simple', i.e. series/parallel combinations, can be assessed by a gradual simplification process. 'Complex' networks require the use of one of the more comprehensive approaches, such as conditional probabilities, star and delta conversions, or path set and cut set evaluations.

One of the difficulties of using a reliability network to evaluate the system performance is that the network structure does not correspond to standard diagrammatical representations of the system, such as P and ID diagrams. As such it is not always a simple matter to obtain the network in the first place.

6.7 Bibliography

Billington, R. and **Allan, R.** (1983) Reliability Evaluation of Engineering Systems. Pitman.

Grosh, D. L. (1989) *A Primer of Reliability Theory*. John Wiley.

Hansler, E. (1972) A fast recursive algorithm to calculate the reliability of a communication network, *IEEE Trans. Communications*, **20** (3), June.

Misra, K. B. (1970) An algorithm for the reliability evaluation of redundant networks, *IEEE Trans. Rel.*, **R-19** (4), November.

Misra, K. B. and **Rao, T. S. M.** (1970) Reliability analysis of redundant networks using flow graphs, *IEEE Trans. Rel.,* **R-19** (1), February.

Nelson, C., James, J., Batts, R., and **Beadles, R.** (1970) A computer program for approximating system reliability, *IEEE Trans. Rel.*, **R-19** (2), May.

Platz, O. and **Olsen, J. V.** (1976) FAUNET: a program package for fault tree and network calculations, Risϕ Report No. 348, September.

Villemeur, A. (1991) *Reliability, Availability, Maintainability and Safety Assessment*. John Wiley.

Chapter 7

Fault Tree Analysis

7.1 The fault tree model

There are two approaches which can be used to analyse the causal relationships between component failures and system failure events. These are inductive, or forward, analysis, and deductive, or backward, analysis. An inductive analysis starts with a set of component failure conditions and proceeds forward, identifying the possible consequences, that is, a 'what happens if' approach. Fault tree analysis is an example of a deductive, 'what can cause this' approach and is used to identify the causal relationships leading to a specific system failure mode.

It is important that the system failure mode and the system boundary are chosen with care so that the analysis will not be too broad or too narrow to satisfy the study objectives.

A fault tree is a structure by which a particular system failure mode can be expressed in terms of combinations of component failure modes and operator actions. The system failure mode to be considered is termed the 'top event' and the fault tree is developed in branches below this event, showing its causes. In this way events represented in the tree are continually redefined in terms of lower resolution events. This development process is terminated when component failure events, termed basic events, are encountered. Analysis of the fault tree can be carried out by providing information on the basic event probabilities.

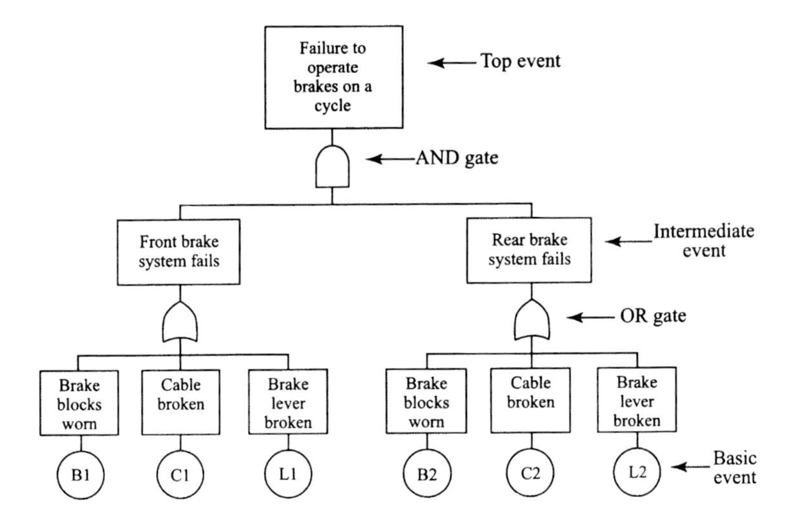

Fig. 7.1 Example fault tree structure

Each fault tree considers only one of the many possible system failure modes and therefore more than one fault tree may be constructed during the assessment of any system. For example, when safety protection systems are being assessed the top event of most concern is that the system fails to carry out its required task when a demand occurs. This top event leads to the development of a fault tree to model causes of this situation. It is possible to add levels of redundancy and diversity to a protection system such that its probability of failing on demand is very small. This course of action can, however, lead to protection system action when no demand arises, that is, a spurious system failure. While this outcome is not as potentially dangerous as the first failure mode it is still of concern, particularly if it results in frequent shutdowns of the production processes. The final solution will be a compromise and may be determined by utilizing a second fault tree for the system which identifies causes of spurious trips.

An example fault tree structure is shown in Fig. 7.1 for the top event 'Failure to operate brakes on a cycle'. The failure logic is developed until basic component failures are encountered.

A fault tree diagram contains two basic elements, 'gates' and 'events'. 'Gates' allow or inhibit the passage of fault logic up the tree and show the relationships between the 'events' needed for the occurrence of a higher level event. The symbols used to represent the causal relationships are the gates and events as shown in Figs 7.2 and 7.3 respectively.

	Gate symbol	Gate name	Causal relation
1		AND gate	Output event occurs if all input events occur simultaneously (number of inputs ≥ 2)
2		OR gate	Output event occurs if at least one of the input events occurs (number of inputs ≥ 2)
3	k n inputs	k-out-of-n gate (voting gate)	Output event occurs if at least k out of the n input events occur
4		Exclusive OR gate	Output event occurs if one, but not both, of the two input events occurs
5		Inhibit gate	Input produces output when conditional event exists
6		Priority AND gate	Output event occurs if all input events occur in the right order from left to right
7		NOT gate	Output event occurs if the input event does not

Fig. 7.2 Gate symbols

	Event symbol	Meaning of symbol
1		Top event or intermediate event description box. These events are further developed by a logic gate
2		Basic event
3		Undeveloped event
4		Conditional event used with inhibit gate
5		House event. Logic event which either occurs or does not occur with certainty (i.e. TRUE or FALSE)
6		Transfer symbol. Represents location of repeated failure logic which is developed elsewhere on the fault tree structure

Fig. 7.3 Event symbols

In Fig. 7.1 the logic gate just below the top event is an AND gate with inputs 'Front brake system fails' and 'Rear brake system fails'. This represents some redundancy in the system in that both of the lower level (intermediate) events would need to have occurred to result in a total failure of the brakes to operate. It is assumed that the brake system on each wheel features brake blocks, cable, and lever, each of which can fail. Failure of any of these elements would render the brake sub-system on a wheel inoperable. This logic is incorporated into the fault tree structure by means of an OR gate. The higher level output event to the OR gate will occur if at least one of the lower level (input) events occurs. This means that as required the wheel brake system will fail if one or more of its elements were to fail. For this simple example, the brake blocks, cable, and lever represent components for which data can be obtained and the development of the failure logic is terminated. Events terminating the branches are known as basic events.

The analysis of the fault tree diagram produces two types of result: qualitative and quantitative. Qualitative analysis identifies the combinations of the basic events which cause the top event. For the fault tree depicted in Fig. 7.1, there are nine failure combinations, which are as follows:

1. B1 B2
2. B1 C2
3. B1 L2
4. C1 B2
5. C1 C2
6. C1 L2
7. L1 B2
8. L1 C2
9. L1 L2

i.e. for each of the three ways that the front brake can fail there are three ways in which the rear brake can also fail.

Fault tree quantification will result in predictions of the system performance in terms of component level performance data (probability of failure or frequency of failure). These system performance indicators are:

(a) top event probability (unavailability or unreliability);
(b) top event unconditional failure intensity;
(c) top event failure rate;
(d) expected number of top event occurrences in a specified time period;
(e) total system downtime in a specified time period

Another very useful piece of information which can be obtained from the fault tree is a set of measures indicating the contribution that each component makes to the system failure. These are known as importance measures.

The remainder of this chapter is concerned with the qualitative and quantitative evaluation of the fault tree structure.

7.2 Examples of the use of fault tree symbols

Basic events, represented by a circle, indicate the limit of resolution of the fault tree. For a quantitative analysis it is these events for which data are required. Consequently it is unnecessary to develop fault tree branches beyond the events for which data are available. The following examples illustrate how the symbols presented in Figs 7.2 and 7.3 are used.

Figure 7.1 illustrates the use of AND gates, OR gates, intermediate events, and basic events. These symbols are the most commonly used and along with the vote gate, are all that are required in the majority of fault tree studies; other symbols are occasionally used.

Example 1 – Vote gate

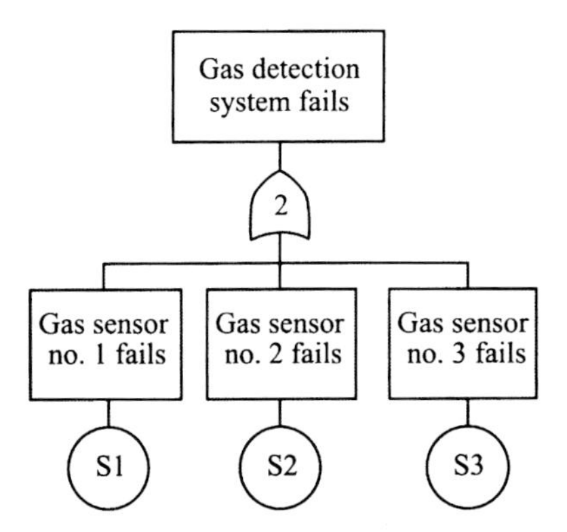

Fig. 7.4 Example 1, vote gate

The vote gate (Fig. 7.4) is a useful shorthand representation of all combinations of k from the n input events. A common application area for this is in the part of systems which monitor the environment. Sensors for temperature, pressure, and gas concentration for example. The more sensors installed, the more likely the system is to monitor correctly for adverse conditions. However, since sensors have more than one failure mode they can also spuriously register the undesired condition (i.e. fail-safe). In order to avoid a single sensor failure tripping the system they are frequently placed in a voting configuration which would require at least k such failures to produce the trip. A vote gate appearing in the fault tree structure can be expanded and re-expressed in terms of AND/OR gates during analysis.

Example 2 – Inhibit gate

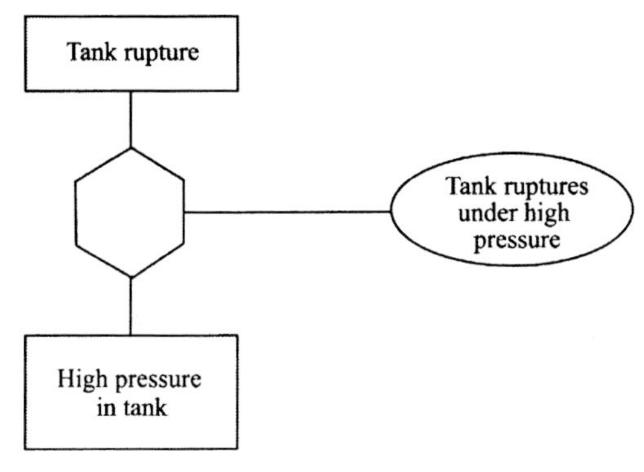

Fig. 7.5 Example 2, inhibit gate

Use of the inhibit gate is shown in Fig. 7.5 and illustrates that in this example, for the tank to rupture, the event 'High pressure in tank' must occur and the condition stated in the oval, 'Tank ruptures under high pressure' must be satisfied. In the inhibit gate the inhibit gate provides a means of incorporating conditional probabilities and is used for convenience in communicating the logic process; however, for the analysis phase it can be replaced by an AND gate.

The difference between the inhibit gate and the normal AND gate is that the inhibit gate is probabilistic. In the inhibit gate the input event will not always cause the output event. The proportion of the time that the output event will result from the input event is indicated by the condition. An AND gate by comparison is deterministic. When all input events to an AND gate occur the output event will definitely result.

Example 3 – Exclusive OR gate

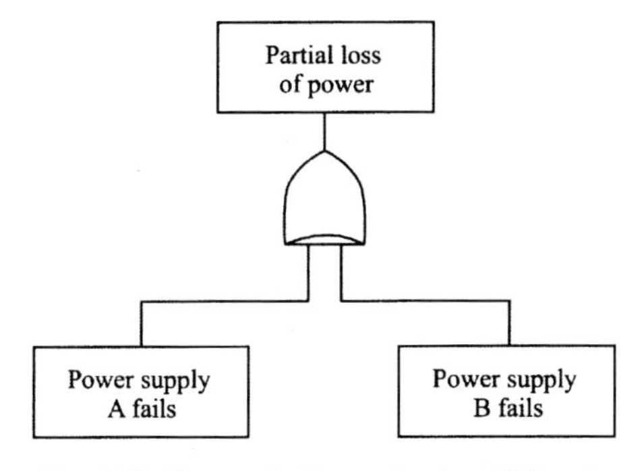

Fig. 7.6 Example 3, exclusive OR gate

Example 3, Fig. 7.6, illustrates the use of the exclusive OR gate or XOR gate. This implies the use of the third logical operator, NOT. For example, consider a plant which has two power sources. A fault tree is to be constructed to represent causes of partial power supply failure. The top event 'Partial loss of power' occurs when either, but not both, of two independent supplies fails. If both supplies fail then this would produce a total loss of power which would not produce the output event defined at the top of the tree. The exclusive OR gate is a shorthand way of representing this logic; it can be replaced by the structure shown in Fig. 7.7 involving AND, OR, and NOT operators.

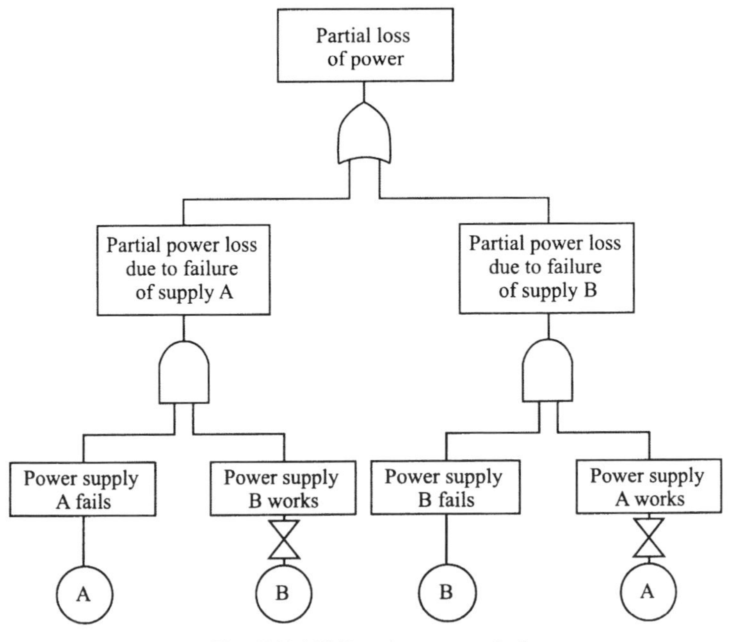

Fig. 7.7 XOR gate expanded

Example 4 – House events

House events are used to model two state events which either occur or do not occur, and so have probabilities of either 1 or 0. They provide a very effective means of turning sections of fault trees on and off. Hence the same fault tree can be used to model several scenarios. The example shown in Fig. 7.8 illustrates how they can be used to indicate the effect on a safety system when part of it is isolated for preventive or corrective maintenance.

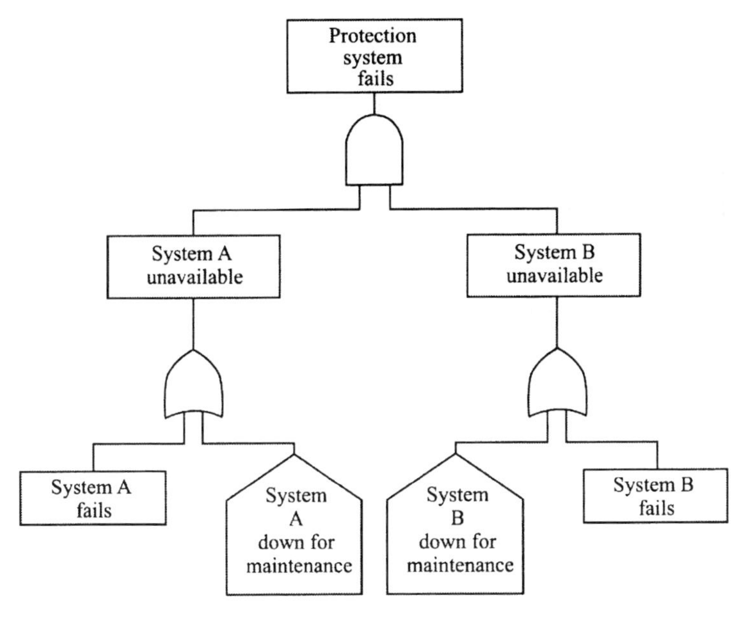

Fig. 7.8 House events

The safety system has two parallel protection systems, A and B. The overall safety system will function provided either of A or B works. Failure to protect can either be caused by a failure on a redundant channel or the channel having been taken out of the system for maintenance. This situation is represented by the fault tree shown in Fig. 7.8. To evaluate the system under normal conditions the two house events used to represent the occurrence of maintenance action on either system are set to OFF, i.e. assigned a probability of zero. To determine the effect of shutting down either system for maintenance the appropriate house event is then turned on and the same fault tree used to look at the degree of system degradation during maintenance.

Example 5 – Transfer symbols

Transfer-in and transfer-out symbols are used to avoid the need for an analyst to duplicate the failure logic for an identical event which occurs more than once in a fault tree. For example, the tree shown in Fig. 7.9 models causes of a protection system failing to function on demand. When a high flammable gas concentration is detected the protection system is designed to fulfil two tasks: inform the operator, and isolate the power to reduce potential ignition sources. It is considered that the system will have failed if it does not complete both tasks. The top section of the tree for this is shown in Fig. 7.9. It can be seen that one cause of both system functions not working is that no alarm signal is received, i.e. the gas has not been detected. Transfer-in and transfer-out symbols are used so that identical failure logic is not reproduced in the tree.

Another task for which transfer symbols are commonly used is paginating fault trees. Frequently the whole of a fault tree diagram will not fit on to a single page. Events which are then developed on other pages are labelled with the transfer-out symbol. An appropriately numbered transfer-in symbol appears just below the event heading the continuation page. In this manner a large fault tree can be split into several sub-trees.

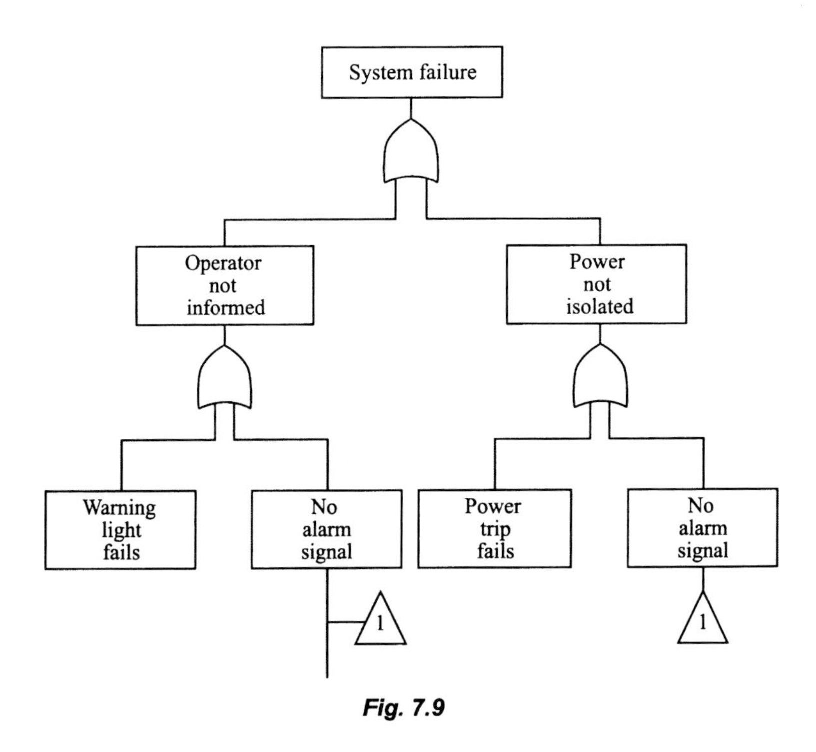

Fig. 7.9

Two symbols, one gate and one event, included in Figs 7.2 and 7.3 have not appeared in any of the examples given. These are the PRIORITY AND gate and the DIAMOND event.

The PRIORITY AND gate is used when the sequence in which events occur is important. Many commercial software packages treat the gate as a simple AND gate during analysis on the assumption that if order is not taken into account, the result will be pessimistic. An alternative approach is to convert this gate into a Markov model which can analyse the sequence of events correctly and then feed the results obtained back into the fault tree structure.

The DIAMOND symbol can be used to represent undeveloped events which terminate fault tree branches. The events may be left undeveloped for a number of reasons, for example it could be beyond the system boundary or the scope of the study. If the system analysis is being performed during the conceptual design phase this symbol can be used to represent failure events which can, at that point, not be further developed until the detail of the design is established

7.3 Boolean representation of a fault tree

The three basic gate types used in a fault tree are the OR gate, the AND gate, and the NOT gate. These gates combine events in exactly the same way as the Boolean operations of 'union', 'intersection', and 'complementation'. There is therefore a one-to-one correspondence between Boolean algebraic expressions and the fault tree structure.

7.4 Component failure categories

Having defined the symbols which are used as the building blocks of a fault tree some of the events which appear in the tree can now be classified as belonging to different categories.

7.4.1 Fault versus failures

For events occurring in a fault tree there is a distinction between events which can be categorized as faults and those which are failures. The difference is best illustrated by means of an example, such as a valve. If the valve closes properly when given the command signal, then this is termed a valve 'success'. If the valve fails to close when it receives the command signal, it is termed a valve **failure**. However, the valve may close at the wrong time due to the improper functioning of an upstream component. This is not a valve failure, even if the resulting valve closure causes the entire system to enter an unsatisfactory state. An occurrence of this type is called a **fault**. In general, all failures are faults but not all faults are failures.

7.4.2 Occurrence versus existence

A fault that is not repairable will continue to exist once it has occurred. For a fault that is repairable, a distinction must be made between the occurrence of a fault and its existence. This distinction is important when performing the quantification of the non-occurrence of the top event over a period of time.

7.4.3 Passive versus active components

It is often convenient to separate components into two types: **passive** and **active**. A passive component can be described as a component that acts as a transmitter of a 'signal'. For example, the signal may be a current and the passive component a wire, or the 'signal' may be fluid flow and the passive component a pipe. The failure of a passive component will result in the non-transmission of its signal.

An active component generally modifies its parent system in some way. For example, a valve modifies a system's fluid flow. Active components originate or modify the signals that passive components transmit. Active components usually require an input signal for the output signal to occur. If an active component fails, there may be no output signal or there may be an incorrect output signal.

The main difference in reliability terms between active and passive components is in their respective failure rates. Passive components generally have failure rates two or three orders of magnitude lower than the failure rates of active components in the same system.

7.5 Fault tree construction

7.5.1 System boundary specification

The specification of the system boundaries is important to the success of the analysis. Many systems have external power supplies and perhaps services such as a water supply. It would not be practical in a system assessment to trace all possible causes of a power failure back through the distribution and generation system. Nor would this extra detail provide any useful information concerning the system being assessed.

The location of the **external boundary** will be partially decided by the aspect of system performance which is of interest. For instance, if the problem is that the bell on a telephone is not loud enough to attract attention in all parts of a house, the external boundary will be within the telephone. If the problem involves a noisy line when the telephone is used, the external boundary will be much more remote and may include the lines into the house and perhaps the local exchange.

Another important consideration in defining the external boundary is the time domain. For example, the start-up or shutdown conditions of a plant can generate different hazards from its steady-state operation and it may be necessary to account for this in tracing the fault.

The **limit of resolution** to which the analysis will develop must also be established. For example, is it necessary to extend the analysis to the sub-system level or beyond this to the component level? The choice of external boundary determines the comprehensiveness of the analysis, whereas the choice of a limit of resolution sets the detail of the analysis.

The system analyst must ensure that the boundaries chosen are feasible and valid with regard to the objectives of the analysis. To reach certain conclusions about a system, it may be necessary to include a large portion of the system within the external boundaries. However, this may involve an extensive and time-consuming analysis. If the resources for such an analysis are not available, the boundaries must be 'restricted' and the amount of information expected to result from the analysis must be reduced.

The limit of resolution (internal boundary) can also be established from considerations of feasibility and from the objective of the analysis. For instance, if the system failure probability is to be calculated, then the limit of resolution should cover component failures for which data are obtainable. Once the limit of resolution has been chosen, it is the interactions at this level with which the analysis will be concerned and no knowledge is needed of interactions at lower levels.

7.5.2 Basic rules for fault tree construction

Once the system has been defined and a particular system failure mode selected as the top event, the fault tree is developed by determining the **immediate**, **necessary**, and **sufficient** causes for its occurrence. It is important to note that these are not the component level causes of the event, but the immediate causes of the event. This is called the 'immediate cause' concept.

The immediate, necessary, and sufficient causes of the top event are then treated as sub-top events and the process then determines their immediate, necessary, and sufficient causes. In this way the tree is developed, continually approaching finer resolution until the limit of resolution is reached and the tree is complete.

Some guidelines that are designed to aid the analyst in following the immediate cause concept are:

1. Classify an event into more elementary events, e.g. 'overpressure of reactor vessel' could be 'overpressure due to overfilling' or 'overpressure due to a runaway reaction'.
2. Identify distinct causes for an event, e.g. 'runaway reaction' could be 'excessive feed' or 'loss of cooling'.
3. Trigger events should be coupled with 'no protective action taken', 'overheating' could be 'loss of cooling' and 'no emergency shutdown'.
4. Find all necessary causes for an event, e.g. 'explosion' could be 'escape of flammable gas' and 'switch sparks'.
5. Identify a component failure event where possible, e.g. 'no current in wire' could be 'electric power supply fails'.

No set of rules can be stated which, if they are adhered to, will guarantee the construction of the correct fault tree in all circumstances. However, certain rules should be followed to ensure that the fault tree is developed in a methodical manner. These are as follows.

Rule 1

Write the statements that are entered in the event boxes as faults; state precisely **what** the fault is and **when** it occurs.

Examples of fault statements are:

1. The door bell fails to sound when the button is pressed.
2. Car fails to start when ignition key is turned.

Rule 2

If the fault in the event box can be caused by a component failure, classify the event as a 'state-of-component' fault. If the fault cannot be caused by a component failure, classify the event as a 'state-of-system' fault.

If the fault event is classified as 'state-of-component' then the event should be developed as in Fig. 7.10. If the fault event is classified as 'state-of-system' then the event is developed to its immediate necessary and sufficient causes.

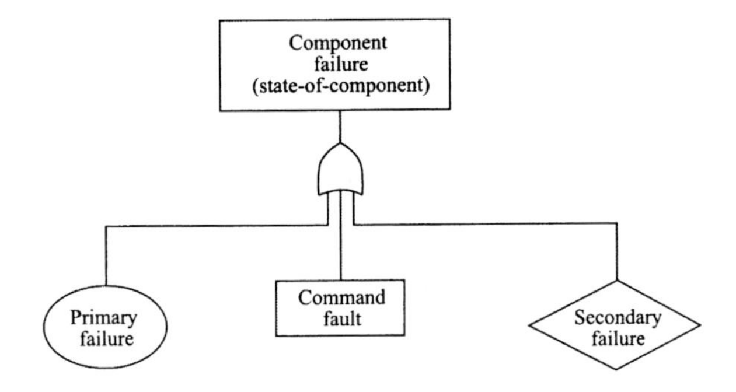

Fig. 7.10 Development of a 'state-of-component' event

A primary failure is defined as any failure of a component that occurs under conditions for which the component was designed to work or that occurs from natural ageing.

A secondary failure is defined as any failure of a component that occurs as a result of the component being under a condition for which it was not designed, either in the past or present, perhaps due to the failure of other components in the system.

A command fault is defined as the component being in the non-working state due to improper control signals or noise.

Rule 3 – No miracles

If the normal functioning of a component propagates a fault sequence, then it is assumed that the component functions normally.

It may be the case that the miraculous failure of some component could block the propagation of a particular fault sequence. For example, consider the situation shown in Fig. 7.11. Gas is passing along a pipe into an area where an ignition source is present. Under normal conditions a fire or explosion could result from the ignition of the gas.

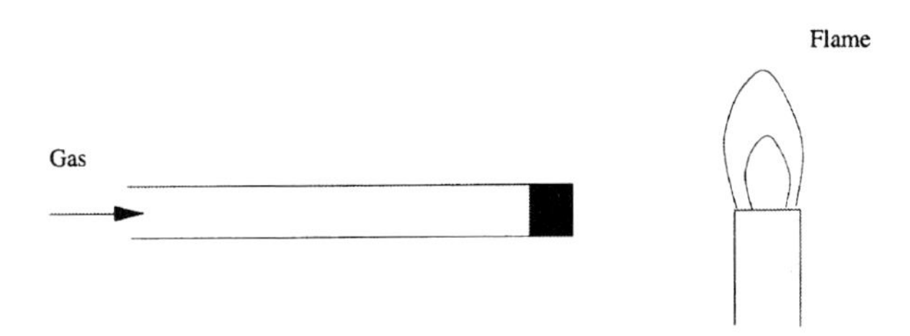

Fig. 7.11

If a failure of the pipe were to happen which caused a blockage of the pipe then the fire would be prevented. This is regarded as a fortuitous or miraculous failure and would not be considered in the fault tree. Its inclusion in a fault tree would require the use of NOT logic since a fire will only occur providing the pipe is 'not failed blocked'.

Rule 4 – Complete-the-gate

All inputs to a particular gate should be completely defined before further development of any one of them is undertaken.

Rule 5 – No gate-to-gate

Gate inputs should be properly defined fault events, described using the rectangular boxes, and gates should not be directly connected to other gates.

Example – fault tree construction

Consider the top event 'Cooling water system fails' for the system shown in Fig. 7.12. The emergency water cooling system is required to work in the event of a high temperature being detected in the pressure vessel. Sensors monitor the vessel temperature and in the event of overheating the water pump P is activated. The temperature monitoring system is assumed to be 100 per cent reliable.

Once started the pump draws water from the pool via two pipes P1 and P2 each with a filter (F1 and F2) placed on the pipe ends. Valves V1 and V2 are normally open and are only closed when maintenance requires the pump to be removed. Water is delivered on to the pressure vessel through a nozzle N fitted to pipe P3.

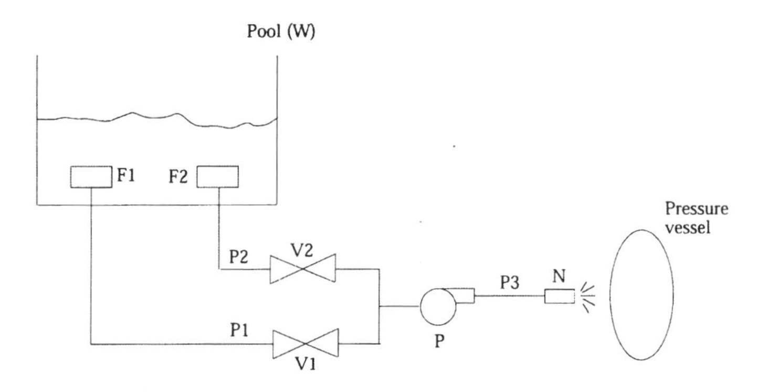

Fig. 7.12 Cooling water system

In accordance with rule 1 the top event is redefined as a fault event, i.e. 'Water cooling system fails on demand'. The fault tree for this event is then developed as shown in Fig. 7.13a. To simplify the procedure only primary failures and command faults have been developed in the tree; secondary failures have been ignored.

Were the system to function as intended, water would be delivered to the pressure vessel from the nozzle N. Hence the immediate cause of the top event in the fault tree is that no water is produced from the nozzle. Hence either the nozzle has failed blocked or if the nozzle is functioning correctly there is no water supply to the nozzle. These two events are added to the fault tree using an OR gate as illustrated in Fig. 7.13a. Since a blockage of the nozzle is a failure of that component and cannot be developed further this branch is terminated with a basic event. Attention is then given to the event 'No water to the nozzle'. Again, from the system diagram it can be seen that no water will reach the nozzle if the pipe P3 is blocked, a basic event, OR if the pipe condition is good and no water is supplied from the pump. As before these events are included using an OR gate and non-basic events developed further. For no water supply from the pump either the pump fails to start, again a primary component failure, OR no water is supplied to the pump. At this point in the system it requires two events to happen simultaneously to prevent water being transported to the pump. Line 1, consisting of components F1, P1, and V1, must fail to supply, as must line 2 (F2, P2, V2). An AND gate is used in the fault tree to represent this situation as shown in Fig. 7.13a. Each of these events is then further developed, considering failures of the valves, pipes, filters, and water reservoir. For line 1 the fault tree is shown in Fig. 7.13b. The tree for line 2 will be of identical structure with different basic event labels.

In the example described here secondary failures have been ignored. A secondary failure occurs as a result of the component operating in conditions for which it was not designed. Illustrations of the types of secondary failure which would appear in the fault tree if they were required can be given by considering the water pump P where the following secondary causes could be appropriate:

1. Pump fails due to out-of-tolerance conditions such as mechanical vibration or thermal stress.
2. Improper maintenance such as inadequate lubrication of motor bearings.

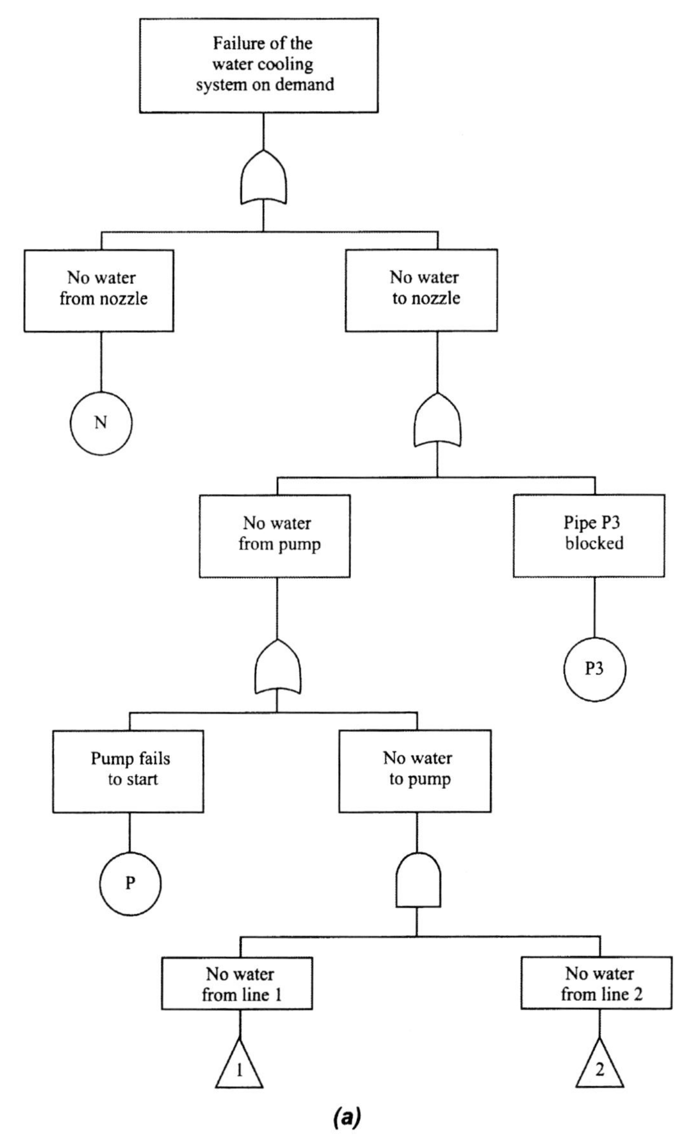

(a)

Fig. 7.13 Fault tree for the cooling water system

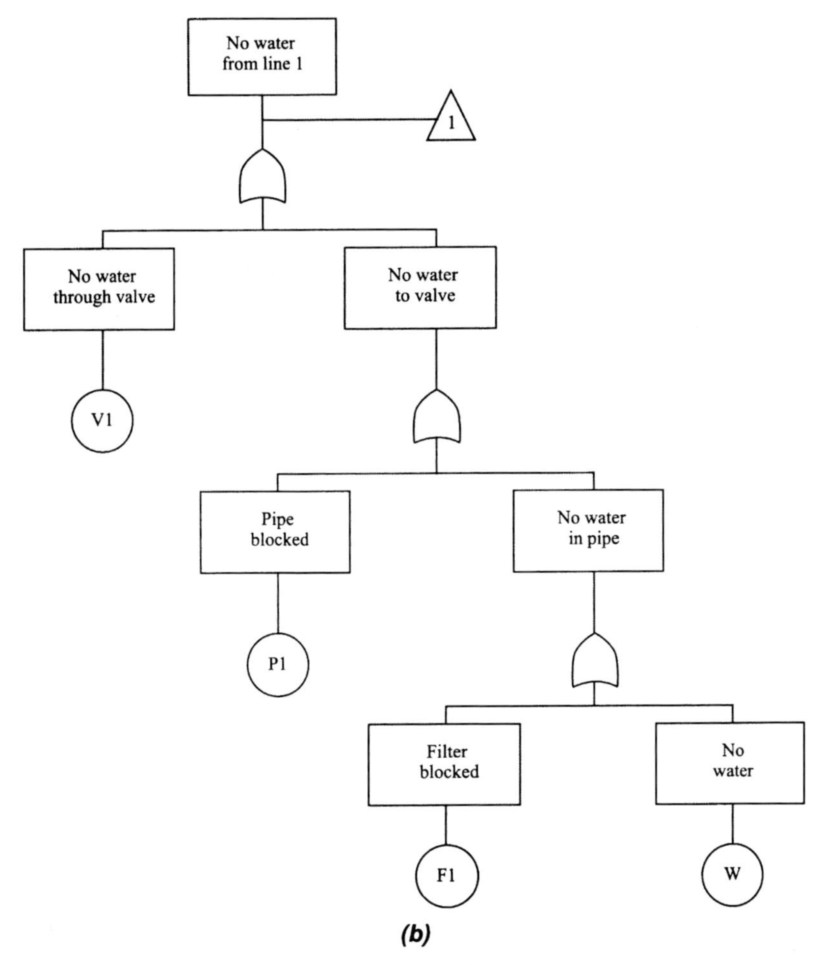

(b)

Fig. 7.13 Continued

7.6 Qualitative fault tree analysis

Each unique way that system failure can occur is a **system failure mode** and will involve the failure of individual components or combinations of components. To analyse a system and to eliminate the most likely causes of failure first requires that each failure mode is identified. One way to identify the system failure modes is to carry out a logical analysis on the system fault tree. The system failure modes are defined by the concept of a **cut set** introduced in Chapter 6 on reliability networks but whose definition is repeated here.

A **cut set** is a collection of basic events such that if they all occur the top event also occurs.

The dual concept of a cut set in the success space is called a **path set.** This is a collection of basic events whose successful operation ensures the non-occurrence of the top event (i.e. a list of components whose functioning ensures successful system operation for the particular top event considered).

For industrial engineering systems there are generally a very large number of cut sets each of which can consist of many component failure events. However, we are only interested in lists of component failure modes which are both necessary and sufficient to produce system failure. For example $\{A, B, C\}$ may be a cut set and the failure of these three components will guarantee system failure. However, if the failure of A and B will alone produce system failure, this means that the state of component C is irrelevant and the system will fail whether C fails or not. This introduces the concept of a **minimal cut set**, which is a cut set such that if any basic event is removed from the set the top event will not occur, i.e. it is the smallest combination of component failures, which if they all occur will cause the top event to occur. Once the minimal cut sets are evaluated the quantification of the fault tree can be carried out.

For the majority of fault trees, the minimal cut sets would be produced using one of the computer codes available for this task. Any fault tree will consist of a finite number of minimal cut sets which are unique for that top event. Two fault trees drawn using different approaches are logically equivalent if they produce identical minimal cut sets. The one-component minimal cut sets (first-order minimal cut sets), if they occur, represent single failures which cause the top event. Two-component minimal cut sets (second order) represent double failures which together will cause the top event to occur. In general the lower-order cut sets contribute most to system failure and effort should be concentrated on the elimination of these in order to improve system performance.

The minimal cut set expression for the top event T can be written in the form

$$T = K_1 + K_2 + K_3 + \cdots + K_n \tag{7.1}$$

where K_i, $i = 1, \ldots, n$, are the minimal cut sets (+ represents logical OR). Each minimal cut set consists of a combination of component failures and hence the general k-component cut set can be expressed as

$$K_i = X_1 . X_2 . \cdots . X_k$$

where $X_1, \ldots, X_k$ are basic component failures on the tree, i.e. if $T = A + B.C + C.D$ (. represents logical AND).

There are three minimal cut sets, one first order, A, and two second order, $B . C$ and $C . D$.

To determine the minimal cut sets of a fault tree either a 'top-down' or a 'bottom-up' approach can be used. Both these methods are straightforward to apply and both involve the expansion of Boolean expressions. They differ in which end of the tree is used to initiate the expansion process. The Boolean algebra laws, particularly the distributive and absorption laws, are then used to remove redundancies in the expressions. Note that the discussions below which describe the qualitative evaluation of fault trees are restricted to trees which contain only AND and OR gates. If the NOT operator occurs or is implied by the use of an exclusive OR gate the combinations of basic events which cause the top event are termed **implicants**. Minimal sets of implicants are called **prime implicants**. The following procedures, while producing a complete list of the minimal cut sets for 'conventional' fault trees, will not necessarily generate a complete list of prime implicants for a fault tree which contains the NOT operator.

Consider the simple fault tree shown in Fig. 7.14. There are three basic events giving a total of $2^3 = 8$ component states. By using the fault tree and determining whether the top event occurs or not from each possible set of component states a table can then be constructed showing all possible system outcomes.

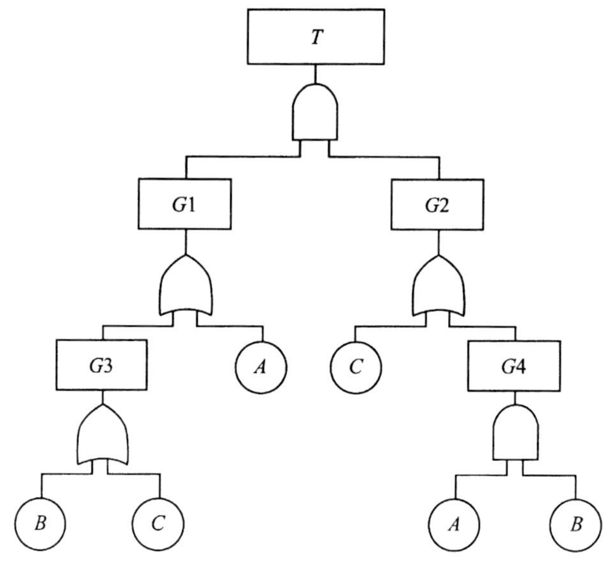

Fig. 7.14 Example fault tree

Table 7.1

	A	B	C	System
1	W	W	W	W
2	W	W	F	F
3	W	F	W	W
4	W	F	F	F
5	F	W	W	W
6	F	W	F	F
7	F	F	W	F
8	F	F	F	F

Of the eight possible system states three represent the system in a working condition (the non-occurrence of the top event) and five in a failed state as represented in the decision table (Table 7.1). By ignoring those component states which give system success and considering only those combinations which give system failure, we are considering the cut sets, i.e. a list of component failures which lead to system failure.

The Boolean nature of the truth table means that half of the rows will contain the failure mode of any given component. By inspection it can be seen that every time component C is in a failed state the system is in a failed state. This is not the case for components A and B. If we introduce a third state in the truth table '–' for 'don't care', then we can replace all rows in which C has failed with the 'don't care' state for components A and B since, whatever their state, if C fails the top event occurs. Half of the rows can be eliminated in this way to give the states shown in Table 7.2

Table 7.2

A	*B*	*C*	System
W	W	W	W
–	–	F	F
W	F	W	W
F	W	W	W
F	F	W	F

Also if A and B both fail, the state of C is irrelevant. The combination of component failures which give system failure is therefore represented in Table 7.3 by just two rows.

Table 7.3

A	*B*	*C*	System
–	–	F	F
F	F	–	F

The minimal cut sets $\{C\}$ and $\{A, B\}$ can now be extracted from this table. This is not, however, a method to be used generally since the table gets too large to manipulate even with relatively few components.

7.6.1 'Top-down' approach

In the 'top-down' method we start with the top event, and expand by substituting in the Boolean events appearing lower down in the tree and simplifying until the expression remaining has only basic component failures.

For example, with the fault tree in Fig. 7.14 substituting $G1 \,.\, G2$ for T is the first operation. $G1$ and $G2$ are then taken in turn and expressions substituted for these. This process is repeated until all references to gates are removed from the expression. Simplifying at each stage gives:

$$
\begin{aligned}
T &= G1 \,.\, G2 \\
&= (G3 + A) \,.\, (C + G4) \\
&= G3 \,.\, C + G3 \,.\, G4 + A \,.\, C + A \,.\, G4
\end{aligned}
$$

Substitute for $G3 \; (= B + C)$:

$$
\begin{aligned}
T &= (B + C) \,.\, C + (B + C) \,.\, G4 + A \,.\, C + A \,.\, G4 \\
&= B \,.\, C + C \,.\, C + B \,.\, G4 + C \,.\, G4 + A \,.\, C + A \,.\, G4
\end{aligned}
$$

as $C \,.\, C = C$

$$T = B \,.\, C + C + B \,.\, G4 + C \,.\, G4 + A \,.\, C + A \,.\, G4$$

By the absorption law ($A + A \,.\, B = A$) this reduces to:

$$T = C + B \,.\, G4 + A \,.\, G4$$

Finally, substitute for $G4$ ($= A \,.\, B$) gives

$$
\begin{aligned}
T &= C + B \,.\, (A \,.\, B) + A \,.\, (A \,.\, B) \\
&= C + B \,.\, A \,.\, B + A \,.\, A \,.\, B \\
&= C + B \,.\, A + A \,.\, B \\
&= C + A \,.\, B
\end{aligned}
$$

As with the decision tables the two minimal cut sets for the top event are C and $A \,.\, B$. The same fault tree could have been represented by a simpler logic diagram, shown in Fig. 7.15, but this may not have been as easy to understand or to construct from the engineering considerations.

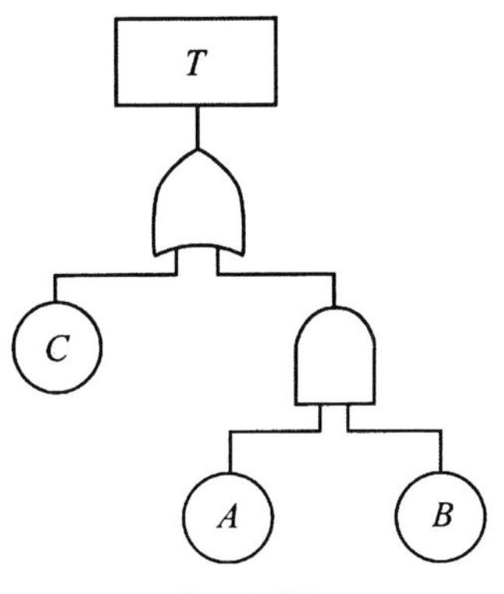

Fig. 7.15

As these two trees have the same minimal cut sets they are equivalent. The fault tree is also logically equivalent to the network shown in Fig. 7.16.

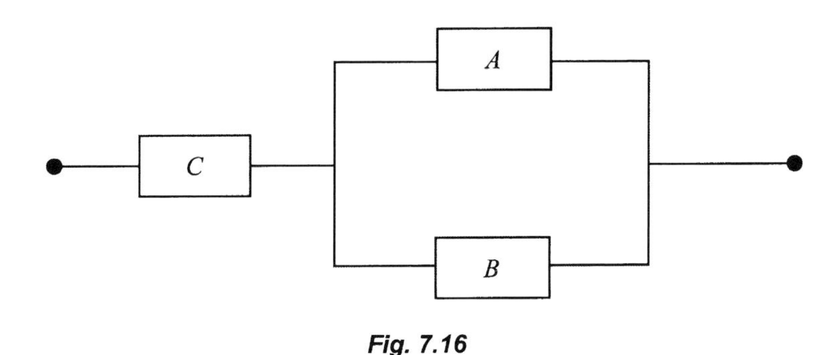

Fig. 7.16

7.6.2 'Bottom-up' approach

The 'bottom-up' method uses the same substitution, expansion, and reduction methods as the 'top-down' approach. The difference with the 'bottom-up' method is that this time the Boolean operations commence at the base of the tree and work towards the top event. Equations containing only basic failures are successively substituted into each gate.

Considering the simple fault tree shown in Fig. 7.14 and beginning at the bottom left of the tree, gate 3: this is equivalent to $B + C$. So $G3$ can be expressed as $B + C$. Proceeding to the next gate up, $G1$, this can then be written as $B + C + A$. Continuing up the tree, the top event is now encountered which is an AND gate combining $B + C + A$ and

$G2$. At this point we need to find an expression for $G2$ in terms of the basic events. This requires going to the base of the right-hand branch and working up.

Gate 4, $G4$ is the last gate on the branch. So $G4 = A \, . \, B$, therefore moving up the tree $G2 = C + A \, . \, B$ which brings us back to the top event gate which is now completely defined in terms of basic events $T = (B + C + A) \, . \, (C + A \, . \, B)$. Applying Boolean reduction gives

$$\begin{aligned} T &= (B + C + A) \, . \, (C + A \, . \, B) \\ &= B \, . \, C + B \, . \, A \, . \, B + C \, . \, C + A \, . \, B \, . \, C + A \, . \, C + A \, . \, A \, . \, B \\ &= B \, . \, C + A \, . \, B + C + A \, . \, B \, . \, C + A \, . \, C + A \, . \, B \\ &= C + A \, . \, B \end{aligned}$$

This example of a fault tree reduction does not reflect the sort of problems which are encountered when analysing real fault trees, which contain large numbers of repeated events and require a good deal of Boolean simplification. For larger problems it is really essential to use computer programs to generate the minimal cut sets.

7.6.3 Computer algorithm

When a fault tree has no mutually exclusive events, an algorithm called MOCUS can be used to obtain the minimal cut sets. This works on the fact that OR gates have the effect of increasing the number of cut sets and AND gates increase the size or order of the cut sets.

This algorithm can be expressed as follows.

1. Provide a unique label for each gate in the fault tree.
2. Label each basic event.
3. Set up a two-dimensional array in the computer and place the top event gate label in the first row, first column.
4. Continually scan the array, replacing:
   (a) each OR gate by a vertical arrangement defining the input events to the gate, increasing the number of cut sets;
   (b) each AND gate by a horizontal arrangement of the inputs to this gate, enlarging the size of the cut sets.

   Repeat this until all gate definitions have disappeared from the array leaving only basic events. When no gate events remain each row in the matrix or array is a cut set.
5. Remove all non-minimal combinations of events such that the array only contains minimal cut sets.

For the fault tree in Fig. 7.14 this would give:

1.

T	

Top event placed 1st row, 1st column

2.

$G1$	$G2$

Top event AND gate – horizontal expansion

3.

$G3$	$G2$
A	$G2$

$G1 = G3 + A$, vertical expansion

4.

$G3$	C
$G3$	$G4$
A	C
A	$G4$

→

A	C
$G3$	C
$G3$	$G4$
A	$G4$

$G2 = C + G4$, vertical expansion [Place fully expanded cut sets at the top of the array]

5.

A	C
B	C
C	C
B	$G4$
C	$G4$
A	$G4$

$G3 = B + C$, vertical expansion

6.

A	C	
B	C	
C	C	
B	A	B
C	A	B
A	A	B

$G4 = A \cdot B$, horizontal expansion

7. Reduction

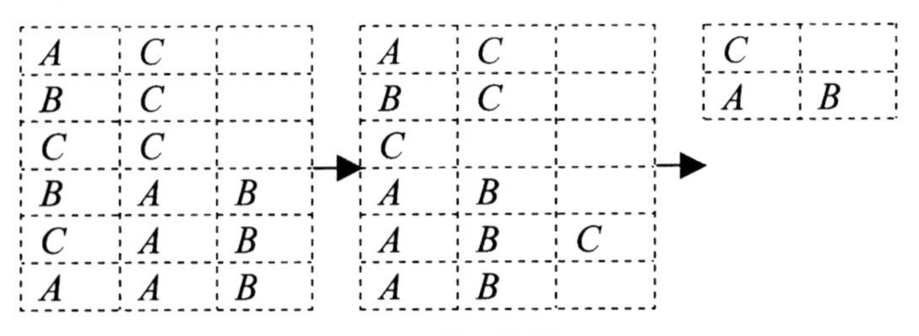

A	C	
B	C	
C	C	
B	A	B
C	A	B
A	A	B

→

A	C	
B	C	
C		
A	B	
A	B	C
A	B	

→

C	
A	B

Fig. 7.17

As described the algorithm is inefficient in its use of computer memory. A simple modification which would improve this is to use a one-dimensional array which contains lists of the cut sets one after the other.

7.6.4 Minimal path sets and dual fault trees

Usually, system failure is represented by the top event of a fault tree. From the point of view of reliability, prevention of the top event is of concern. The top event of a fault tree can be represented by a Boolean equation. Complementation of this expression then gives the non-occurrence of the top event, i.e. system success. This complemented tree or **dual tree** can be obtained from the original fault tree by complementing all basic events and interchanging OR gates for AND gates and vice versa. The Boolean reduction of the dual tree gives the **minimal path sets**.

7.7 Fault tree quantification

7.7.1 Top event probability

Fault trees without repeated events

In the event that the fault tree for a top event T contains independent basic events which appear only once in the tree structure, then the top event probability can be obtained by working the basic event probabilities up through the tree. In doing this, intermediate gate event probabilities are calculated starting at the base of the tree and working upwards until the top event probability is obtained.

For example consider the fault tree in Fig. 7.18 which has four non-repeated basic events A, B, C, and D which occur independently of each other with probability 0.1.

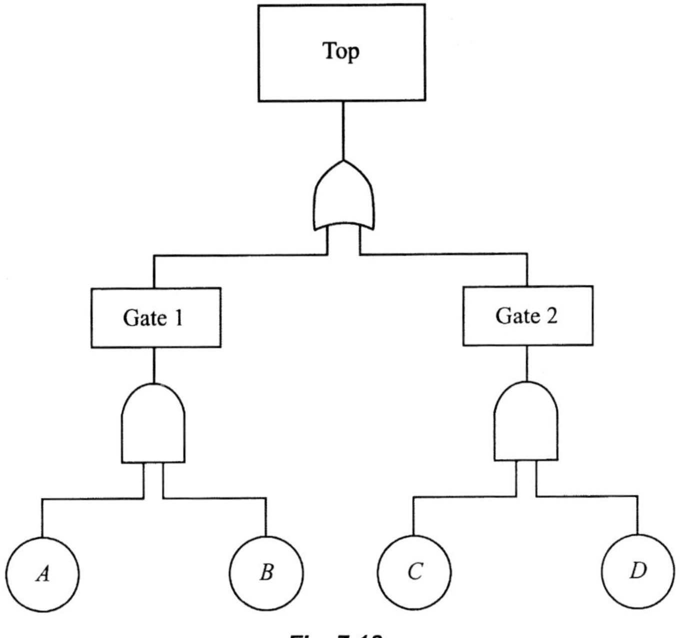

Fig. 7.18

Working from the bottom of the fault tree up we get

$$P(\text{Gate 1}) = P(A)\ .\ P(B) = 0.01$$
$$P(\text{Gate 2}) = P(C)\ .\ P(D) = 0.01$$

and

$$\begin{aligned} P(\text{Top}) &= P(\text{Gate 1 OR Gate 2}) \\ &= P(\text{Gate 1}) + P(\text{Gate 2}) - P(\text{Gate 1})\ .\ P(\text{Gate 2}) \\ &= 0.01 + 0.01 - 0.0001 \\ &= 0.0199 \end{aligned}$$

This approach is simple and accurate but unfortunately it can only be applied to the small category of fault trees which do not contain repeated events. If we refer back to the fault tree constructed for the failure of the cooling water system (Fig. 7.13) it can be seen that even for this simple fault tree such an approach cannot be used since event W, failure of the water supply, occurs twice.

When trees with repeated events are to be analysed, this method is not appropriate since intermediate, gate events will no longer occur independently. If this method is used it is entirely dependent upon the tree structure whether an overestimate or an underestimate of the top event probability is obtained, as will be shown in the following section.

Fault trees with repeated events

When fault trees have basic events which appear more than once, the methods most often used to obtain the top event probability utilize the minimal cut sets produced in the qualitative analysis. Since this approach is appropriate in all circumstances whether basic events are repeated or not, it is a good general approach which will yield the correct result providing the assumption that the basic events are independent is true.

If a fault tree has n_C minimal cut sets K_i, $i = 1, \ldots, n_C$ then the top event exists if at least one minimal cut set exists, i.e.

$$T = K_1 + K_2 + \cdots + K_{n_C}$$

$$= \bigcup_{i=1}^{n_C} K_i$$

and $P(T) = P\left(\bigcup_{i=1}^{n_C} K_i\right)$

Using the general expansion in equation (2.6) gives

$$P(T) = \sum_{i=1}^{n_C} P(K_i) - \sum_{i=2}^{n_C}\sum_{j=1}^{i-1} P(K_i \cap K_j) + \cdots + (-1)^{n_C - 1} P(K_1 \cap K_2 \cap \cdots \cap K_{n_C}) \tag{7.2}$$

This expansion is known as the **inclusion–exclusion expansion**.

Example 1

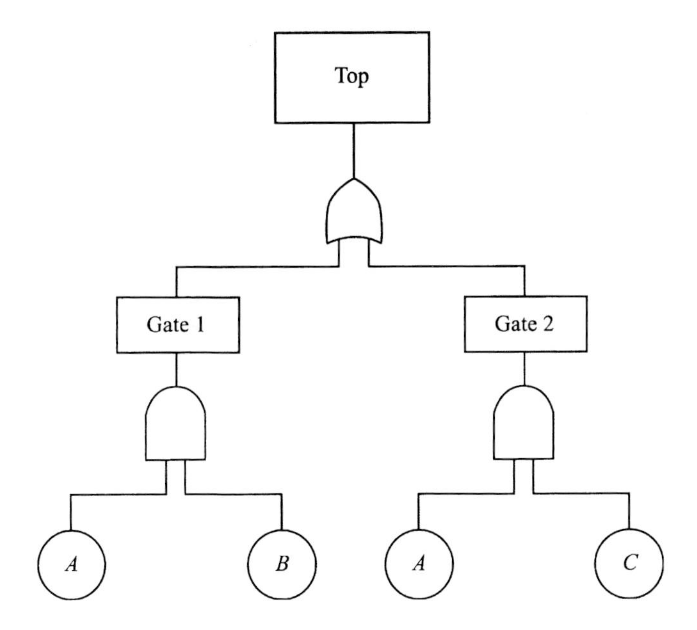

All basic events are independent with $q_A = q_B = q_C = 0.1$

Fig. 7.19

For the first example the fault tree shown in Fig. 7.19 has three basic events A, B, and C with probabilities of occurrence q_A, q_B, and q_C respectively. Event A appears twice in the tree structure.

By applying Boolean reduction the two minimal cut sets for this fault tree are found to be:

$$AB$$
$$AC$$

So

$$T = AB + AC$$

and

$$Q_s(t) = P(T) = P(AB + AC)$$

Using two terms of the inclusion–exclusion expansion gives

$$= P(AB) + P(AC) - P(AB \text{ AND } AC)$$
$$= q_A q_B + q_A q_C - q_A q_B q_C$$
$$= 0.01 + 0.01 - 0.001$$
$$= 0.019$$

[Note: working probabilities up through the fault tree would in this case give P(Gate 1) = P(Gate 2) = 0.01, $Q_s(t)$ = 0.01 + 0.01 – 0.0001 = 0.0199, which is a pessimistic approximation that is reasonably accurate!]

Example 2

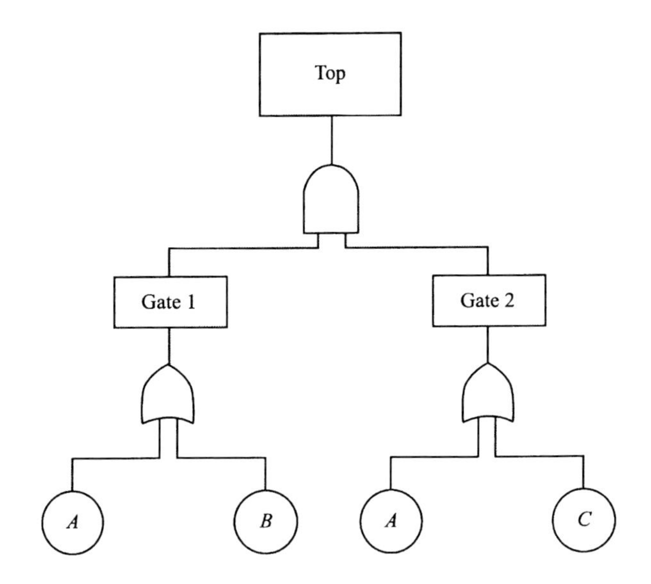

All basic events independent with $q_A = q_B = q_C = 0.1$

Fig. 7.20

For the second example (Fig. 7.20) Boolean reduction of the fault tree structure gives one first-order and one second-order minimal cut set:

A

BC

So

$$T = A + B \,.\, C$$
$$Q_s(t) = P(T) = P(A + B \,.\, C)$$
$$= P(A) + P(B \,.\, C) - P(ABC)$$
$$= 0.1 + 0.01 - 0.001$$
$$= 0.109$$

[Note: working probabilities up through the fault tree would give P(Gate 1) = P(Gate 2) = 0.1 + 0.1 – 0.01 = 0.19, $Q_s(t)$ = 0.19 × 0.19 = 0.0361, which is neither pessimistic nor even reasonably accurate!]

Example 3

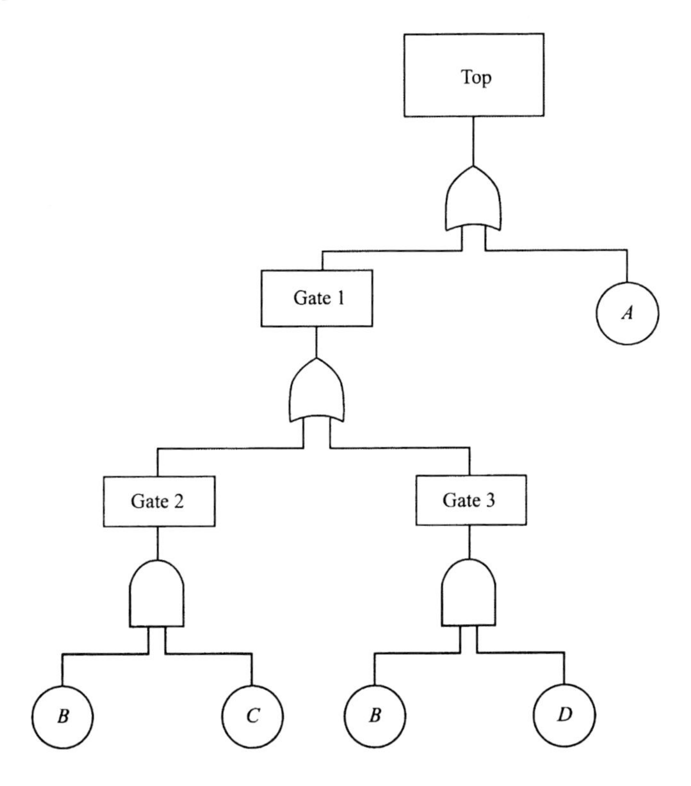

All basic events are independent and $q_A = q_B = q_C = q_D = 0.1$

Fig. 7.21

The minimal cut sets of the fault tree shown in Fig. 7.21 are

$$A$$
$$BC$$
$$BD$$

So

$$T = A + BC + BD$$
$$Q_s(t) = P(T) = P(A + BC + BD)$$

Using three terms of the inclusion–exclusion expansion gives:

$$\underbrace{\left[P(A)+P(BC)+P(BD)\right]}_{\text{1st term}}-\underbrace{\left[P(ABC)+P(ABD)+P(BCD)\right]}_{\text{2nd term}}$$
$$+\underbrace{\left[P(ABCD)\right]}_{\text{3rd term}}$$
$$=[0.1+0.01+0.01]-[0.001+0.001+0.001]+[0.0001]$$
$$=[0.12]-[0.003]+[0.0001]$$
$$=0.1171$$

Example 3 shows that calculating each term in the inclusion–exclusion expansion is tedious and time consuming. In this example it is only necessary to deal with three minimal cut sets. It is not unusual to have fault trees for engineering systems which result in tens of thousands of minimal cut sets. It is impractical in these situations to calculate all terms in the complete expansion! Therefore, approximations which yield accurate top event probabilities with less effort are required.

The probabilities produced for the full expansion in Example 3 typify the trends which result when analysing fault trees. The first term is numerically more significant than the second term. The second term is in turn more significant than the third, and so on. The series converges with each term providing a less significant contribution to the final probability as shown in Fig. 7.22.

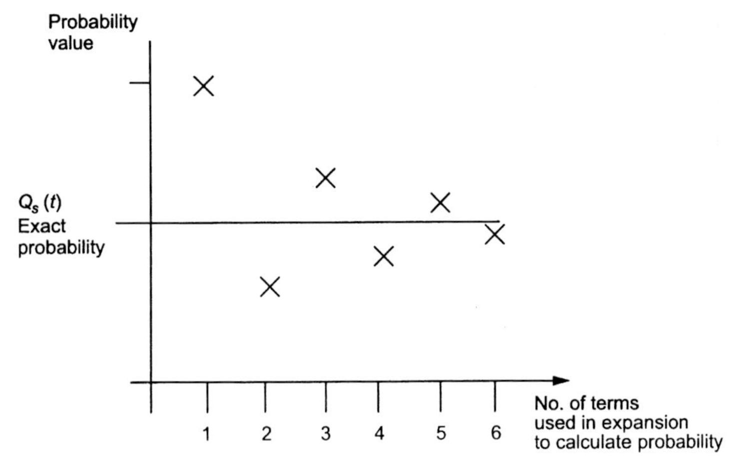

Fig. 7.22 Convergence of inclusion–exclusion expansion

The series evaluation adds successive odd-numbered terms and subtracts successive even-numbered terms and in addition each term is numerically less significant. Therefore truncating the series at an odd-numbered term will provide an upper bound and truncating the series after an even-numbered term provides a lower bound for the exact probability.

Upper and lower bounds for system unavailability

Considering the first two terms in the inclusion–exclusion expansion gives

$$\underbrace{\sum_{i=1}^{n_C} P(K_i) - \sum_{i=2}^{n_C}\sum_{j=1}^{i-1} P(K_i \cap K_j)}_{\text{lower bound}} \le \underbrace{Q_s(t)}_{\text{exact}} \le \underbrace{\sum_{i=1}^{n_C} P(K_i)}_{\text{upper bound}} \tag{7.3}$$

From Example 3 we obtain the following bounds on the exact top event probability:

$$\begin{aligned}\text{Upper bound} &= 0.12\\ \text{Lower bound} &= 0.12 - 0.003\\ &= 0.117\end{aligned}$$

This enables the estimation of the top event probability with a maximum error of 0.0015 by producing a figure mid-way between the upper and lower bounds. For Example 3 basic event probabilities were 0.1; the smaller this value becomes the faster the convergence rate for the series. The upper bound for the top event probability used here is known as the **rare event approximation** since it is itself accurate if the component failure events are rare.

Minimal cut set upper bound

A more accurate upper bound is the **minimal cut set upper bound**

$$\begin{aligned} P(\text{system failure}) &= P(\text{at least 1 minimal cut set occurs}) \\ &= 1 - P(\text{no minimal cut set occurs}) \end{aligned}$$

Since

$$P(\text{no minimal cut set occurs}) \geq \prod_{i=1}^{n_C} P(\text{minimal cut set } i \text{ does not occur})$$

(equality being when no event appears in more than one minimal cut set), therefore

$$P(\text{system failure}) \leq 1 - \prod_{i=1}^{n_C} P(\text{minimal cut set } i \text{ does not occur})$$

i.e.

$$Q_s(t) \leq 1 - \prod_{i=1}^{n_C} \left[1 - P(K_i)\right] \qquad (7.4)$$

It can be shown that

$$\underset{\text{exact}}{Q_s(t)} \leq \underset{\text{minimal cut set upper bound}}{1 - \prod_{i=1}^{n_C} \left[1 - P(K_i)\right]} \leq \underset{\text{rare event approximation}}{\sum_{i=1}^{n_C} P(K_i)}$$

In Example 3 the minimal cut set upper bound approximation is:

$$\begin{aligned} Q_{\text{MCSU}} &= 1 - \prod_{i=1}^{n_C} \left[1 - P(K_i)\right] \\ &= 1 - \left[1 - P(A)\right]\left[1 - P(BC)\right]\left[1 - P(BD)\right] \\ &= 1 - (1 - 0.1)(1 - 0.01)(1 - 0.01) \\ &= 0.117\,91 \end{aligned}$$

The inequalities hold as stated.

$$\begin{array}{ccc} Q_s(t) \le 1 - \prod_{i=1}^{n_C} \left[1 - P(K_i)\right] \le \sum_{i=1}^{n_C} P(K_i) \\ 0.1171 \le \quad 0.117\,91 \quad \le 0.12 \end{array}$$

7.7.2 *Top event failure intensity*

The procedure to calculate the top event failure intensity or failure rate progresses in the same way as that to calculate the top event probability. First the component probabilities and failure intensities are calculated, then the minimal cut set parameters, and finally the system or top event failure intensities. Chapter 5 describes how component failure parameters are calculated; the remainder of this section gives the details of how these are used to determine the minimal cut set and then the top event failure characteristics.

7.7.3 *Minimal cut set parameters*

Minimal cut set reliability characteristics will be denoted by the same symbols as the component reliability parameters defined in Chapter 5 and will carry the subscript *C*. The calculation of minimal cut set probabilities has already been dealt with. Calculation of the conditional failure intensity $\lambda_C(t)$ of a minimal cut set *C* requires consideration of the order in which components in the set fail.

$\lambda_C(t)$ is the probability of occurrence of a cut set per unit time at *t*, given that no minimal cut set failure existed at time *t*. Therefore

$$\lambda_C(t)\,\mathrm{d}t = P\left[C[t, t + \mathrm{d}t) \middle| \bar{C}(t)\right] \qquad (7.5)$$

where $C[t, t + \mathrm{d}t)$ denotes the **occurrence** of the minimal cut set in time interval *t* to $t + \mathrm{d}t$. $C(t)$ denotes the **existence** of the minimal cut set at time *t*.

$$\lambda_C(t)\,\mathrm{d}t = \frac{P\left[C[t,\ t+\mathrm{d}t)\cap\bar{C}(t)\right]}{P\left[\bar{C}(t)\right]}$$
$$= \frac{P\left[C[t,\ t+\mathrm{d}t)\right]}{P\left[\bar{C}(t)\right]} \quad \text{as } C[t,\ t+\mathrm{d}t) \text{ implies } \bar{C}(t) \tag{7.6}$$

Now $P[C[t,\ t+\mathrm{d}t)]$ is the probability that the minimal cut set occurs during $[t,\ t+\mathrm{d}t)$. Since only one component failure event can take place in a small time element $\mathrm{d}t$ this is the probability of an event i from the set occurring during $[t,\ t+\mathrm{d}t)$ given that all other basic events in the minimal cut set have already occurred. Considering each of the n elements of the minimal cut set in turn gives

$$P\left[C[t,\ t+\mathrm{d}t)\right] = \sum_{i=1}^{n}\left(w_i(t)\,\mathrm{d}t\prod_{\substack{j\neq i\\ j=1}}^{n}Q_j(t)\right) \tag{7.7}$$

and also

$$P\left[\bar{C}(t)\right] = 1 - Q_C(t) \tag{7.8}$$

The unconditional failure intensity of the minimal cut set, w_C, the probability of the minimal cut set occurring per unit time at t, is defined as the quantity expressed in equation (7.7):

$$w_C\,\mathrm{d}t = \sum_{i=1}^{n}\left(w_i(t)\,\mathrm{d}t\prod_{\substack{j\neq i\\ j=1}}^{n}Q_j(t)\right) \tag{7.9}$$

If all the minimal cut sets for a particular top event are indexed 1 to n_C, then minimal cut set i occurs at rate $w_{C_i}(t)$. If components fail independently and there are n_i components in a minimal cut set i then using equation (7.9) gives:

$$
\begin{aligned}
w_{C_i}(t)\,\mathrm{d}t = {} & q_2(t)q_3(t)\ldots q_{n_i}(t)w_1(t)\,\mathrm{d}t \\
& + q_1(t)q_3(t)\ldots q_{n_i}(t)w_2(t)\,\mathrm{d}t \\
& + q_1(t)q_2(t)\ldots q_{n_i}(t)w_3(t)\,\mathrm{d}t \\
& + \ldots \\
& + q_1(t)q_2(t)\ldots q_{n_i-1}(t)w_{n_i}(t)\,\mathrm{d}t
\end{aligned}
\tag{7.10}
$$

Cancelling throughout by dt gives an expression for the minimal cut set occurrence rate.

Therefore from equations (7.6), (7.7), (7.8), and (7.9)

$$\lambda_C(t) = \frac{w_C(t)}{\left[1 - Q_C(t)\right]} \tag{7.11}$$

For reliable systems $w_C(t)$ can be approximated by $\lambda_C(t)$ since $1 - Q_C(t) \approx 1.0$. Similarly for the repair process

$$\mu_C(t) = \frac{v_C(t)}{Q_C(t)} \tag{7.12}$$

and the expected number of cut set failures and repairs W_C and V_C respectively are

$$W_C(0,t) = \int_0^t w_C(u)\,\mathrm{d}u \tag{7.13}$$

$$V_C(0,t) = \int_0^t v_C(u)\,\mathrm{d}u \tag{7.14}$$

If steady-state parameters are being used then $w_C(t) = w_C$ and

$$W_C(t_1, t_2) = \int_{t_1}^{t_2} w_C\,\mathrm{d}t = w_C(t_2 - t_1) \tag{7.15}$$

$W_{C_i}(t_1, t_2)$ is the expected number of times the system fails in time period t_1 to t_2 due to minimal cut set i. $W_{C_i}(t_1, t_2)$ is strictly the number of cut set failures in the time interval but it is also a good conservative approximation to the probability of the minimal cut set failure in the time period. The probability of a minimal cut set failure in time interval (t_1, t_2) is the **minimal cut set unreliability**. Therefore

$$W_{C_i}(t_1, t_2) \approx \text{minimal cut set unreliability in time } t_1 \text{ to } t_2$$

System parameters

For the top event intensity or system failure intensity $w_s(t)$ is defined as the probability that the top event occurs at t per unit time. Therefore $w_s(t)\,dt$ is the probability that the top event occurs in the time interval $[t, t + dt)$.

For the top event to occur between t and $t + dt$ all the minimal cut sets must not exist at t and then one or more occur during t to $t + dt$. More than one minimal cut set can occur in a small time element dt since component failure events can be common to more than one minimal cut set.

Hence

$$w_S(t)\,dt = P\left[A \bigcup_{i=1}^{n_C} C_i\right] \tag{7.16}$$

where A is the event that all minimal cut sets do not **exist** at time t and $\bigcup_{i=1}^{n_C} C_i$ is the event that one or more C_i **occur** in time t to $t + dt$.

Now $A = u_1 \cap u_2 \cap \cdots \cap u_{n_C}$ where u_i means the ith minimal cut set does not exist at t. Also $P(A) = 1 - P(\bar{A})$ and so equation (7.16) can be written

$$P\left[A \bigcup_{i=1}^{n_C} C_i\right] = P\left[\bigcup_{i=1}^{n_C} C_i\right] - P\left[\bar{A} \bigcup_{i=1}^{n_C} C_i\right] \tag{7.17}$$

i.e. $\bar{A}$ means at least one minimal cut set exists at t

$$\bar{A} = \bigcup_{i=1}^{n_C} \bar{u}_i$$

Replacing the left-hand term of equation (7.17) gives a two-term expression for the top event or system unconditional failure intensity:

$$w_S(t)\,\mathrm{d}t = \underbrace{P\left[\bigcup_{i=1}^{n_C} C_i\right]}_{\substack{\text{contribution from}\\ \text{the occurrence of}\\ \text{at least one}\\ \text{minimal cut set}}} - \underbrace{P\left[\bar{A}\bigcup_{i=1}^{n_C} C_i\right]}_{\substack{\text{contribution of minimal}\\ \text{cut sets occurring while}\\ \text{other minimal cut sets}\\ \text{already exist}\\ \text{[i.e. system already failed]}}}$$

We can denote these two terms by $w_S^{(1)}(t)$ and $w_S^{(2)}(t)$ respectively so

$$w_S(t)\,\mathrm{d}t = w_S^{(1)}(t)\,\mathrm{d}t - w_S^{(2)}(t)\,\mathrm{d}t \tag{7.18}$$

Each of these two failure intensity terms can now be developed separately. Expanding for the first term, the occurrence of at least one minimal cut set. This gives the following series expression:

$$\begin{aligned} w_S^{(1)}\,\mathrm{d}t &= \sum_{i=1}^{n_C} P(C_i) - \sum_{i=2}^{n_C}\sum_{j=1}^{i-1} P(C_i \cap C_j) + \cdots \\ &\quad + (-1)^{n_C-1} P\left[C_1 \cap C_2 \cap \cdots \cap C_{n_C}\right] \end{aligned} \tag{7.19}$$

where $\sum_{i=1}^{n_C} P(C_i)$ is the sum of the probabilities that minimal cut set i fails in t to t + dt. All other terms in equation (7.19) involve the simultaneous occurrence of two or more minimal cut sets.

Since only one basic event can fail in time interval dt the occurrence of more than one minimal cut set in t to t + dt must result from the failure of a component common to all failed minimal cut sets.

For the general term which requires m minimal cut sets to occur in t to t + dt which have k common basic events:

$$\text{if } k = 0 \quad \text{then} \quad P\left[C_1 \cap C_2 \cap \cdots \cap C_m\right] = 0$$

otherwise if $k > 0$ then

$$P[C_1 \cap C_2 \cap \cdots \cap C_m] = w_A(t, B_1, \ldots, B_k)\,\mathrm{d}t \prod Q_A(\underline{B}) \tag{7.20}$$

where $\prod Q_A(\underline{B})$ is the product of the probabilities of the component failures which exist in at least one of the *m* minimal cut sets but are not common to them all, and $w_A(t, B_1, \ldots, B_k)$ is the failure intensity for a set of components which consists of the *k* common components.

The second term of equation (7.18), $w_S^{(2)}\,\mathrm{d}t$, has a more involved expansion:

$$\begin{aligned} w_S^{(2)}\,\mathrm{d}t &= P\left[\bar{A} \bigcup_{i=1}^{n_C} C_i\right] \\ &= \sum_{i=1}^{n_C} P(C_i \cap \bar{A}) - \sum_{i=2}^{n_C} \sum_{j=1}^{i-1} P(C_i \cap C_j \cap \bar{A}) + \cdots \\ &\quad + (-1)^{n_C - 1} P(C_1 \cap C_2 \cap \cdots \cap C_{n_C} \cap \bar{A}) \end{aligned} \tag{7.21}$$

Now since

$$\bar{A} = \bigcup_{i=1}^{n_C} \bar{u}_i$$

then each term above can be expanded again, so for a general term in equation (7.21):

$$\begin{aligned} P(C_1 \cap C_2 \cap \cdots \cap C_m \cap \bar{A}) &= \sum_{i=1}^{n_C} P(C_1 \cap C_2 \cap \cdots \cap C_m \cap \bar{u}_i) \\ &\quad - \sum_{i=2}^{n_C} \sum_{j=1}^{i-1} P(C_1 \cap C_2 \cap \cdots \cap C_m \cap \bar{u}_i \cap \bar{u}_j) + \cdots \\ &\quad + (-1)^{n_C - 1} P(C_1 \cap C_2 \cap \cdots \cap C_m \cap \bar{u}_1 \cap \bar{u}_2 \cap \cdots \cap \bar{u}_{n_C}) \end{aligned} \tag{7.22}$$

where C_i means the minimal cut set *i* **occurs** in *t* to *t* + d*t* and $\bar{u}_i$ means the minimal cut set *i* **exists** at *t*.

For a general term in this expansion, $P(C_1, C_2, \ldots, C_m, \bar{u}_1, \bar{u}_2, \ldots, \bar{u}_k)$ is the probability that minimal cut sets 1,...,*k* exist at time *t* and minimal cut sets $C_1, \ldots, C_m$ occur in *t* to *t* + d*t*.

Therefore all basic events contained in $\overline{u}_1, \overline{u}_2, \ldots, \overline{u}_k$ exist at time t and we can evaluate the general term from

$$P\left(C_1, C_2, \ldots, C_m, \overline{u}_1, \overline{u}_2, \ldots, \overline{u}_k\right) = w_B\left(t, B_1, \ldots, B_r\right)\ \mathrm{d}t \prod Q_B\left(\underline{B}\right)$$

where $\prod Q_B\left(\underline{B}\right)$ is the product of probabilities of component failures which exist in any minimal cut set $\overline{u}_j$ together with component failures which exist in at least one of the C_m minimal cut sets but are not common to them all, and $w_B(t,B_1,\ldots,B_r)$ is the failure intensity for a set of components consisting of all component failures which are common to all of the minimal cut sets $C_1,C_2,\ldots,C_m$ but not in minimal cut sets $\overline{u}_1, \overline{u}_2, \ldots, \overline{u}_m$.

Example
Find the unconditional failure intensity and expected number of system failures for the fault tree shown in Fig. 7.23.

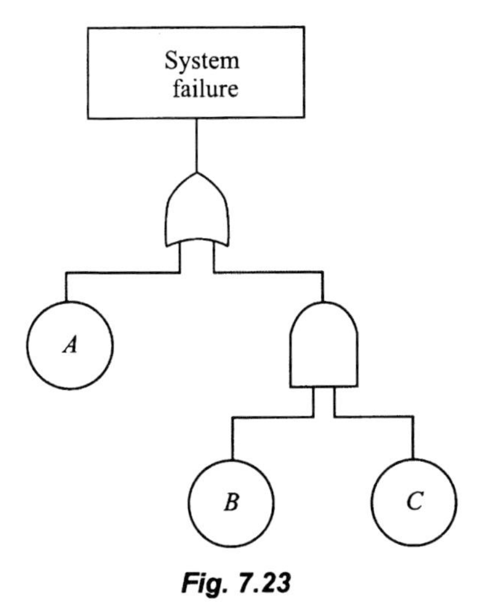

Fig. 7.23

Boolean reduction gives minimal cut sets: $\{A\}$, $\{B,C\}$.
The component failure data are given in Fig. 7.24.

	Failure rate λ/h	Mean time to repair τ(h)
A	1.0×10^{-6}	4
B	4.0×10^{-6}	24
C	2.0×10^{-4}	2

Fig. 7.24

From these data and using the steady-state unavailability approximation we get the following component unavailabilities:

$$q_A = \lambda_A \tau_A = 1.0 \times 10^{-6} \times 4 = 4.0 \times 10^{-6}$$
$$q_B = \lambda_B \tau_B = 4.0 \times 10^{-6} \times 24 = 9.6 \times 10^{-5}$$
$$q_C = \lambda_C \tau_C = 2.0 \times 10^{-4} \times 2 = 4.0 \times 10^{-4}$$

The component conditional failure intensities w can be derived from the expression given in Chapter 5, equation (5.16):

$$w = \lambda(1 - q)$$

This gives

$$w_A = 9.999\ 96 \times 10^{-7}$$
$$w_B = 3.999\ 916 \times 10^{-6}$$
$$w_C = 1.9992 \times 10^{-4}$$

Minimal cut set parameters. Minimal cut set 1 = $\{A\}$. From equation (7.9) we get the minimal cut set unconditional failure intensity:

$$w_{C_1} = \sum_{j=1}^{n} \left(w_j \prod_{\substack{i=1 \\ i \neq j}}^{n} Q_i \right)$$

Therefore

$$w_{C_1} = w_A = 9.999\ 96 \times 10^{-7}$$

and the probability that minimal cut set 1 has occurred is

$$q_1 = q_A = 4 \times 10^{-6}$$

Minimal cut set 2 = $\{B, C\}$. From equation (7.9) we get

$$\begin{aligned} w_{C_2} &= w_B q_C + w_C q_B \\ &= 3.999\ 916 \times 10^{-6} \times 4 \times 10^{-4} + 1.9992 \times 10^{-4} \times 9.6 \times 10^{-5} \\ &= 2.079 \times 10^{-8} \end{aligned}$$

and also

$$q_{C_1} = q_B q_C = 3.84 \times 10^{-8}$$

System parameters. From equation (7.18)

$$w_S = w_S^{(1)} - w_S^{(2)}$$

$w_S^{(1)}\,\mathrm{d}t$ is the probability that at least one minimal cut set occurs in time interval dt. We can evaluate this using equation (7.19):

$$\begin{aligned} w_S^{(1)}\,\mathrm{d}t &= \sum_{i=1}^{2} P(C_i) - \sum_{i=2}^{2}\sum_{j=1}^{i-1} P(C_i \cap C_j) \\ &= P(C_1) + P(C_2) - P(C_1 \cap C_2) \\ &= w_{C_1}\,\mathrm{d}t + w_{C_2}\,\mathrm{d}t - 0 \end{aligned}$$

(Note that C_1 and C_2 cannot both occur in time interval dt since they have no common components.)

$$\begin{aligned} w_S^{(1)} &= 9.999\ 96 \times 10^{-7} + 2.079 \times 10^{-8} \\ &= 1.02 \times 10^{-6} \end{aligned}$$

For the second term $w_S^{(2)}$ we are interested in the occurrence of minimal cut sets which do not fail the system since other minimal cut sets already exist. Using equation (7.21) gives

$$\begin{aligned} w_S^{(2)}\,\mathrm{d}t &= \sum_{i=1}^{2} P\left(C_i,\bar{A}\right) - P\left(C_1,C_2,\bar{A}\right) \\ &= P\left(C_1,\bar{A}\right) + P\left(C_2,\bar{A}\right) - P\left(C_1,C_2,\bar{A}\right) \end{aligned}$$

Taking each of these terms in turn and expanding we get

$$\begin{aligned} P\left(C_1,\bar{A}\right) &= P\left(C_1,\bar{u}_1\right) + P\left(C_1,\bar{u}_2\right) - P\left(C_1,\bar{u}_1,\bar{u}_2\right) \\ &= 0 + w_{C_1} q_{C_2}\,\mathrm{d}t - 0 \\ &= w_{C_1} q_{C_2}\,\mathrm{d}t \end{aligned}$$

(The first and third terms equal zero since minimal cut sets cannot exist and occur at the same time.)

Similarly for the second term:

$$\begin{aligned} P\left(C_2,\bar{A}\right) &= P\left(C_2,\bar{u}_1\right) + P\left(C_2,\bar{u}_2\right) - P\left(C_2,\bar{u}_1,\bar{u}_2\right) \\ &= w_{C_2} q_{C_1}\,\mathrm{d}t + 0 - 0 \end{aligned}$$

(as above minimal cut set 2 cannot occur and exist at the same time) and the third term:

$$P\left(C_1,C_2\bar{A}\right) = 0$$

Therefore combining these results

$$\begin{aligned} w_S &= w_S^{(1)} - w_S^{(2)} \\ &= \left(w_{C_1} + w_{C_2}\right) - \left(w_{C_1} q_{C_2} + w_{C_2} q_{C_1}\right) \\ &= w_{C_1}\left(1 - q_{C_2}\right) + w_{C_2}\left(1 - q_{C_1}\right) \\ &= 1.02 \times 10^{-6} \end{aligned}$$

The expected number of system failures in 10 years is

$$\begin{aligned} W\left(0,\ 87\,600\right) &= \int_0^{87\,600} w_S\,\mathrm{d}t = w_S\left(87\,600\right) \\ &= 0.089\,42 \end{aligned}$$

and the system failure probability is

$$\begin{aligned} Q_S &= q_{C_1} + q_{C_2} - q_{C_1} q_{C_2} \\ &= 4.038 \times 10^{-6} \end{aligned}$$

Approximation for the system unconditional failure intensity

From the previous example it is evident that the full expansion and evaluation of all terms in the equation $w_S(t) = w_S^{(1)}(t) - w_S^{(2)}(t)$ is a tedious and time-consuming task. Like the series expansion for the fault-tree top event probability it would be an advantage to find some approximation for $w_S(t)$ which is acceptably accurate and easier to calculate. In the circumstance that component failures are rare then minimal cut set failures will also be rare events. The second term $w_S^{(2)}(t)$ requires minimal cut sets to exist and occur at the same time. When component failures are unlikely this occurrence rate is also very small and since we are subtracting $w_S^{(2)}(t)$ from $w_S^{(1)}(t)$ to get $w_S(t)$, approximating $w_S^{(2)}(t)$ by zero leaves an upper bound for $w_S(t)$. That is

$$w_{S_{\mathrm{MAX}}}(t)\,\mathrm{d}t = w_S^{(1)}(t)\,\mathrm{d}t \tag{7.23}$$

Also, since

$$w_S^{(1)}(t)\,\mathrm{d}t = \bigcup_{i=1}^{n_C} P(C_i)$$

results in a series expansion we can truncate this at the first term as we did for the top event probability expression to derive the rare event approximation, so

$$\begin{aligned} w_{S_{\mathrm{MAX}}}(t)\,\mathrm{d}t &\le \sum_{i=1}^{n_C} P(C_i) \\ &\le \sum_{k=1}^{n_C} w_{C_i}(t)\,\mathrm{d}t \end{aligned} \tag{7.24}$$

As with the top event probability calculation where equation (7.24) is equivalent to the rare event approximation there is an alternative which is equivalent to the minimal cut set upper bound. This is given in equation (7.25).

$$w_{S_{\mathrm{MAX}}}(t) \le \sum_{i=1}^{n_C} w_{C_i}(t) \prod_{\substack{j=1 \\ j \ne i}}^{n_C} \left[1 - Q_{C_j}(t)\right] \tag{7.25}$$

Each term in the summation is the frequency of minimal cut set i occurring and none of the other minimal cut sets existing. So minimal cut set i causes system failure. Equality will hold if each minimal cut set is independent (contains no events common to other minimal cut sets).

Example 1
From the fault tree given in the previous example we had minimal cut sets $\{A\}$, $\{B, C\}$:

$$w_{C_1} = 9.999\,96 \times 10^{-7}$$

$$w_{C_2} = 2.079 \times 10^{-8}$$

using the above approximation, equation (7.24):

$$w_{S_{\mathrm{MAX}}} = \sum_{i=1}^{n_C} w_{C_i} = 1.0384 \times 10^{-6}$$

Using the above approximation, equation (7.25)

$$\begin{aligned} w_{S_{\mathrm{MAX}}} &= 9.999\,96 \times 10^{-7}\left(1 - 3.84 \times 10^{-8}\right) + 2.079 \times 10^{-8}\left(1 - 4 \times 10^{-6}\right) \\ &= 1.02 \times 10^{-6} \end{aligned}$$

(This compares with the exact answer of 1.02×10^{-6}.)

Example 2
Consider the system shown below in Fig. 7.25. Liquid is pumped from a tank into a butane vaporizer where it is heated to form a gas. In the event of a pump surge the pressure in the vaporizer exceeds the

rating of the vaporizer tubes. To prevent the tubes rupturing, safety systems have been incorporated at several locations on the inlet to the vaporizer, which will shut down the process on the pump surge. In total three protective systems have been used: two trip loops which close a valve halting the butane flow and a vent valve which opens allowing the butane to return to the tank if the pressure exceeds the preset limit. The component failure modes which need to be considered are given in Fig. 7.25 together with the failure and repair data.

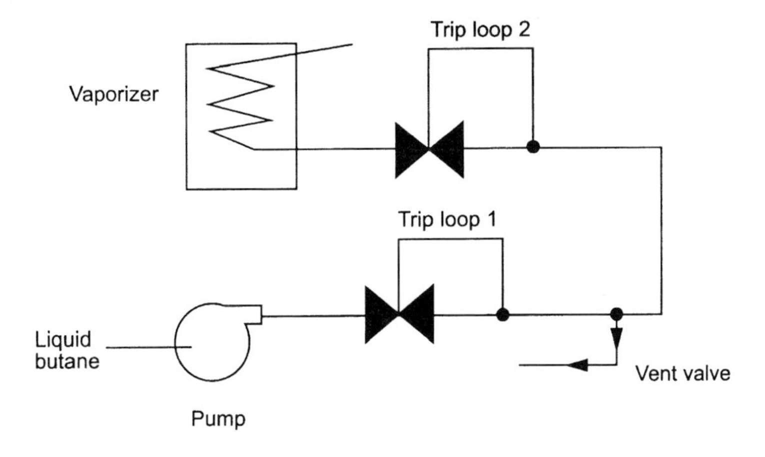

Component failure modes		λ/h	τ(h)	$q = \dfrac{\lambda\tau}{\lambda\tau + 1}$	$w = \lambda(1-q)$
Pump surge	*A*	0.01	50	0.3333	6.667×10^{-3}
Trip loop 1 fails to act	*B*	0.01	10	0.0909	9.091×10^{-3}
Trip loop 2 fails to act	*C*	0.01	10	0.0909	9.091×10^{-3}
Vent valve fails to act	*D*	0.01	10	0.0909	9.091×10^{-3}

Fig. 7.25 Vaporizer system

The fault tree for the top event 'Vaporizer coil ruptures under high pressure' is shown in Fig. 7.26. It must be noted that the fault tree has been constructed assuming that the occurrence of a high pressure will definitely rupture the tank.

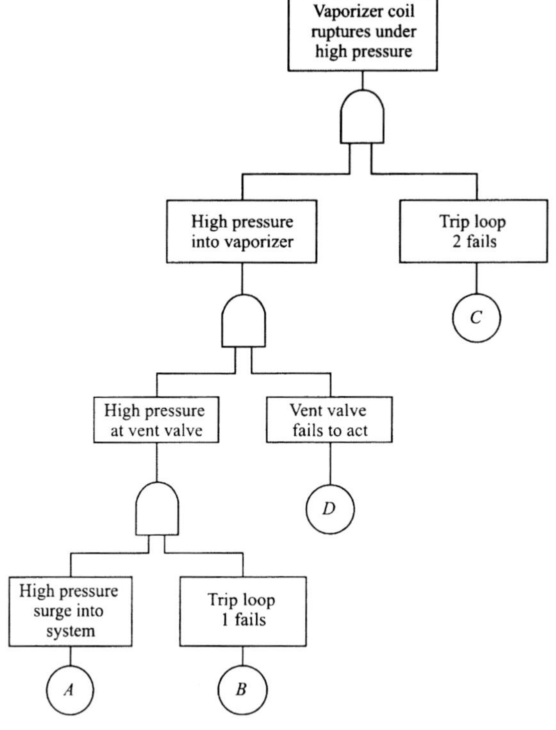

Fig. 7.26 Vaporizer fault tree

Boolean reduction produces one minimal cut set of order 4:

$\{A, B, C, D\}$

Since there is only one minimal cut set

$$
\begin{aligned}
w_S(t) &= w_{C_i}(t) \\
&= \sum_{i=1}^{4}\left(w_i(t) \prod_{\substack{j=1 \\ j \neq i}}^{4} q_i(t) \right) \\
&= w_A q_B q_C q_D + w_B q_A q_C q_D + w_C q_A q_B q_D + w_D q_A q_B q_C \\
&= 5.0075 \times 10^{-6} + 2.5037 \times 10^{-5} + 2.5037 \times 10^{-5} \\
&\quad + 2.5037 \times 10^{-5} \\
&= 8.012 \times 10^{-5}
\end{aligned} \tag{7.26}
$$

Over a 10-year operating period (i.e. 87 600 h) the expected number of top event occurrences is

$$\begin{aligned} W(0,\ 87\,600) &= \int_0^{87\,600} 8.012\times10^{-5}\,\mathrm{d}t \\ &= 8.012\times10^{-5}\times 87\,600 \\ &= 7.02 \end{aligned}$$

The window for initiating events

The previous treatment of the calculation of system failure intensity assumed that the order in which the events in any minimal cut set occur is unimportant. In some analyses the order of basic event failures happening is of vital importance to the occurrence of the fault tree top event. This is particularly true when the analysis of a safety protection system is being carried out. For example if a hazardous event occurs and the protection systems have failed the outcome will be a dangerous system failure. However, if failures occur in another sequence where the hazardous event occurs prior to the protection system's failing then a shutdown will have resulted. This type of situation is modelled by considering the failures as either initiating or enabling events. In the diagram shown in Fig. 7.27 the safety systems are shown to be inactive between time t_0 and t_1. When the safety systems are unable to respond to a hazardous condition the system is in a critical state due to the occurrence of enabling events. If the initiating event occurs between t_0 and t_1 the hazardous system failure will happen. However, if the initiating event occurs prior to t_0 or after t_1 the safety systems will respond as designed. Therefore the order of component failures is important. Initiating and enabling events are formally defined as follows:

Initiating events perturb system variables and place a demand on control/protective systems to respond.

Enabling events are inactive control/protective systems which permit initiating events to cause the top event.

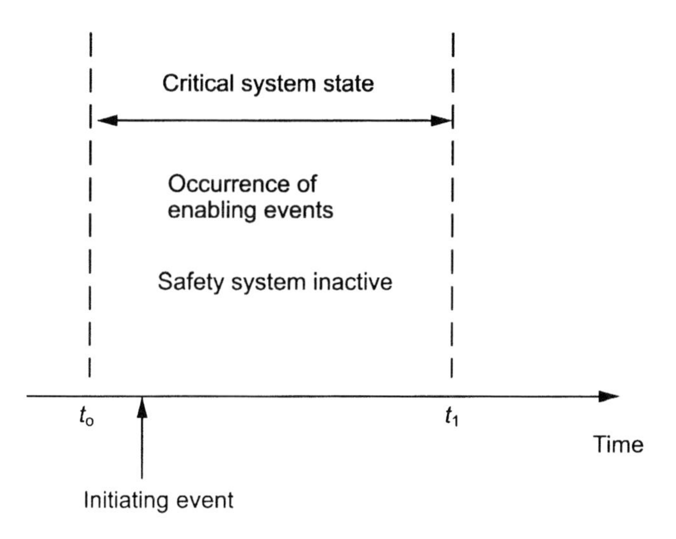

Fig. 7.27 Safety system example

7.7.4 Calculating system unconditional failure intensity using initiator/enabler events

$$w_S(t)\ \mathrm{d}t = P\left[\bigcup_{i=1}^{n_C} C_i\right] - P\left[\bar{A}\bigcup_{i=1}^{n_C} C_i\right]$$
$$= w_S^{(1)}(t)\ dt - w_S^{(2)}(t)\ \mathrm{d}t$$

For general terms in $w_S^{(1)}(t)\ \mathrm{d}t$

$$P[C_1 \cap C_2 \cap \cdots \cap C_m] = w_A(t, B_1, \ldots, B_k)\ \mathrm{d}t \prod Q_A(B)$$

where $\prod Q_A(B)$ is the product of the probabilities of component failures which exist in at least one of the m minimal cut sets but are not common to all of them OR are enabling events common to all minimal cut sets; $w_A(t, B_1, \ldots, B_k)$ is the failure intensity of a set of components which contains the k initiating events common to all m minimal cut sets.

For general terms in $w_S^{(2)}(t)\,\mathrm{d}t$

$$P\left[C_1 \cap C_2 \cap \cdots \cap C_m \cap \bar{u}_1 \cap \bar{u}_2 \cap \cdots \cap \bar{u}_k\right]$$
$$= w_B\left(t, B_1, \ldots, B_r\right) \, dt \prod Q_B(B)$$

where $\prod Q_B(B)$ is the product of the probabilities of component failures which exist in any minimal cut set $\bar{u}_1$ OR components which exist in at least one of the m minimal cut sets but are not common to all of them OR are enabling events which are common to all m minimal cut sets; $w_B(t,B_1,\ldots,B_r)$ is the failure intensity for a set of components which contains initiating events common to all minimal cut sets $C_1, C_2, \ldots, C_m$ but are not in minimal cut sets $\bar{u}_1, \ldots, \bar{u}_k$.

Example

Let us return to the previous example of the butane vaporizer and consider what changes the distinction between initiating and enabling events makes to the predicted number of top event occurrences in a 10-year period. Without considering initiator/enabler events we had from before that

$$w_S = w_A q_B q_C q_D + w_B q_A q_C q_D + w_C q_A q_B q_D + w_D q_A q_B q_C \tag{7.26}$$

However, examining each of the four terms which give w_S it can be seen that only the first term $w_A q_B q_C q_D$ actually causes the top event of the fault tree. This term represents the sequence of events whereby the safety systems have failed (events B, C, D) and then the pump surge occurs (event A). Each of the remaining three terms does not cause the top event since the pump surge occurs prior to all the safety systems failing. This is because event A, the pump surge, places a demand on the safety systems to respond – it is an initiating event. Events B, C, and D are failures of the protective systems which allow the initiating event to cause the top event – they are enabling events.

Therefore, accounting for initiating and enabling events, the system failure intensity is

$$\begin{aligned} w_S &= w_A q_B q_C q_D \\ &= 5.0075 \times 10^{-6} \end{aligned} \tag{7.27}$$

Over a 10-year period the expected number of system failures is

$$w(0,\ 87\ 600) = \int_0^{87\ 600} 5.0075 \times 10^{-6}\,\mathrm{d}t$$
$$= 0.4387$$

Note that if account is not made of initiating and enabling events then an overestimate of the expected number of system failures can be predicted. In this case 7.02 failures as compared with 0.4387 failures; a factor of 16 overestimate. This could result in very costly system modifications if the predicted system performance is deemed inadequate. The modifications may not have been considered necessary had the correct calculation procedure been adopted. This result is important to note since the majority of computer codes which perform fault tree analysis do not distinguish between initiating and enabling events.

7.8 Importance measures

A very useful piece of information which can be derived from a system reliability assessment is the **importance** measure for each component or minimal cut set. An importance analysis is a sensitivity analysis which identifies weak areas of the system and can be very valuable at the design stage. For each component its importance signifies the role that it plays in either causing or contributing to the occurrence of the top event. In general a numerical value is assigned to each basic event which allows them to be ranked according to the extent of their contribution to the occurrence of the top event.

Importance measures can be categorized in two ways:

(a) Deterministic
(b) Probabilistic

The probabilistic measures can themselves be categorized into those which are appropriate for system availability assessment (top event probability) and those which are concerned with system reliability assessment (expected number of top event occurrences).

The importance measures commonly used or referred to are discussed separately, below, and their application demonstrated using: a simple series system, a simple parallel system, and a two-out-of-three voting system.

7.8.1 Deterministic measures

Deterministic measures assess the importance of a component to the system operation without considering the component's probability of occurrence. One such measure is the **structural measure of importance** which is defined for a component i as

$$I = \frac{\text{number of critical system states for component } i}{\text{total number of states for the } (n-1) \text{ remaining components}}$$

A **critical system state** for component i is a state for the remaining $n - 1$ components such that failure of component i causes the system to go from a working to a failed state. This system criticality is illustrated using the three example systems shown in Figs 7.28–7.30.

Series system

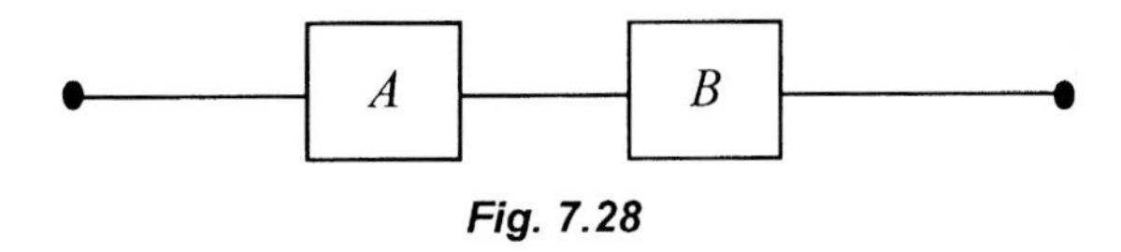

Fig. 7.28

Component A:

States for other components	Critical state for A
$(. , B)$	No
$(. , \bar{B})$	Yes

where B denotes B failed and $\bar{B}$ denotes B working

$$I_A = \frac{1}{2}$$

Component B:

States for other components	Critical state for B
$(A, .)$	No
$(\bar{A}, .)$	Yes

$$I_B = \frac{1}{2}$$

Parallel system

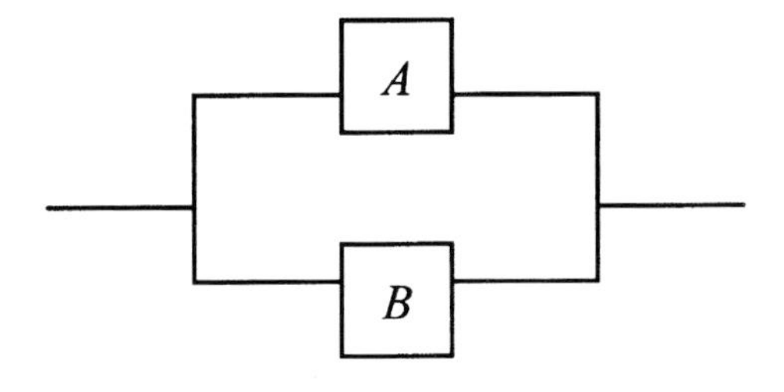

Fig. 7.29

Component A:

States for other components	Critical state for A
$(\,.\,, B)$	Yes
$(\,.\,, \overline{B})$	No

$$I_A = \frac{1}{2}$$

Component B:

States for other components	Critical state for B
$(A, \,.\,)$	Yes
$(\overline{A}, \,.\,)$	No

$$I_B = \frac{1}{2}$$

Two-out-of-three system

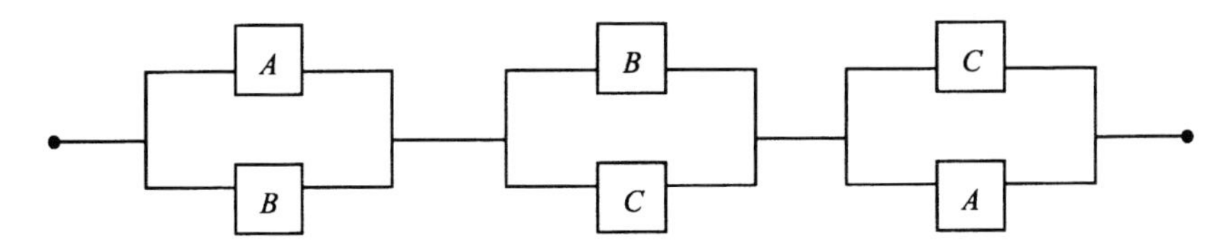

Fig. 7.30

Component A:

States for other components	Critical state for A
$(.\,,B,C)$	No
$(.\,,B,\bar{C})$	Yes
$(.\,,\bar{B},C)$	Yes
$(.\,,\bar{B},\bar{C})$	No

$$I_A = \frac{2}{4} = \frac{1}{2}$$

As all components play the same part in the structure

$$I_A = I_B = I_C = \frac{1}{2}$$

Since this measure of importance takes no account of the component probabilities it is of little use in practical reliability problems.

7.8.2 Probabilistic measures (systems availability)

Several probabilistic measures for importance assessment can be used. Those used when assessing system availability or top event probability are listed and defined in Table 7.4 and are discussed separately below.

Table 7.4

Measure	Probabilistic expression	Interpretation
Criticality measure of basic event importance	$\dfrac{G_i(\boldsymbol{q}(t))q_i(t)}{Q_{SYS}(\boldsymbol{q}(t))}$	Probability that the system is in a critical state for component *i*, and *i* has failed*
Fussell–Vesely measure of basic event importance	$I_i = \dfrac{P\left(\bigcup_{k\mid i\in k} C_k\right)}{Q_{SYS}(\boldsymbol{q}(t))}$	Probability of the union of the minimal cut sets containing basic event *i* *
Fussell–Vesely measure of minimal cut set importance	$\dfrac{P(C_k)}{Q_{SYS}(\boldsymbol{q}(t))}$	Probability of occurrence of minimal cut set C_k *

* Given system is in a failed state at time *t*.

7.8.3 Birnbaum's measure of importance

Birnbaum's measure of importance is also known as the criticality function. The criticality function for a component i is denoted by $G_i(\boldsymbol{q})$ and is defined as the probability that the system is in a critical system state for component i. Thus it is the sum of the probabilities of occurrence of the critical system states for component i.

Table 7.5 summarizes the states and gives the probabilities of their occurrence for the three simple configurations considered in Figs 7.28–7.30. Criticality being considered for component A, since the three systems are structurally symmetrical it can be used to obtain importance measures for A, B, and C. q_i denotes the probability that component i fails.

Table 7.5

System	State	Probability	Critical system state for component *A*
Series	$(\,.\,, B)$	q_B	No
	$(\,.\,, \bar{B})$	$(1-q_B)$	Yes
Parallel	$(\,.\,, B)$	q_B	Yes
	$(\,.\,, \bar{B})$	$(1-q_B)$	No
Two out of three	$(\,.\,, B, C)$	$q_B q_C$	No
	$(\,.\,, B, \bar{C})$	$q_B(1-q_C)$	Yes
	$(\,.\,, \bar{B}, C)$	$(1-q_B)q_C$	Yes
	$(\,.\,, \bar{B}, \bar{C})$	$(1-q_B)(1-q_C)$	No

Therefore the criticality function for component A is:

Series system $\quad G_A = 1 - q_B$

Parallel system $\quad G_A = q_B$

Two-out-of-three system $\quad G_A = (1 - q_B)q_C + q_B(1 - q_C)$

$$= q_C + q_B - 2q_B q_C$$

Two expressions for the criticality function are:

1. $G_i(\boldsymbol{q}) = Q(1_i, \boldsymbol{q}) - Q(0_i, \boldsymbol{q})$ $\quad$ (7.28)

where

$Q(\boldsymbol{q})$ = probability that the system fails
$(1_i, \boldsymbol{q}) = (q_1, \ldots, q_{i-1}, 1, q_{i+1}, \ldots, q_n)$
$(0_i, \boldsymbol{q}) = (q_1, \ldots, q_{i-1}, 0, q_{i+1}, \ldots, q_n)$

This is the probability that the system fails with component i failed minus the probability that the system fails with component i working. It is therefore the probability that the system fails only if component i fails.

For example, for the two-out-of-three system we have:

$$Q(\boldsymbol{q}) = q_A q_B + q_B q_C + q_C q_A - 2q_A q_B q_C$$

Therefore

$$Q(1_A, \boldsymbol{q}) = q_B + q_C - q_B q_C$$

and

$$Q(0_A, \boldsymbol{q}) = q_B q_C$$

giving

$$G_A(\boldsymbol{q}) = Q(1_A, \boldsymbol{q}) - Q(0_A, \boldsymbol{q}) = q_B + q_C - 2q_B q_C$$

as obtained from Table 7.5.

2. $$G_i(\boldsymbol{q}) = \frac{\partial Q(\boldsymbol{q})}{\partial q_i} \tag{7.29}$$

This is defining the criticality function as a partial derivative which is the same as the first expression since

$$\frac{\partial Q(\boldsymbol{q})}{\partial q_i} = \frac{Q(1_i, \boldsymbol{q}) - Q(0_i, \boldsymbol{q})}{1 - 0}$$

again using the two-out-of-three system as an example.

Since $Q(\boldsymbol{q}) = q_A q_B + q_B q_C + q_C q_A - 2q_A q_B q_C$

$$G_A(\boldsymbol{q}) = \frac{\partial Q}{\partial q_A} = q_B + q_C - 2q_B q_C$$

It should be noted that this expression is not itself a function of a component's own failure probability. Many of the remaining importance measures are defined using Birnbaum's measure.

7.8.4 Criticality measure of importance

This importance measure is defined as the probability that the system is in a critical state for component i, and i has failed (weighted by the system unavailability):

$$I_i = \frac{G_i(\boldsymbol{q}(t))\, q_i(t)}{Q_{\text{SYS}}(\boldsymbol{q}(t))} \tag{7.30}$$

Series system:

$$I_A = \frac{(1 - q_B)q_A}{q_A + q_B - q_A q_B}$$

Parallel system:

$$I_A = \frac{q_B q_A}{q_A q_B} = 1$$

Two-out-of-three system:

$$I_A = \frac{(q_B + q_C - 2q_B q_C)q_A}{q_A q_B + q_B q_C + q_C q_A - 2q_A q_B q_C}$$

7.8.5 Fussell–Vesely measure of importance

This measure is defined as the probability of the union of the minimal cut sets containing i given that the system has failed:

$$I_i = \frac{P\left(\bigcup_{k|i\in k} C_k\right)}{Q_{\text{SYS}}\left(\boldsymbol{q}(t)\right)} \tag{7.31}$$

Series system:

$$I_A = \frac{q_A}{q_A + q_B - q_A q_B}$$

Parallel system:

$$I_A = \frac{q_A q_B}{q_A q_B} = 1$$

Two-out-of-three system:

$$I_A = \frac{q_A\left(q_B + q_C - q_B q_C\right)}{q_A q_B + q_B q_C + q_C q_A - 2q_A q_B q_C}$$

The importance rankings produced by the Fussell–Vesely method are very similar to those produced by the criticality measure.

7.8.6 Fussell–Vesely measure of minimal cut set importance

The previously defined importance measures ranked component failures in the order of their contribution to the top event. This measure provides a similar function except that the minimal cut sets are themselves ranked. The importance measure is defined simply as the probability of occurrence of cut set i given that the system has failed:

$$I_i = \frac{P(C_i)}{Q_{\text{SYS}}\left(\boldsymbol{q}(t)\right)} \tag{7.32}$$

Series system:

minimal cut sets $1 = \{A\}$
$2 = \{B\}$

$$I_1 = \frac{q_A}{q_A + q_B - q_A q_B}$$

$$I_2 = \frac{q_B}{q_A + q_B - q_A q_B}$$

Parallel system:

minimal cut sets 1 = {A, B}

$$I_1 = \frac{q_A q_B}{q_A q_B} = 1$$

Two-out-of-three system:

minimal cut sets 1 – {A, B}
2 – {B, C}
3 – {C, A}

$$I_1 = \frac{q_A q_B}{q_A q_B + q_B q_C + q_C q_A - 2q_A q_B q_C}$$

$$I_2 = \frac{q_B q_C}{q_A q_B + q_B q_C + q_C q_A - 2q_A q_B q_C}$$

$$I_3 = \frac{q_C q_A}{q_A q_B + q_B q_C + q_C q_A - 2q_A q_B q_C}$$

7.8.7 Probabilistic measures (systems reliability)

The remaining measures of importance, two for basic events and one for cut sets, are shown in Table 7.6. These measures are appropriate for systems where the interval reliability is being assessed and the sequence in which components fail matters. All measures in Table 7.6 are weighted according to the expected number of system failures, $W(0, t)$.

7.8.8 Barlow–Proschan measure of initiator importance

This is the probability that initiating event i causes the system failure over the interval $[0, t)$. It is defined in terms of the criticality function and the unconditional failure intensity of the component:

$$I_i = \frac{\int_0^t \{Q(1_i, \boldsymbol{q}(t)) - Q(0_i, \boldsymbol{q}(t))\} w_i(t)\, \mathrm{d}t}{W(0, t)} \tag{7.33}$$

7.8.9 Sequential contributory measure of enabler importance

This measure is the probability that enabling event i permits an initiating event to cause system failure over $[0, t)$. The index j in the expression below runs over each initiating event which is contained in the same minimal cut set as enabling event i:

$$I_i = \frac{\sum\limits_{\substack{j \\ i \neq j \\ i \text{ and } j \in C_k \\ \text{for some } k}} \int_0^t \{Q(1_i, 1_j, \boldsymbol{q}(t)) - Q(1_i, 0_j, \boldsymbol{q}(t))\} q_i(t) w_j(t)\, \mathrm{d}t}{W(0, t)} \tag{7.34}$$

This expression is an approximation.

7.8.10 Barlow–Proschan measure of minimal cut set importance

This measure of cut set importance is the probability that a minimal cut set causes the system failure in interval $[0, t)$ given that the system has failed:

$$I_i = \frac{\sum\limits_{j \in i} \int_0^t \left[1 - Q\left(0_j, 1^{i-\{j\}}, \boldsymbol{q}(t')\right)\right] \prod\limits_{\substack{k \neq j \\ k \in i}} q_k(t') w_j(t')\, \mathrm{d}t'}{W(0, t)} \tag{7.35}$$

(j is each initiating event in the minimal cut set $\{i\}$.)

To illustrate how these three measures of importance for interval reliability problems are used consider the two-out-of-three system which has reached steady state. The minimal cut sets are:

$$\{A, B\} \quad \{B, C\} \quad \text{and} \quad \{C, A\}$$

Assume that C is an enabling event and that A and B are initiating events with steady-state component unavailabilities and unconditional failure frequencies q and w respectively. (For reliable systems the unconditional failure rate, w, can be approximated by the conditional failure rate λ.)

$$W(0, t) = \int_0^t \sum_{\substack{i=1 \\ i \text{ initiator}}}^{n} G_i(q(t)) w_i \, \mathrm{d}t$$

$$= \int_0^t \left[(q_B + q_C - 2q_B q_C) w_A + (q_A + q_C - 2q_A q_C) w_B \right] \mathrm{d}t$$

$$= \left[q_B w_A + q_C w_A + q_A w_B + q_C w_B - 2q_C (q_B w_A + q_A w_B) \right] t$$

Note that there are two ways in which the first minimal cut set can fail but only one way for the remaining two minimal cut sets to fail. The Barlow–Proschan measures of basic event importance for initiating events A and B are:

$$I_A = \frac{(q_B + q_C - 2q_B q_C) w_A t}{W(0, t)}$$

$$I_B = \frac{(q_A + q_C - 2q_A q_C) w_B t}{W(0, t)}$$

Both of the above importance measures could be improved for a component by decreasing its failure intensity w.

The sequential contributory importance measures for the components are

$$I_A = \frac{(1 - q_C) q_A w_B t}{W(0, t)}$$

$$I_B = \frac{(1 - q_C) q_B w_A t}{W(0, t)}$$

$$I_C = \frac{((1 - q_A) w_B + (1 - q_B) w_A) q_C t}{W(0, t)}$$

Note that these measures are increasing functions of q_i, the unavailability of the enabling event. Thus to decrease an enabler ranking, its unavailability should be decreased. This can be achieved by decreasing its failure rate or decreasing its mean time to repair (i.e. increasing its repair rate).

The Barlow–Proschan measures of cut set importance are approximated by:

$$I_{AB} = \frac{(w_A q_B + w_B q_A)t}{W(0,\ t)}$$

$$I_{BC} = \frac{q_C w_B t}{W(0,\ t)}$$

$$I_{CA} = \frac{q_C w_A t}{W(0,\ t)}$$

The above calculations assume that when a minimal cut set occurs, it causes system failure. This assumption may not be true if the system is not reliable and other minimal cut sets have failed.

7.9 Expected number of system failures as a bound for systems unreliability

The expected number of system failures $W(0, t)$ is an upper bound for the system unreliability. If system failure is rare this approximation is a close upper bound, i.e.

$$F(t) \quad \leq \quad W(0, t)$$

Unreliability Expected number of system failures

Let

$$P_i(t) = P(\text{exactly } i \text{ system failures in } [0,\ t))$$

The unreliability $F(t)$ is therefore given by

$$F(t) = \sum_{i=1}^{\infty} P_i(t)$$

The expected number of system failures is

$$W(0,t) = \sum_{i=1}^{\infty} iP_i(t)$$

Therefore we have

$$\sum_{i=1}^{\infty} P_i(t) \le \sum_{i=1}^{\infty} iP_i(t)$$

and hence

$$F(t) \le W(0,\ t)$$

The equality condition holds if P(two or more failures) = 0, i.e. for reliable systems the unreliability can be accurately approximated by the expected number of system failures.

7.10 Use of system performance measures

1. For continuously operating systems for which the top event cannot be tolerated (e.g. top events such as 'fire' or 'explosion') $W(0, t)$ is the relevant measure of performance.
2. When system failure can be tolerated, $Q(t)$, $F(t)$, and $W(0, t)$ all have relevance. For moderately large systems $F(t)$ is difficult to calculate exactly, but if the system is reliable then

$$F(t) \approx W(0,\ t)$$

3. For standby or safety systems $Q(t)$, the probability that the system will fail on demand, is the relevant measure.

7.11 Benefits to be gained from fault tree analysis

1. The fault tree construction focuses the attention of the analyst on one particular undesired system failure mode which is usually that

identified as the most critical with respect to the desired system function.

2. The fault tree diagram can be used to help communicate the results of the analysis to peers, supervisors, and subordinates. It is particularly useful with multidisciplinary teams with some members who may be unfamiliar with the numerical performance measures.
3. Qualitative analysis often reveals important system features.
4. Using component failure data the fault tree can be quantified.
5. The qualitative and quantitative results together provide the decision makers with an objective means of assessing the adequacy of the system design.

7.12 Summary

Fault tree analysis is the most important and most frequently used of the methods available to quantify system performance. Not only does it provide a means for system quantification but also a diagrammatic representation of the system failure causes, which is ideal for communicating the failure logic to other personnel.

The chapter has covered the means by which the fault trees are constructed. Once constructed the methods available to analyse the tree both qualitatively (through minimal cut sets) and quantitatively (through top event probabilities, top event frequencies, and component importance measures) have been described.

7.13 Bibliography

Apostolakis, G., Garribba, S., and **Volta, G.** (Eds) (1978) Synthesis and analysis methods for safety and reliability studies. In Proceedings of NATO Conference, Urbino, Italy, pp. 3–14, Plenum Press.

Barlow, R. E. and **Chatterjee, P.** (1973) *An Introduction to Fault Tree Analysis*. ORC.

Barlow, R. E., Fussell, J. B., and **Singpurwalla, N. D.** (1975) Reliability and fault tree analysis. In SIAM, Philadelphia.

Hassl, D. F., Roberts, N. H., Vesely, W. E., and **Goldberg, F. F.** (1981) *Fault Tree Handbook*. US Nuclear Regulatory Commission, NUREG-0492.

Henley, E. J. and **Kumamato, H.** (1980) *Reliability Engineering and Risk Assessment*. Prentice-Hall.

Lapp, S. A. and **Powers, G. J.** (1977) Computer-aided synthesis of fault-trees. *IEEE Trans. on Rel.*, April.

Vesely, W. E. (1970) A time-dependent methodology for fault tree evaluation. *Nucl. Engng Des.*, **13,** 337–360.

Vesely, W. E. and **Narum, R. E.** (1970) PREP and KITT: computer codes for the automatic evaluation of a fault tree. IN-1349.

Villemeur, A. (1991) *Reliability, Availability, Maintainability and Safety Assessment.* John Wiley.

Chapter 8

Common Cause Failures

8.1 Introduction

Safety systems commonly feature redundancy, incorporated to provide a high likelihood of the protective system success. When quantifying the success probability, the redundant sub-systems will not always fail independently. A single common cause can affect all redundant channels at the same time. In safety studies account must be taken of these common cause failures (CCF). A procedure known as the beta factor method is commonly used. The beta factor method assumes that common cause effects can be represented in the system reliability model as a proportion of the failure probability of any single channel of the multi-channel redundant system. Where the channels have different levels of complexity the beta factor is applied to the predicted failure rate or failure probability of the highest reliability channel.

8.2 Common mode and common cause failures

If a system consists of two identical components A and B in a parallel structure then the system will fail only if both components are in a failed state. The probability of this event, providing the components fail independently of each other, is $p \times p = p^2$ where p is the probability of either component failing.

This parallel system will be far more reliable than a system with a single component. However, if the components do not fail independently and can both fail under the same condition then the parallel system may only be slightly more reliable than each single component. A condition which can cause more than one component to fail is known as a **common cause**. (A common cause may result in more than one failure mode.)

One- and two-event cut sets are generally the most significant contributors to the top event of a fault tree and it is these which must be eliminated to increase system reliability. However, if there is a common mode failure influence on the system then higher order cut sets may also have comparable probabilities and contribute to the system failure in a significant way, and they cannot be neglected.

Potential common mode failures can be caused by the following:

(a) components contain identical manufacturing faults;
(b) components are maintained by the same maintenance engineer;
(c) components are situated in the same location and are subject to the same environmental hazards such as impacts or vibration;
(d) components have a common power source;
(e) components are subject to the same operator.

Common cause failures can generally be placed into one of the following three categories:

(a) ageing
(b) plant personnel
(c) system environment.

A more detailed list of potential common causes is given in Table 8.1.

8.2.1 Common mode cut sets

Once a fault tree has been constructed and the minimal cut sets obtained, then providing no common mode failures exist in the system, these cut sets can be directly used in the quantification process. If any of the possible common mode causes are present for the system then a list of each common cause should be made.

For each common cause on this list all basic events which could be influenced by this event should be identified; these are called **common mode events**. If a basic event is unaffected by the common cause then it is a **neutral event**. Where common mode events exist in the minimal cut sets these can be replaced by a complex event representing the common mode failure. The cut sets can now be re-minimized and by providing a probability or frequency for the common mode event the resulting cut sets can be analysed using conventional fault tree analysis programs.

Table 8.1 Categories and examples of common causes

Source	Symbol	Category	Examples
Environment, system components, or sub-systems	I	Impact	Pipe whip, water hammer, missiles, earthquake, structural failure
	V	Vibration	Machinery in motion, earthquake
	P	Pressure	Explosion, out-of-tolerance system changes (pump overspeed, flow blockage)
	G	Grit	Airborne dust, metal fragments generated by moving parts with inadequate tolerances
	S	Stress	Thermal stress at welds of dissimilar metals, thermal stresses and bending moments caused by high conductivity and density
	T	Temperature	Fire, lightning, welding equipment, cooling system faults, electrical short circuits
	E	Loss of energy source	Common drive shaft, same power supply
	C	Calibration	Misprinted calibration instruction
	F	Manufacturer	Repeated fabrication error, such as neglect to properly coat relay contacts; poor workmanship; damage during transportation
Plant personnel	IN	Installation contractor	Same subcontractor or crew
	M	Maintenance	Incorrect procedure, inadequately trained personnel
	O	Operator or operation	Operator disabled or overstressed, faulty operating procedures
	TS	Test procedure	Faulty test procedures which may affect all components normally tested together
	A	Ageing	Components of same materials

For large fault trees, the list of minimal cut sets may be extremely large and it would be prohibitive for the common mode events to be manipulated manually. Computer codes do, however, exist which will carry out the task of identifying common cause events, replacing them in the cut sets with a single complex event and re-minimizing the resulting cut set list ready for quantification.

Where common causes provide an influence on a system it is not advisable that the qualitative analysis should be carried out using truncation or culling of the minimal cut sets generated, based on cut set order. With a common cause influence a cut set of order 4 may be more likely to occur than one with say only two events.

Strictly, a common cause event is one whose occurrence causes the failure of more than one component with near certainty. An event which only increases the likelihood of components failing is a common influence event.

Common cause failures are defined by Edwards and Watson (**1**) as follows.

> A common cause failure is the result of an event or events which, because of dependencies, cause a coincidence of failure states of components in two or more separate channels of a redundancy system leading to the system failing to perform its intended function.

This definition aims to exclude situations which can be predicted from existing failure statistics such as loss of mains electric supply. Common cause failures thus defined generally fall within the areas shown in Fig. 8.1, which Bourne *et al.* (**2**) proposed for the classification of common cause failures. These areas are also used by some analysts to estimate modifying factors which may be applied to the recommended mean value of β.

8.2.2 The beta factor method

The beta factor relates the CCF rate of redundant channels to the total failure rate of the highest reliability channel. If the total failure rate of the channel is λ and the independent and common cause failure rates for that channel are λ_i and λ_{cc} then:

$$\lambda = \lambda_i + \lambda_{cc}$$

The beta (β) factor is defined as the ratio of the common cause failure rate to the total failure rate. Therefore

$$\beta = \frac{\lambda_{cc}}{\lambda} \quad \text{or} \quad \lambda_{cc} = \beta\lambda \tag{8.1}$$

and

$$\lambda_i = \lambda - \lambda_{cc} = (1-\beta)\lambda \tag{8.2}$$

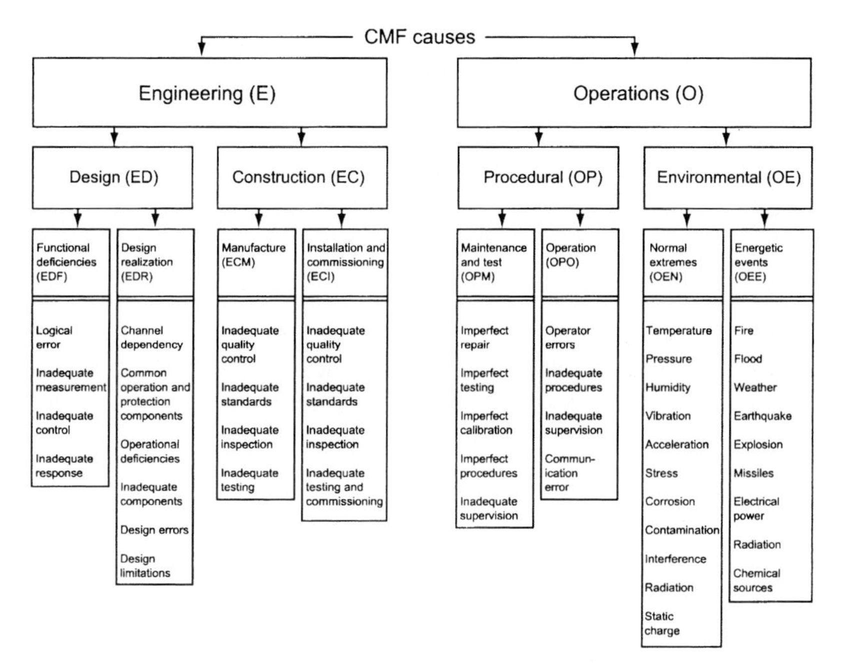

Fig. 8.1 Classification of common mode failures

Example

A simple active-parallel redundant system is shown in Fig. 8.2.

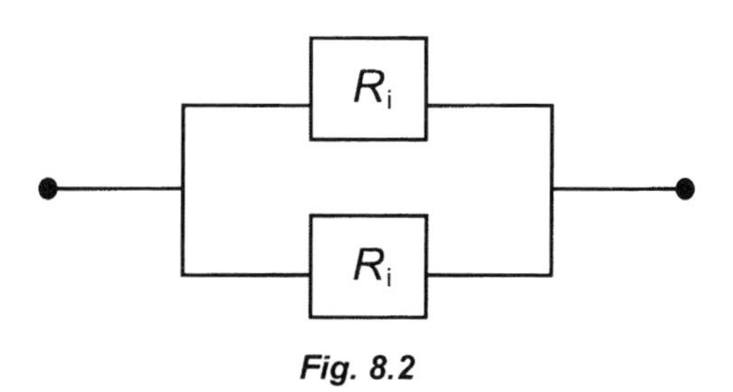

Fig. 8.2

Considering only independent failures the reliability of the system R_S is given by

$$R_S = 1 - [(1 - R_i)(1 - R_i)]$$

where R_i is the probability of survival of a sub-system with independent failures.

For a period of time t

$$\begin{aligned} R_S(t) &= 1 - \{[1 - \exp(-\lambda_i t)][1 - \exp(-\lambda_i t)]\} \\ &= 1 - [1 - 2\exp(-\lambda_i t) + \exp(-2\lambda_i t)] \\ &= 2\exp(-\lambda_i t) - \exp(-2\lambda_i t) \end{aligned} \tag{8.3}$$

With common cause failures included the system network is modified as illustrated in Fig. 8.3.

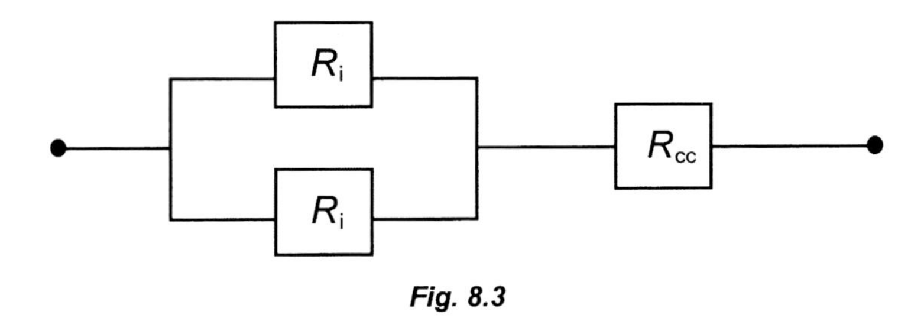

Fig. 8.3

The system reliability is then

$$R'_S = R_S R_{cc} \tag{8.4}$$

i.e. the probability of system survival with independent failures multiplied by the probability of survival with common cause failures.

For a period of time t

$$\begin{aligned} R'_S(t) &= R_S(t) R_{cc}(t) \\ &= [2\exp(-\lambda_i t) - \exp(-2\lambda_i t)]\exp(-\lambda_{cc} t) \end{aligned} \tag{8.5}$$

Substituting from equation (8.2) for λ_i and equation (8.1) for λ_{cc} gives

$$\begin{aligned} R'_S(t) &= \{2\exp[-(1-\beta)\lambda t] - \exp[-2(1-\beta)\lambda t]\}\exp(-\beta\lambda t) \\ &= 2\exp(-\lambda t) - \exp[-(2-\beta)\lambda t] \end{aligned}$$

If $\beta = 1$(i.e. **all** common cause failures)

$$
\begin{aligned}
R'_S(t) &= 2\exp(-\lambda t) - \exp[-(2-1)\lambda t] \\
&= 2\exp(-\lambda t) - \exp(-\lambda t) \\
&= \exp(-\lambda t)
\end{aligned}
$$

and, since $\lambda = \lambda_{cc}$ in this case,

$$R'_S(t) = \exp\left(-\lambda_{cc} t\right) \tag{8.6}$$

If $\beta = 0$ (i.e. **no** common cause failures)

$$
\begin{aligned}
R'_S(t) &= 2\exp(-\lambda t) - \exp[-(2-0)\lambda t] \\
&= 2\exp(-\lambda t) - \exp(-2\lambda t)
\end{aligned}
$$

and, since $\lambda = \lambda_i$ in this case,

$$R'_S(t) = 2\exp(-\lambda_i t) - \exp(-2\lambda_i t) \tag{8.7}$$

Thus it can be seen that the β factor method gives consistent results in both cases. A base value of 0.2 for β has generally been assumed. This base figure is then adjusted by modifying factors. It should be noted that modifying factors need to be based on engineering judgement since common cause failures are relatively rare events.

The analysis given above corresponds to the interpretation of Edwards and Watson (**1**), and is based on the original paper from the General Atomic Company in the USA (**3**).

8.3 Other common cause failure models

Attention has been focused by Atwood (**4**) on the problems of predicting common cause failures. He proposed refinements to the simple β factor model. His approach, the binomial failure rate (BFR) model, assumes that a component in a system has a constant independent failure rate and a susceptibility to common cause shocks which are assumed to hit the system at a constant rate. Vesely (**5**) and others have extended the BFR model to include common cause shocks which are not lethal to the component. The BFR model is equivalent to the β factor model where all common cause shocks are considered to be lethal.

Martin and Wright (**6**) have also proposed a modified β factor model which takes account of the levels of redundancy and diversity in the systems. A similar approach is evident in the so-called multiple Greek letter model (**7**) which includes an additional factor in the basic β factor expression to account for the level of redundancy.

There are, however, some reservations of the validity of models employing global β factors. Obviously the independent failure rate of a channel will increase as the number of components increases and it may not be unreasonable to assume that the common cause failure rate will also increase as the number of components increases. To assume that the common cause failure rate increases in direct proportion to the independent failure rate is, however, less obvious unless the parameters have been estimated from event data specifically collected and classified in a manner consistent with the assumptions built into the models. The practice of applying a global β factor, estimated from applications as diverse as a nuclear reactor and aircraft systems to chemical plant systems, would appear to be questionable. Common cause failure studies conducted for the United States Atomic Energy Commission by Fleming *et al.* (**8**) are more component specific and, as shown in Table 8.2, β factors significantly lower than 0.2 have been observed.

A significantly different approach to predicting common cause failure probabilities proposed by Hughes (**9**) is also worthy of note. The Hughes method does not require a specific model since predictions are made directly from event data; the approach is based on identifying the root causes of failure and the likelihood of generating simultaneous failures in similar equipment. The basis of the Hughes method is to represent the variability of a component's failure probability by distributions which can be estimated directly from relatively small data on multiple failures. The 'environment' distributions reflect the component response to the different root causes, which include factors such as the quality of maintenance as well as more obvious environmental influences such as temperature. The estimation of these distributions requires the systematic collection and analysis of data on environments as well as failures. Currently these data are sparse, which limits the application of the Hughes method.

Limiting values of failure probability have also been applied to account for the effect of common cause failures, particularly in reliability studies of protective instrumentation systems. The

Table 8.2

Component	Reactor years	Number of events classified	Event distribution: Independent	Dependent	Generic, common cause events	Estimated β factors
Reactor trip breakers	563	72	56	16	11	0.15
Diesel generators	394	674	639	35	22	0.03
Motor-operated valves	394	947	842	105	42	0.04
Safety/relief valves:						
PWR	318	54	30	24	0	—
BWR	245	172	136	36	14	0.08
Pumps:						
Safety injection	394	112	77	35	8	0.07
RHR	394	117	67	50	5	0.04
Containment spray	394	48	32	16	2	0.04
Auxiliary feedwater	394	255	194	61	5	0.02
Service water	394	203	159	44	4	0.02
Total	—	2654	2232	422	113	0.04

Events classified include those having one or more actual or potential component failures or functionally unavailable states.

assumption is that, because of common cause failures, the system reliability can never exceed an upper limit determined by the configuration. The minimum failure probability levels proposed by Bourne *et al.* (**2**) are:

Configuration	*Minimum failure probability*
Single instrument	1×10^{-2}
Redundant system	1×10^{-3}
Partially diverse system	1×10^{-4}
Fully diverse system	1×10^{-5}
Two diverse systems	1×10^{-6}

These limiting values were employed by Martin and Wright (**6**) to modify the value of the recommended β factor for different levels of redundancy and diversity.

8.4 Choice of CCF model

In all the treatments of common causes described above there are limitations to the use of the different models either because of the scarcity of common cause failure data or because of problems in their application. The β factor model scores by its simplicity and its ability to cope with situations where failure rates and failure probabilities can both be involved.

Table 8.3 Estimates of β factors from Sellafield data

Sample number	Description	CCF/total failures	Estimated β factor
1	Electrical distribution systems	2/23	0.09
2	Ventilation fans (running failures)	2/166	0.01
3	Ventilation fans (standby failures)	2/47	0.04
4	Data loggers (processors)	2/66	0.03
5	Data loggers (disk drives)	2/34	0.06

Tables 8.2 and 8.3 show the results from studies by EPRI (**10**) and BNFL (**11**) respectively from which estimates of β for a number of different types of equipment have been derived. These data are not comprehensive but it is considered reasonable to estimate β factors for other types of equipment by applying engineering judgement to this basic information. From these data it can be seen that equipment-specific β factors can often be a factor of 10 lower than the recommended value of 0.2 suggested by Edwards and Watson (**1**). The exception appears to be equipment which performs some safety or standby function, particularly when the piece of equipment is operated intermittently or when energetic events such as electrical trips or pressure relief are involved. In these cases the β factor is seen to more closely approach the recommended value of 0.2.

Care needs to be exercised when evaluating any fault tree to ensure that common cause events are recognized as repeated events and are not counted twice. Boolean reduction must be carried out before calculating the top event probability. To this end fault tree computer programs which automatically carry out Boolean reduction before evaluating the minimal cut sets are clearly a useful aid which will ensure that the effect of common cause failures is correctly included in a reliability analysis.

Proposals for modifying the value of β for specific situations have been made by a number of organizations. In this respect the most important factors which warrant consideration are: redundancy and diversity; system complexity; the extent to which the design, operation, and maintenance procedures provide defences against common cause effects; and the probability of the failure remaining unrevealed in normal operation.

8.4.1 Redundancy and diversity

A partially diverse system is defined as a system where the measured parameters (temperature, pressure, etc.) are processed by channels employing different physical principles. Full diversity is when the measured parameters are also different and the diversity extends to full independence of power supplies, maintenance, etc. Increasingly the degree of diversity should reduce the probability of common cause failure and hence the β factor may be modified to reflect this.

8.4.2 System complexity

Much of the research into common cause failures from which the recommended base value of β was derived concentrated on complex control and protection systems. Less complex systems or individual components may not be prone to some of the potential causes of common cause failures such as design and maintenance errors experienced by protective systems, so lower values of β may apply. On the other hand, systems incorporating computers where software validation poses significant problems may attract higher β factors.

8.4.3 Defences against CCF

Edwards and Watson (1) evaluated data from a number of different sources to estimate the proportion of common cause failures attributable to the different areas shown in Fig. 8.1. The results are shown in the last column of Table 8.4. From these data it can be seen that the defences against common cause failure inherent in the design, and the maintenance and test procedures, are likely to have the biggest impact on the probability of common cause failure. If these defences are systematically evaluated during a reliability assessment it may be appropriate to adopt a lower value for the β factor.

Table 8.4 Classification of CCF [Martin and Wright (6)]

Engineering (E)				Operations (O)			
Design (ED)		Construction (EC)		Procedural (OP)		Environmental (OE)	
Funct. def.	Design real.	Manufac.	Instal. and commis.	Maint. and test	Operation	Normal extremes	Energetic events
(EFD)	(EDR)	(ECM)	(ECI)	(OPM)	(OPO)	(OEN)	(OEE)

CCF cause	Proportion of failures on dual running systems: Simultaneous – Assigned values	Simultaneous – Weighted values	Non-simultaneous – Assigned values	Non-simultaneous – Weighted values	Proportion of CCF failures in each category
EFD	1	0.026	—	—	0.026
EDR	0.5	0.145	0.5	0.145	0.29
ECM	—	—	1	0.031	0.031
ECI	—	—	1	0.028	0.028
OPM	—	—	1	0.228	0.228
OPO	0.5	0.060	0.5	0.069	0.138
OEN	—	—	1	0.090	0.090
OEE	1	0.082	—	—	0.082
Total	3	0.322	5	0.591	
Proportion of	0.375	0.35	0.625	0.65	

Note: each CCF cause has been assigned to the category of simultaneous or non-simultaneous failure (or split equally between them) of running (rather than standby) systems. It is seen that approximately one-third of failures would be expected to affect running systems simultaneously.

8.4.4 Unrevealed failures

Protective systems and equipment on standby duty will mainly experience unrevealed failures; that is, failures will only become evident when a demand or test is made on the system. When failure of one channel is revealed (for example, failure of the running pump in a pumping system) then action can frequently be taken to ensure that the system is quickly restored to its original operating state. Thus only simultaneous equipment failures will cause change of the operating state at system level and a lower value of β may be adopted in the reliability assessment. Martin and Wright (**6**) estimate that approximately one-third of all failures would affect running systems simultaneously.

In practice the most significant effect on common cause failure frequency will be the degree of redundancy and diversity applied in the system design. Redundancy and diversity can, therefore, be expected to have the greatest effect on β factors. The following modifying factors have been proposed (**6**):

Configuration	*Modifying factor*
Redundant channel system	1.0
Partially diverse system	0.1
Fully diverse system	0.01
Two diverse systems	0.001

The other factors such as system complexity and the defences against CCF inherent in the design, operating, and maintenance of the system will have less effect on the β factor. This also applies to the unrevealed failure rates of dormant protective and standby equipment; however, their impact on common cause failures should be systematically evaluated during the system reliability assessment, and reductions in β factor applied where appropriate.

Proper consideration of redundancy, diversity, and these other factors will generally lead to smaller β factors and thus lower common cause failure rates in system reliability studies. Unless realistic values are employed the common cause failure probability is likely to be dominant so that the impact of specific combinations of independent failures may not be accorded the importance they actually warrant.

Data input to a fault tree needs careful consideration since it can influence the importance ranking of the minimal cut sets. Common cause failure rates, and thus the β factor to be applied, need as much if not more consideration than the data for independent failures to ensure that realistic minimal cut set probabilities are generated by the analysis.

8.5 Fault tree analysis with CCF

It is not generally acceptable to incorporate CCF effects as a single input immediately below the top event gate of a fault tree. Figure 8.4 shows the preferred way to incorporate common cause failures. This fault tree illustrates how the four component failures $A1$, $A2$, B, and C can combine to produce system failure. Both $A1$ and $A2$ have an associated common cause event CCFA. Boolean reduction is applied to this fault tree to obtain the minimal cut sets as follows:

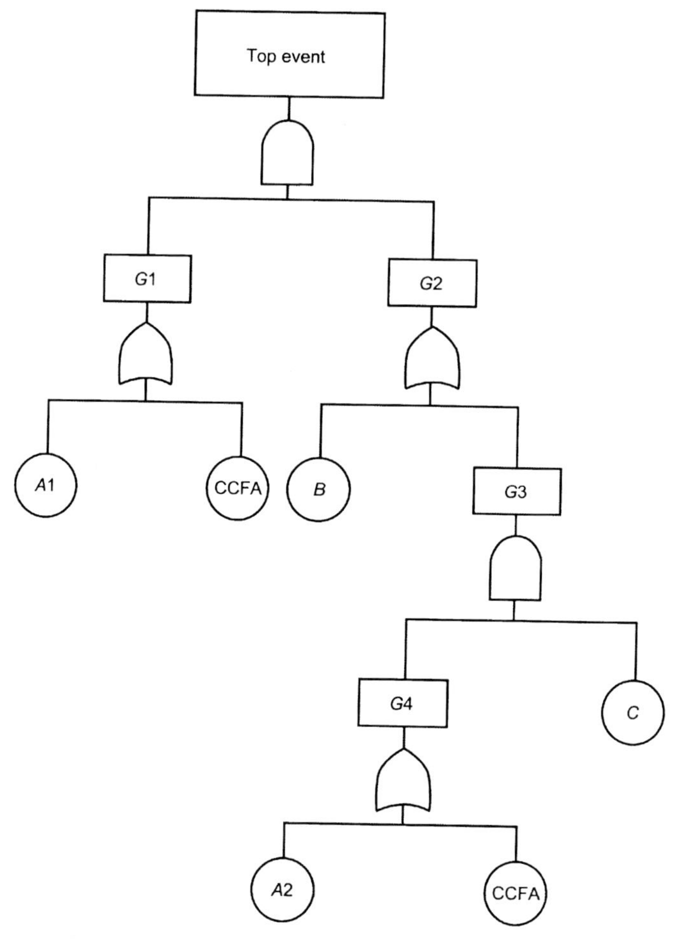

Fig. 8.4 CCF incorporated at primary event level

$$T = G1.G2$$

Since

$$G1 = A1 + \text{CCFA} \quad \text{and} \quad G2 = B + G3$$

$$T = (A1 + \text{CFFA}).(B + G3)$$
$$= A1.B + A1.G3 + \text{CCFA}.B + \text{CCFA}.G3$$

Substituting

$$G3 = G4.C$$

$$T = A1.B + A1.(G4.C) + \mathrm{CCFA}.B + \mathrm{CCFA}.(G4.C)$$

Finally substituting

$$G4 = A2 + \mathrm{CCFA}$$

$$T = A1.B + A1.C.(A2 + \mathrm{CCFA}) + \mathrm{CCFA}.B + \mathrm{CCFA}.C.(A2 + \mathrm{CCFA})$$

Expanding and reducing we get

$$T = A1.B + A1.C.A2 + \mathrm{CCFA}.B + \mathrm{CCFA}.C$$

Minimal cut sets are then:

$B.A1$
$B.\mathrm{CCFA}$
$C.\mathrm{CCFA}$
$C.A1.A2$

If the basic events have the following probabilities:

$P(A1) = P(A2) = 0.1$
$P(B) = 0.2$
$P(C) = 0.3$

and

$$\beta = 0.2$$

then

$$\mathrm{CCFA} = \beta \times 0.1 = 0.02$$

Using the rare event approximation to obtain the top event probability:

$$P(T) = (0.2 \times 0.1) + (0.2 \times 0.02) + (0.3 \times 0.02) + (0.3 \times 0.1 \times 0.1)$$
$$= 0.02 + 0.004 + 0.006 + 0.003$$
$$= 0.033$$

The effect of diversity on the top event probability for the fault tree in Fig. 8.4 can be seen from the example in Table 8.5, which contains results from redundant, partially- and fully-diverse systems with components $A1$ and $A2$.

Table 8.5 Effect on diversity of system failure probability

Minimal cut sets	Redundant $\beta = 0.2$, CCFA = 0.02	Partially diverse $\beta = 0.02$, CCFA = 0.002	Fully diverse $\beta = 0.002$, CCFA = 0.0002
B . *A*1	0.02	0.02	0.02
B . CCFA	0.004	0.0004	0.000 04
C . CCFA	0.006	0.0006	0.000 06
C . *A*1 . *A*2	0.003	0.003	0.003
Top event probability	0.033	0.024	0.0231

The contribution from common cause failures might be further reduced if consideration had been given to the defences against CCF incorporated into design and operation. If failures were also revealed then the contribution to CCF could perhaps be a factor of 4 or 5 lower in all cases than the figures shown above. However, it can be seen that the major impact on the CCF contribution will be through the incorporation of diversity in the redundant channels.

8.6 Summary

Common cause failure effects need to be considered in reliability studies. The β factor model provides a practical means of incorporating these effects in fault trees. The recommended base value of $\beta = 0.2$ should, however, be used with caution. Lower values of β have been experienced with certain equipment. These values should be incorporated in the fault tree at component level whenever possible.

The effect of diversity is particularly important in limiting the effect of CCF. Attention to defences against CCF during design, construction, operation, and maintenance can also significantly reduce this effect.

8.7 References

(1) **Edwards, G. T.** and **Watson, I. A.** (1979) A study of common-mode failures. UKAEA report SRD R146, July.

(2) **Bourne, A. J., *et al*.** (1981) Defences against common-mode failures in redundancy systems. UKAEA report SRD R195, January.

(3) **Fleming, K. N.** (1974) A reliability model for common-mode failures in redundant safety systems. General Atomic Report, GA 13284, December.

(4) **Atwood, C. L.** (1980) Estimates for the binomial failure rate common-cause model: NUREG/CR-1401, EFF-EA-5112.

(5) **Vesely, W. E.** (1977) Estimating common-cause failure probability in reliability and risk analysis. Marshall Olkin specialisations. In Proceedings of the International Conference on *Nuclear Systems Reliability and Risk Assessment*, Gattinburg.

(6) **Martin, B. R.** and **Wright, R. I.** (1987) A practical method of common-cause failure modelling. *Proceedings Reliability* **87**, 14–16 April.

(7) **Fleming, K. N.** and **Kalinowski, A. M.** (1983) An extension of the beta factor to systems with high levels of redundancy. Picard, Lowe, and Garrick, Inc., PLG-0289, June.

(8) **Fleming, K. N., *et al.*** (1985) Classification and analysis of reactor operating experience involving dependent failures. EPRI NP-3967, June.

(9) **Hughes, R. P.** (1987) A new approach to common-cause failure. *Reliability Engng*, **17**, 211–236.

(10) **Crellin, G. L., *et al.*** (1985) A study of common-cause failures. Phase 2: a comprehensive classification system for component fault analysis. EPRI NP-3837, June.

(11) **Moss, T. R.** and **Sheppard, G. T.** (1989) Common-cause failures – evidence from Sellafield. Euredata Conference, Sienna, Italy.

Chapter 9

Maintainability

9.1 Introduction

The maintainability of a system can be defined as the probability that the system will be restored to a fully operational condition within a specified period of time. Clearly maintainability is related to the design features of the system which impact on the maintenance function. In industrial systems, the upkeep costs we incur today are the outcome of design decisions taken 10, 15, or even 20 years ago. Improvements in maintenance efficiency can only have a small percentage effect – the major cost parameters are inseparable from the design and only major design changes will affect them.

Figure 9.1 shows the stages in a typical project and the life-cycle costs (LCC) committed against actual spend. Note how quickly the LCC commitment builds up to the stage where the conceptual design phase is completed. At this stage the cumulative spend is a very small percentage of the projected total. It follows that for maximum impact, maintainability must be given serious consideration during the early stages of a project.

9.2 Maintainability analysis

Maintainability analysis is carried out to provide information for maintenance planning, test and inspection scheduling, and logistic support. During the engineering design phases of a project the objective of the maintainability analysis is to support the achievement of the most cost-effective design within the constraints of the required

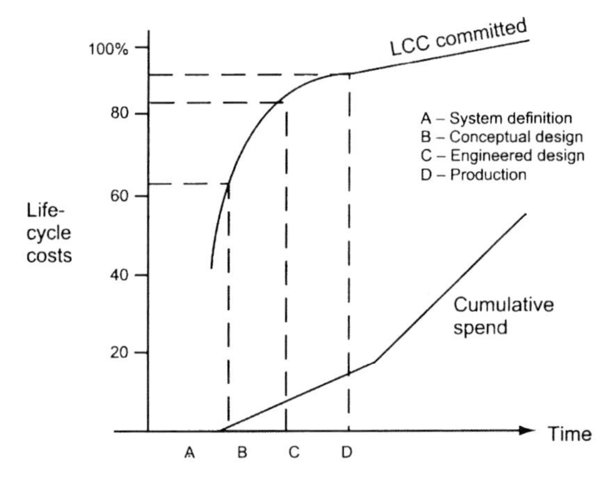

Fig. 9.1 LCC commitment per project phase

operational availability and the selected maintenance policy. Taking inputs from other studies such as reliability prediction and failure mode and effects analysis trade-off studies will be made to evaluate the overall effect in terms of cost and availability of the different design options.

The procedure for the maintainability analysis of mechanical systems described in MIL-HDBK 472 (**1**) entails:

(a) determining the relative frequencies of failure modes
(b) obtaining estimates for elemental activities
(c) summing times of elemental activities
(d) factoring by relative contribution
(e) summing factored times to give mean corrective/preventive times.

Data required from a maintainability analysis for availability prediction and logistic support analysis are:

(a) mean and maximum repair times
(b) scheduled maintenance requirements
(c) consumption rates of limited life items (seals, filters, bearings, etc.)
(d) range and scale of spares necessary to support a given stock-out risk
(e) requirement for special tools and test equipment at all levels of maintenance
(f) maintenance actions requiring special attention (for example because of potential safety hazards)

(g) maintenance crew skill levels and training requirements
(h) maintenance support documentation.

Maintainability prediction, to determine the repair time distribution or expected mean time to repair (MTTR) for specified equipment failure modes is a necessary input to system availability models.

9.3 The maintainability model

For repairable systems the characteristics of the repair process need to be identified in addition to failure characteristics. The general term for the repair process is maintenance, which includes both scheduled and unscheduled repair operations. A probabilistic concept of this characteristic is maintainability where, by definition:

> Maintainability is the probability $M(t)$ that a failed item is repaired in the time interval $(0,t)$.

(Maintainability is clearly dependent on many conditions not explicitly identified in this definition – for example, the work environment and training given to maintenance staff both of which can have a very marked effect on the repair process.)

Maintainability has a probability density function $m(t)$ and a conditional repair intensity $\mu(t)$ given by

$$\mu(t) = \frac{m(t)}{1 - M(t)}$$

[analogous to the hazard rate function $z(t)$]. Intuitively it can be seen that $\mu(t)\,\mathrm{d}t$ describes, to the first order of magnitude in dt, the probability that a system will be repaired within the interval $(t, t+\mathrm{d}t)$ under the conditions that it was in a failed state at time t. If the repair rate $\mu(t) = \mu$, a constant, then the maintainability function

$$M(t) = \int_0^\infty \mathrm{e}^{-\mu t}\,\mathrm{d}t$$

which is similar in form to the equation for unreliability $F(t)$ of a non-repairable system. The MTTR may be defined as

$$\text{MTTR} = \int_0^\infty t \cdot m(t)\,\mathrm{d}t$$

This characteristic describes the average time taken to repair a system with maintainability $M(t)$.

The maintainability function of many systems has been shown to fit a log-normal distribution but often an exponential distribution is employed in practice because the use of constant repair rates leads to simple and manageable formulae. Its use is dictated in part by the limitations in available data – nevertheless it should be noted that as log-normal distributions become increasingly skewed to the left (variances > 1) the distribution of t is quite well represented by an exponential distribution except in the extremes of the tails. Since repair times are frequently truncated to ensure that the 2σ limit is not unduly exceeded the assumption of exponentiality does not seem unreasonable.

For an exponential distribution of repair times maintainability is given by the expression

$$M(t) = 1 - \mathrm{e}^{-\mu t}$$

and the mean time to repair

$$\mathrm{MTTR} = \frac{1}{\mu}$$

MTTR can be estimated by analysing field data or by synthesis using a modified parts-count technique.

9.4 Maintainability prediction

Predicting maintenance times is a key task in maintainability analysis. The system downtime determines system availability, and maintenance cost – particularly the cost of labour and spares – is a major factor in the total life-cycle cost. It is important that predicted maintenance times have a high confidence level. Scheduled maintenance is usually well documented, with the maintenance actions and their frequency well defined. The unscheduled maintenance is subject to much wider variability. A system MTTR is required for calculating system availability but in many cases the maximum allowable system downtime is just as important.

9.4.1 Field data analysis

As with reliability prediction, field data provide the soundest base for prediction of MTTRs, but the data need careful examination before use. Of course it is important that the systems on which the analysis is based are similar to those for which the prediction is being made and are operating in nominally similar conditions. However, there are other pitfalls which will become evident from the following example.

Centrifugal compressor data

The objective of this analysis was to determine the mean time between failures (MTBF) and MTTR of a centrifugal compressor set. The data were required for estimating the unavailability of a gas processing train on an offshore platform.

Table 9.1 Repair times for four compressor sets (6-month period)

Failure	Compressor restoration times (h)			
number	C1	C2	C3	C4
1	1.2	9.5	75.0	1.7
2	6.2	7.8	5.1	2.2
3	2.7	1.5	1.0	2.8
4	51.0	1.7	2.0	10.0
5	4.2	205.0	720.0	135.0
6	1.9	3.6	2.1	14.0
7	1.9	1.4	18.5	6.3
8	1.5		2.2	7.8
9	12.2		8.6	
10	1.3		4.6	
11	3.4			

The repair times from four similar gas compressors recorded over a period of 6 months are shown in Table 9.1. It can be seen that the 36 failures produced a total aggregated downtime of 1336.9 h. Assuming exponential distributions of times to failure and times to repair, the MTBF and MTTR of the compressor set can be calculated respectively as

$$\text{MTBF} = \frac{\Sigma t_i}{n} = \frac{17\,520 - 1336.9}{36} = 450 \text{ h}$$

$$\text{MTTR} = \frac{\sum t_{\text{r}}}{n} = \frac{1336.9}{36} = 37.1 \text{ h}$$

where

$\sum t_i$ = total operating times
n = number of failures
$\sum t_{\text{r}}$ = total repair times

Thus, for this compressor set a first estimate of the steady-state unavailability would be

$$Q = \frac{\text{MTTR}}{\text{MTBF} + \text{MTTR}} = \frac{37.1}{450 + 37.1} = 0.076 \text{ or } 7.6 \text{ per cent}$$

On this basis an estimate of the downtime (DT) due to the compressor set is

$$\text{DT} = 0.076 \times 8760 = 666 \text{ h/year}$$

Closer inspection of Table 9.1, however, shows a very wide range of recorded repair times and the assumption of an exponential distribution of times to repair was called into question. The individual times to repair were, therefore, ordered by magnitude as shown in Table 9.2 and plotted on Weibull probability paper. Three distinct modes of failure for the compressors were evident, namely: instrument failures, external mechanical failures, and internal mechanical failures. The distributions of repair times for these different modes of failure were subsequently shown to be approximately log-normal. However, for this example where only first-order estimates are required it is reasonable to assume exponentiality. The estimates of MTBF and MTTR for the three failure sub-sets are shown in Table 9.3 and indicate the following MTTRs:

1. Instrumentation failures – $\text{MTTR}_1 = 1.7$ h
2. External mechanical failures – $\text{MTTR}_2 = 7.5$ h
3. Internal mechanical failures – $\text{MTTR}_3 = 117$ h.

Table 9.2 Combined set – ranked repair times

Order number (i)	Ranked repair times (t_r)	Mean rank $R_i = i/(n+1)$
Combined set $n = 36$		
1	1.0	0.03
2	1.2	0.05
3	1.3	0.08
4	1.4	0.11
5	1.5	0.14
6	1.5	0.16
7	1.7	0.19
8	1.7	0.22
9	1.9	0.24
10	1.9	0.27
11	2.0	0.30
12	2.1	0.32
13	2.2	0.35
14	2.2	0.38
15	2.7	0.41
16	2.8	0.43
17	2.8	0.43
18	3.6	0.49
19	4.2	0.51
20	4.6	0.54
21	5.1	0.57
22	6.2	0.59
23	6.3	0.62
24	7.8	0.65
25	7.8	0.68
26	8.6	0.70
27	9.5	0.73
28	10.0	0.76
29	12.2	0.78
30	14.0	0.81
31	18.5	0.84
32	51.0	0.86
33	75.0	0.89
34	135.0	0.92
35	205.0	0.95
36	720.0	0.97
$n = 36$	$\sum t_r = 1336.9$	

Table 9.3 Mode 1, 2, and 3 sub-sets – ranked repair times

Order number (i)	Ranked repair times (t_r)	Mean rank $MR_i = i/(n + 1)$
Mode 1 sub-set n = 14		
1	1.0	0.07
2	1.2	0.13
3	1.3	0.20
4	1.4	0.27
5	1.5	0.33
6	1.5	0.40
7	1.7	0.47
8	1.7	0.53
9	1.9	0.60
10	1.9	0.67
11	2.0	0.73
12	2.1	0.80
13	2.2	0.87
14	2.2	0.93
$\text{MTBF}_1 = \frac{17\,520 - 1336.9}{14} = 1156\text{ h}$		$\text{MTTR}_1 = \frac{\sum t_{ri}}{n} = \frac{23.6}{14} = 1.7\text{ h}$
Mode 2 sub-set n = 17		
1	2.7	0.06
2	2.8	0.11
3	3.4	0.17
4	3.6	0.22
5	4.2	0.28
6	4.6	0.33
7	5.1	0.39
8	6.2	0.44
9	6.3	0.50
10	7.8	0.56
11	7.8	0.61
12	8.6	0.67
13	9.5	0.72
14	10.6	0.78
15	12.2	0.83
16	14.6	0.89
17	18.5	0.94
$\text{MTBF}_2 = \frac{17\,520 - 1336.9}{17} = 952\text{ h}$		$\text{MTTR}_2 = \frac{\sum t_{ri}}{n} = \frac{127.3}{17} = 7.5\text{ h}$
Mode 3 sub-set n = 4		
1	51.0	0.17
2	75.0	0.33
3	135.0	0.50
4	205.0	0.67
$\text{MTBF}_3 = \frac{17\,520 - 1336.9}{4} = 3237\text{ h}$		$\text{MTTR}_3 = \frac{\sum t_{ri}}{n} = \frac{466}{4} = 117\text{ h}$

Using these data the compressor set unavailability can be calculated as follows:

$$U_c = U_1 + U_2 + U_3$$

where

$$U_1 = \frac{1.7}{1156 + 1.7} = 0.0015$$

$$U_2 = \frac{7.5}{952 + 7.5} = 0.0078$$

$$U_3 = \frac{117}{3237 + 117} = 0.0349$$

Hence

$$U_c = 0.0015 + 0.0078 + 0.0349 = 0.0442 \text{ or } 4.42 \text{ per cent}$$

In this case the estimated downtime per year due to the compressor set is

$$\text{DT} = 0.0442 \times 8760 = 387 \text{ h}$$

(cf. 666 h previously.)

It can be seen that care must be taken with field data. The biggest contribution to downtime was shown to be due to internal failures which require the compressor casing to be split. These failures occur on average two or three times a year and keep the plant unavailable for about 4 days at a time. Most of the other repairs will be completed within a normal shift of 8 h.

The benefit of this level of information is obvious for availability studies and for developing an overall maintenance strategy for the plant.

9.5 MTTR synthesis

Where field data are not available an alternative method is to synthesize the system MTTR by techniques similar to failure mode and effect analysis (FMEA). As with FMEA the system is broken down into a number of discrete components – generally known as maintainable items. Each maintainable item is then considered in turn for each failure mode in the following seven categories:

(a) failure recognition
(b) failure diagnosis
(c) dismantle
(d) repair/replace
(e) reassemble
(f) test
(g) restore system.

Times to perform each activity are allocated by engineering judgement to the appropriate column on the worksheet and the column times then aggregated to give the total estimated time to perform the repair action. The sum of these estimates for all failure modes represents the MTTR for the maintainable item. This procedure is then repeated for each maintainable item in the system. A typical maintainability analysis worksheet is shown in Fig. 9.2.

To obtain the system MTTR the relative frequency of the maintainable item (MI) failures need to be taken into account. For each MI, therefore, the MTTR is multiplied by the MI failure rate and entered into the final column of the worksheet – this effectively shows the expected downtime per year due to each maintainable item. The sum of the MI downtimes gives the average downtime per year of the system. Hence, to obtain the system MTTR this figure must be divided by the system failure rate ($\sum\lambda$). The system MTBF is the reciprocal of the system failure rate.

For availability studies involving large mechanical process equipment a less rigorous approach is frequently used. The same tabular format is applied but in this case the different types of maintainable item are grouped together. For each type of MI only failures causing a forced outage are considered for two categories of restoration action, namely, active repair time and waiting time. The former category covers the calendar time interval to execute the repair (irrespective of the number of men employed) the latter category accounts for activities such as arranging the permit to work, assembling the necessary tools and spares, isolation of equipment, testing, ramping up to full output after repair, etc.

Maintainability Analysis Example

MAIN EQUIPMENT Surveillance Radar
SYSTEM Transmitter

Date:
Sheet: 6 0f 26

Maintainable item	Failure rate λ(f/y)	Failure mode	Average time to perform corrective maintenance action							Mean time to repair tr (min)	Downtime/y λtr
			Failure recognition	Failure diagnosis	Dismantle	Repair/ replace	Reassemble	Test	Restore sytsem		
Travelling wave tube	0.42									44	
Modulator	0.28									28	
L V Power	0.12									18	2.2
EHT supply	0.22									18	4.0
Start/stop sequencer	0.03									32	1.0
BITE board	0.02									12	0.2
	0.06									22	1.3
Filter	0.01	Blocked	(a) System S/D lamp lights (b) Over-temp lamp lights 2	Inspection (Maint Hdbk 42) 10	Release 4 fixing screws Withdraw filter 5	Insert new filter 1	Tighten 4 fixing screws 3	Power up - check full air flow 3	EHT on	32	0.3
$\Sigma\lambda$	1.15									$\Sigma\lambda$tr	35.3

NOT SHOWN IN THIS EXAMPLE

Transmitter system MTTR = $\frac{\Sigma\lambda tr}{1.15}$ = 31 minutes

Fig. 9.2 Maintainability analysis worksheet

The procedure is then identical to that described previously as shown by the example in Fig. 9.3. From this example it can be seen that rotating equipment is the main contributor to downtime, with about five failures on average each year and outage times ranging from 3 to 48 h. The contribution from the different types of equipment are as follows:

Type of maintainable item	*Average downtime/year*
Rotating equipment	175.2 h
Lube oil system	27.6 h
Vessels	19.2 h
Valves	16.8 h
Instrumentation	12.3 h

Clearly this analysis does not give the same degree of information or the level of confidence obtained from the analysis of field data. Nevertheless, it can take significantly less time (given that suitable generic data are available), it produces well-documented records of the analysis and it provides a systematic and consistent approach to the estimation of MTTRs.

MAINTAINABILITY ANALYSIS WORKSHEET

SYSTEM -- GAS COMPRESSOR Date Sheet 1 of 1

Maintainable item	Failure rate (f/year)	Failure mode	FOFR (λ)	Average time to perform maintenance action		Mean time to repair t_r	λt_r
				Active repair time	Waiting time		
Compressor casing (2)	1.5×2	Forced outage	3.0	32	16	48	144
Gearbox	0.3	"	0.3	8	4	12	3.6
Drive couplings (3)	0.1×3	"	0.3	8	4	12	3.6
El motor	1.0	"	1.0	16	8	24	24
Sub-total			**4.6**				
Lube oil system	2.3	"	2.3	8	4	12	27.6
Sub-total			**2.3**				
Suction scrubbers (4)	0.1×4	"	0.4	8	4	12	4.8
Recycle/suction coolers (4)	0.3×4	"	1.2	8	4	12	14.4
Control valves (9)	0.2×9	"	1.8	4	2	12	10.8
Blowdown valves (4)	0.1×4	"	0.4	4	2	6	2.4
Non-return valves (3)	0.1×3	"	0.3	4	2	6	1.8
Pressure relief valves (3)	0.1×3	"	0.3	4	2	6	1.8
Surge controllers (3)	0.4×3	"	1.2	2	1	6	3.6
Temp controllers (2)	0.2×4	"	0.8	2	1	3	2.4
Pressure controllers (2)	0.2×2	"	0.4	2	1	3	1.2
Level controllers (4)	0.2×4	"	0.8	2	1	3	2.4
D P transmitters (3)	0.1×3	"	0.3	2	1	3	0.9
Flow transmitters (3)	0.3×3	"	0.9	2	1	3	2.7
Sub-total			**8.8**			3	
Pipework connections, etc.		"	0.9	8	4	12	10.8

$\sum \lambda = 16.6$ f/year $\qquad \sum \lambda t_r = 262.8$

$$\text{System mean time to repair } = \frac{\sum \lambda t_r}{\sum \lambda} = \frac{262.8}{16.6} = 15.8 \text{ h}$$

Fig. 9.3 Process equipment maintainability analysis example

9.6 Summary

Maintainability analysis is a necessary part of reliability studies concerned with repairable systems. It should be carried out in the early stages of design to obtain the maximum benefit since procurement and plant layout decisions made at this stage can significantly influence the through-life costs on a project.

Generally, estimates of the system mean time to repair derived from field data are to be preferred since they provide the highest degree of information and confidence in the results. Nevertheless, alternative methods based on a systematic evaluation of the repair process for each failure mode can be worthwhile and can provide valuable input to availability studies.

9.7 Reference

(1) MIL-HDBK 472 (1980). Maintainability Analysis. US Department of Defense.

Chapter 10

Markov Analysis

10.1 Introduction

Situations may arise in availability and reliability modelling which do not conform to the assumptions required by techniques such as fault tree analysis. For example, the kinetic tree theory utilized in fault tree methods requires the statistical independence of the basic events. Standby redundancy, common cause failures, secondary failures, and multiple-component states are all commonly encountered while assessing systems availability. All of these situations violate the independence assumption.

10.1.1 Standby redundancy

When an operating component fails, a standby component is brought into operation and the unit continues to function. The probability of failure of the standby component may change when it starts to function and experience load. Its failure probability is therefore dependent upon the failure of the operating component. The unit fails when both components are in the failed state but the failure of the two components is not statistically independent.

10.1.2 Common causes

A common cause event can render more than one component failed or their performance degraded. Such component failures are no longer independent.

10.1.3 Secondary failures

Secondary failures occur due to loads or stresses which lie outside the bounds within which the component is intended to function. When one component failure causes the secondary failure of another component its failure does not occur independently. This circumstance can occur when the failure of a component results in other components experiencing an increased load. The increased load may change the failure likelihood of the functioning component violating independence.

10.1.4 Multiple-state component failure modes

If a component can exist in more states than simply working and failed then mutually exclusive basic events may need to be considered in an analysis. For example, a valve can be considered to fail open, closed, or stuck. A valve not failing in the open mode does not therefore imply that it works successfully.

When these conditions of dependence are encountered a solution can be obtained by applying Markov modelling. The Markov method is not one which should be used to assess all systems. It should only be used when necessary. It has the capability of modelling dependencies in systems but the trade-off for this is that the model size can grow very quickly with the number of components. The advantages and disadvantages of the approach will be discussed later. The Markov approach can be used for systems that vary discretely or continuously with respect to time and space. Reliability and availability problems are generally discrete in space and continuous in time. That is, systems or components can be considered to have a fixed number of identifiable discrete states but transitions between states can occur at any time point.

A Markov model comprises two elements: states and transitions. To help understand the discrete state continuous time Markov model, consider the simple analogy with a traveller moving from city to city. The traveller can reside in one of the discrete set of possible cities (**states**). He can move between cities that have roads which link them. The movement between one city and another is known as a **transition**. Transitions take negligible time. The traveller's location at any point in time represents the system's current **status** and the time the traveller spends in a city before moving on is the **state holding time**. Due to the road system it will not be possible to move directly to all other cities from the current city. It is only possible to move to those which are linked. It may be necessary to move to one or more intermediate cities before reaching a required destination. If for some reason it is not possible to leave a city it is known as an **absorbing** city.

For a reliability problem the model states represent the possible states of a system. This is the complete set of mutually exclusive states in which a system can reside. The states are usually described by defining which components are working or failed. Transitions between states then represent component failures and repairs. Following the formulation of a Markov model its analysis will yield the probability of being in any of the states. The system failure probability is then determined by summing the probabilities of residing in those states which represent a system failure condition.

For the basic Markov approach to be applicable the system must be characterized by a lack of memory, that is, the future states of the system are independent of all past states except the immediately preceding one. So the future behaviour depends only on its present state, not the history of what has happened in the past. This is represented by the equation below where X_t is a random variable representing thestate in which the system resides at time t.

$$\begin{aligned} P(X_{t+\mathrm{d}t} = k \mid X_t = j,\ X_{t-\mathrm{d}t} = i,\ X_{t-2\mathrm{d}t} = h, \ldots, X_0 = a) \\ = P(X_{t+\mathrm{d}t} = k \mid X_t = j) \end{aligned}$$

Also, for the methods described in this section the system behaviour must not vary with time. The probability of making a transition from one state to another must be constant. This type of process is called stationary or homogeneous. If the transition probabilities are functions of time then the process is not stationary and is known as non-Markovian.

With the lack-of-memory property, the likelihood of a component failure is only dependent on the fact that it is currently working. It does not matter that it may have been perfectly reliable up to this time or that it may have failed several times before.

The homogeneous property means that the transitions between states are not dependent on time. They are, therefore, governed by a constant rate, and times between transition are governed by the negative exponential distribution.

For the Markov approach to be valid the system must be stationary, the process must lack memory, and the states of the system must be identifiable. If these conditions are satisfied then the first stage of the Markov analysis is to draw the state transition diagram. This takes the form of a directed graph where each node represents one of the discrete system states, and the edges indicate the transition probabilities or frequencies between the states in the direction indicated by an arrow drawn on the edge.

10.2 Example – single-component failure/repair process

As an example consider the simple case of a single repairable component. The component undergoes an alternating process of failure and repair with constant rates, i.e. characterized by the negative exponential distribution. The component can exist in one of two states, working or failed. The state transition diagram for this situation is given in Fig. 10.1.

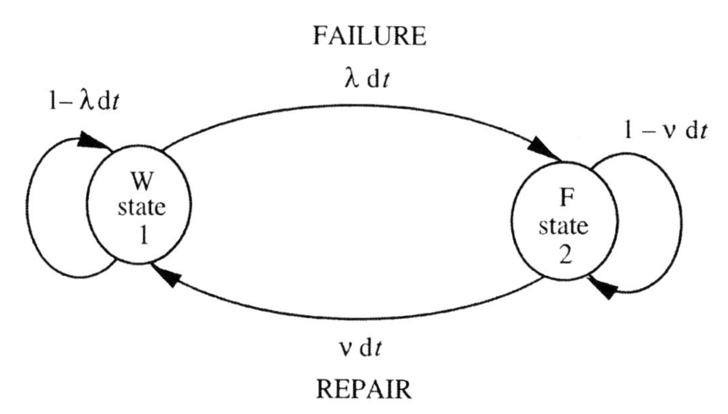

Fig. 10.1 Repairable component state transition diagram

The component can be considered to start in the working state at time $t = 0$. Transition from the working state, 1, to the failed state, 2, occurs with constant rate λ. Therefore the probability of failure at $t + dt$ given that the component is working at t is $\lambda\,dt$. Failure is immediately revealed and transition back to the working state, the repair process, occurs with constant rate ν. For components in state 1, they can either move to state 2 with probability $\lambda\,dt$ or remain in state 1 with probability $1 - \lambda\,dt$ as indicated on the diagram. The time interval dt must be small enough such that two or more transitions cannot occur in dt.

Let $x(t)$ be an indicator variable denoting the state of the component at time t, i.e.

$$x(t) = \begin{cases} 1 & \textit{Failed} \\ 0 & \textit{Working} \end{cases}$$

The probability that the component exists in the failed state (unavailability) after time increment dt depends only on the state of the component at present, i.e.

$P[x(t + \mathrm{d}t) = 1] = P$ (the component was working at t and undergoes failure in time dt OR the component was failed at time t and remained in the failed state during dt)

$$P[x(t+\mathrm{d}t)=1] = P[x(t)=0]\,\lambda\,\mathrm{d}t + P[x(t)=1]\,(1-\nu\,\mathrm{d}t) \quad (10.1)$$

If $P_w(t)$ denotes the probability of the component working at time t, i.e. $P[x(t) = 0]$, and $P_f(t)$ denotes the probability of the component being in the failed state at t, i.e. ($P[x(t) = 1]$), then

$$P_f(t+\mathrm{d}t) = P_w(t)\lambda\,\mathrm{d}t + P_f(t)\,(1-\nu\,\mathrm{d}t)$$

Rearranging gives

$$\frac{P_f(t+\mathrm{d}t) - P_f(t)}{\mathrm{d}t} = P_w(t)\lambda - P_f(t)\nu \quad (10.2)$$

As d$t \to 0$

$$\frac{P_f(t+\mathrm{d}t) - P_f(t)}{\mathrm{d}t} \to \frac{\mathrm{d}P_f(t)}{\mathrm{d}t}$$

so

$$\frac{\mathrm{d}P_f(t)}{\mathrm{d}t} = \lambda P_w(t) - \nu P_f(t) \quad (10.3)$$

Since $P_w(t) + P_f(t) = 1$ then equation (10.3) becomes

$$\frac{dP_f(t)}{dt} = \lambda - (\lambda + \nu)P_f(t) \tag{10.4}$$

Solving this linear first-order differential equation with initial condition $P_f(0) = 0$ gives the unavailability:

$$P_f(t) = \frac{\lambda}{\lambda + \nu}\left(1 - e^{-(\lambda+\nu)t}\right) \tag{10.5}$$

Considering availability, $P[x(t + dt) = 0]$, in the same way gives

$$P_w(t + dt) = (1 - \lambda\, dt)P_w(t) + \nu\, dt\, P_f(t)$$

So

$$\frac{dP_w(t)}{dt} = -\lambda P_w(t) + \nu P_f(t) \tag{10.6}$$

or

$$\frac{dP_w(t)}{dt} = \nu - (\lambda + \nu)P_w(t) \tag{10.7}$$

The solution of equation (10.7) with initial condition $P_w(0) = 1$ is

$$P_w(t) = \frac{\nu}{\lambda + \nu} + \frac{\lambda e^{-(\lambda+\nu)t}}{\lambda + \nu} \tag{10.8}$$

The differential equations and hence their solutions for the probabilities of being in the failed state [equation (10.5)] and the working state [equation (10.8)] are identical to those obtained in Chapter 5 [equation (5.30)]. These solutions are illustrated in Fig. 10.2.

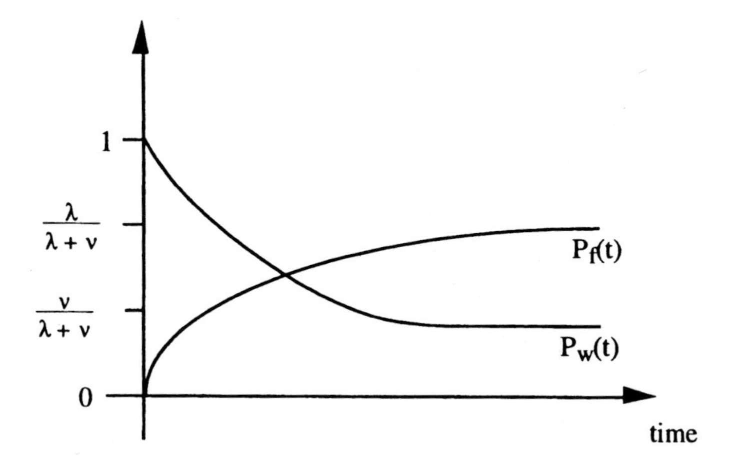

Fig. 10.2 Time-dependent solution for the two-state component

Another way to obtain this solution is to write down the differential equations direct from the state transition diagram:

$$\frac{dP_{state}}{dt} = \text{(rate of entering state)} - \text{(rate of leaving state)}$$

Therefore

$$\frac{dP_w(t)}{dt} = -\lambda P_w(t) + \nu P_f(t) \tag{10.9a}$$

$$\frac{dP_f(t)}{dt} = \lambda P_w(t) - \nu P_f(t) \tag{10.9b}$$

In matrix form this is

$$[\dot{P}_w(t),\ \dot{P}_f(t)] = [P_w(t),\ P_f(t)] \begin{bmatrix} -\lambda & \lambda \\ \nu & -\nu \end{bmatrix} \tag{10.10}$$

i.e.

$$\dot{\boldsymbol{P}} = \mathbf{P}\ \mathbf{A}$$

10.3 General Markov state transition model construction

The two-state Markov model representing a single component described in the previous section is the most simple model possible. The method is more commonly used to model more complex systems comprising several components. States on the Markov model are required to be mutually exclusive and exhaustive. That is, they must be non-overlapping and represent every possible state in which the system can reside. A possible way to generate the system states is to identify the functionality or failure mode for each component in the system and list all possible combinations. Sometimes it is easy to write down all possible states; for example, for a system with two components which either work or fail, the Markov state transition diagram is shown in Fig. 10.3. For components A and B, either both components work, state 1, or both components fail, state 4, or there are two ways in which one component works and one fails, states 2 and 3. The transition rates λ_A, λ_B and repair rates ν_A, ν_B are incorporated on the diagram.

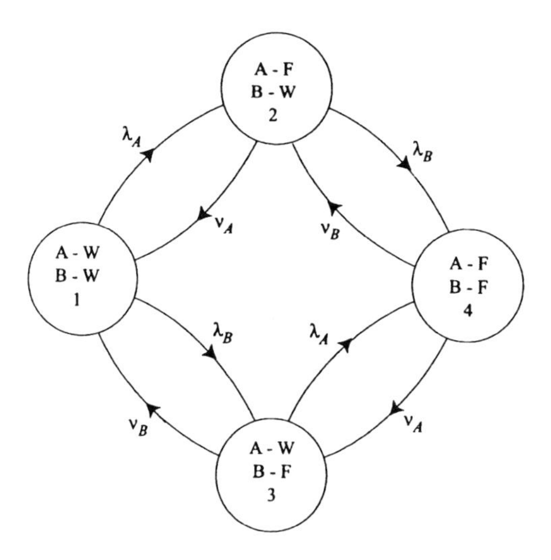

Fig. 10.3 Two-component Markov model

It is possible to consider components in states other than simply working and failed. In these circumstances it is not always as simple to construct the state transition diagram. A procedure which will be successful regardless of the system is to write down all system states in which the system would ideally reside. For the two-component models this would be state 1, components A and B both work. Then consider all possible transitions from this state. There are two possibilities: A fails which results in state 2, or B fails which results in state 3. This completes all possible transitions from state 1. Now consider one of the new states generated – say state 2. Again two possibilities exist: A which is failed can be repaired and return the system to state 1, B which is working could fail, which generates a new state with both components failed (state 4). Continuing this process for system states 3 and 4 will not generate any new possible states but it completes the full set of transitions on the diagram (Fig. 10.3 results).

As yet no consideration of the system structure has been made. Analysis of the Markov model shown in Fig. 10.3 will yield the probabilities of being in each state, i.e. Q_1, Q_2, Q_3, and Q_4.

If the system structure is parallel, i.e. all components must fail in order for the system to fail, then the system failure probability Q_{SYS} is

$$Q_{SYS} = Q_4$$

For a series system which fails when at least one component fails then

$$Q_{SYS} = Q_2 + Q_3 + Q_4$$

10.4 Markov state equations

Consider a general Markov model which consists of a maximum of p states. Let these be numbered from 1 to p with the system working states numbered first 1 to k. So the system failed states are $k + 1$ to p.

Let a_{ij} denote the transition rate from state i to state j, i.e.

$$P(\text{system in state } j \text{ at } t + \mathrm{d}t \mid \text{system in state } i \text{ at } t) = a_{ij}\,\mathrm{d}t$$

10.4.1 State equations

Consider the probability that the system is failed at t + dt. This is given by:

$$Q_i(t+\mathrm{d}t) = P\text{ (system in state } i \text{ at } t \text{ and remains in state } i \text{ after } \mathrm{d}t)$$

$$+\sum_{\substack{j=1\\ j\neq i}}^{p} P(\text{system in state } j \text{ at } t \text{ and makes a transition to state } i \text{ in the next } \mathrm{d}t)$$

i.e.

$$Q_i(t+\mathrm{d}t) = Q_i(t)\left[1-\sum_{\substack{j=1\\ j\neq i}}^{p} a_{ij}\ \mathrm{d}t\right] + \sum_{\substack{j=1\\ j\neq i}}^{p} Q_j(t)a_{ji}\ \mathrm{d}t \tag{10.11}$$

Rearranging gives

$$\frac{Q_i(t+\mathrm{d}t)-Q_i(t)}{\mathrm{d}t} = -Q_i(t)\sum_{\substack{j=1\\ j\neq i}}^{p} a_{ij} + \sum_{\substack{j=1\\ j\neq i}}^{p} Q_j(t)a_{ji} \tag{10.12}$$

As $\mathrm{d}t \to 0$

$$\frac{Q_i(t+\mathrm{d}t)-Q_i(t)}{\mathrm{d}t} \to \frac{\mathrm{d}Q_i}{\mathrm{d}t}$$

$$\frac{\mathrm{d}Q_i(t)}{\mathrm{d}t} = -Q_i(t)\sum_{\substack{j=1\\ j\neq i}}^{p} a_{ij} + \sum_{\substack{j=1\\ j\neq i}}^{p} Q_j(t)a_{ji} \tag{10.13}$$

If we define a_{ii} as

$$a_{ii} = -\sum_{\substack{j=1\\ j\neq i}}^{p} a_{ij}$$

equation (10.13) becomes

$$\frac{\mathrm{d}Q_i(t)}{\mathrm{d}t} = \sum_{j=1}^{p} Q_j(t)a_{ji} \qquad \text{for } i = 1,2,\ldots,p \tag{10.14}$$

In matrix form:

$$\left[\frac{dQ_1(t)}{dt}, \frac{dQ_2(t)}{dt}, \ldots, \frac{dQ_p(t)}{dt}\right] = [Q_1(t) \cdots Q_p(t)][\mathbf{A}] \tag{10.15}$$

where [**A**] is the transition rate matrix ($p \times p$)

$$[\mathbf{A}] = \begin{cases} a_{ij} & \text{Transition rate from } i \rightarrow j \\ a_{ii} & -\sum_{\substack{j=1 \\ j \neq i}}^{p} a_{ij} \end{cases}$$

i.e. the elements on any row sum to zero. **A** is a singular matrix (i.e. Det **A** = 0).

The square matrix **A** above is called the state transition matrix and can be formulated directly from the transition diagram using the following rules:

(a) the dimensions of the matrix are equal to the number of states in the model;
(b) all rows summate to zero;
(c) an off-diagonal element in row i column j represents the transition rate from state i to state j;
(d) a diagonal element row i, column i is the transition rate out of state i (always negative).

Equation (10.15) is dependent and so when solving the corresponding matrix equation additional information is required. This is

$$\sum_{i=1}^{p} Q_i = 1 \tag{10.16}$$

Using these rules to form the state transition matrix for the two-component system whose Markov diagram is illustrated in Fig. 10.3 gives

$$\mathbf{A} = \begin{bmatrix} -(\lambda_A + \lambda_B) & \lambda_A & \lambda_B & 0 \\ \nu_A & -(\nu_A + \lambda_B) & 0 & \lambda_B \\ \nu_B & 0 & -(\nu_B + \lambda_A) & \lambda_A \\ 0 & \nu_B & \nu_A & -(\nu_A + \nu_B) \end{bmatrix}$$

and the system equations are

$$\begin{aligned} \dot{Q}_1 &= -(\lambda_A + \lambda_B)Q_1 + \nu_A Q_2 + \nu_B Q_3 \\ \dot{Q}_2 &= \lambda_A Q_1 - (\nu_A + \lambda_B)Q_2 + \nu_B Q_4 \\ \dot{Q}_3 &= \lambda_B Q_1 - (\nu_B + \lambda_A)Q_3 + \nu_A Q_4 \\ \dot{Q}_4 &= \lambda_B Q_2 + \lambda_A Q_3 - (\nu_A + \nu_B)Q_4 \end{aligned} \qquad (10.17)$$

10.5 Dynamic solutions

The solution to the set of linear first-order differential equations developed for the Markov models will give the time-dependent solution. Laplace transforms are the most widely used method of solving these sets of equations. They transform the differential equations to a set of algebraic equations.

Transforming equation (10.15) gives

$$\begin{bmatrix} \frac{dQ_1(t)}{dt} \\ \frac{dQ_2(t)}{dt} \\ \vdots \\ \frac{dQ_p(t)}{dt} \end{bmatrix} = [\mathbf{A}]^T \begin{bmatrix} Q_1 \\ Q_2 \\ \vdots \\ Q_p \end{bmatrix} \qquad (10.18)$$

Let the Laplace transform of Q_i be

$$\mathcal{L}[Q_i(t)] = Q_i^*$$

Then since

$$\mathcal{L}\left[\dot{Q}_i(t)\right] = sQ_i^* - Q_i(0) \quad \text{for } i = 1, \ldots, p$$

the Laplace transform of the state equation (10.18) becomes:

$$s\begin{bmatrix} Q_1^* \\ Q_2^* \\ \vdots \\ Q_p^* \end{bmatrix} - \begin{bmatrix} Q_1(0) \\ Q_2(0) \\ \vdots \\ Q_p(0) \end{bmatrix} = [\mathbf{A}]^{\mathrm{T}} \begin{bmatrix} Q_1^* \\ Q_2^* \\ \vdots \\ Q_p^* \end{bmatrix} \tag{10.19}$$

By substituting $s\sum_{i=1}^{p} Q_i^* = 1$ into equation (10.19), which is equivalent to $\sum_{i=1}^{p} Q_i(t) = 1$, the equation can be solved.

10.6 Steady-state probabilities

A process is **ergodic** if the limiting probabilities are not dependent upon the initial conditions. So it is unimportant which state the system starts in at time zero; when steady-state conditions are achieved the probabilities of being in any of the states will not be affected. The steady-state probabilities of an ergodic Markov process, such as the single repairable component discussed in the previous section, will be non-zero. The time-dependent probability of being in the failed and working states are given by equations (10.5) and (10.8) respectively. As time gets very large, i.e. $t \to \infty$, the limiting values become

$$P_{\mathrm{f}} = \frac{\lambda}{\lambda + \nu}$$

$$P_{\mathrm{w}} = \frac{\nu}{\lambda + \nu} \tag{10.20}$$

These steady-state values can be produced without solving for the full time-dependent solution since at steady state

$$\frac{\mathrm{d}P_\mathrm{f}(t)}{\mathrm{d}t} = 0 \quad \text{and} \quad \frac{\mathrm{d}P_\mathrm{w}(t)}{\mathrm{d}t} = 0 \tag{10.20a}$$

Equations (10.3) and (10.6) then become

$$\lambda P_\mathrm{w}(t) - \nu P_\mathrm{f}(t) = 0$$

and

$$-\lambda P_\mathrm{w}(t) + \nu P_\mathrm{f}(t) = 0 \tag{10.21}$$

Both of these equations are the same so $P_\mathrm{w}(t) + P_\mathrm{f}(t) = 1$ is needed to obtain the solution to equation (10.20a).

The solution method for these sets of equations can be obtained by Gaussian elimination. This method is presented in Section 10.7.3. Alternatively it can be shown that the asymptotic probabilities can be obtained from

$$Q_i(\infty) = \frac{\begin{vmatrix} a_{11} & \cdots & a_{1,p-1} & 0 \\ \vdots & \vdots & \vdots & \vdots \\ \cdots & \cdots & \cdots & 0 \\ a_{i1} & \cdots & a_{i,p-1} & 1 \\ \cdots & \cdots & \cdots & 0 \\ \vdots & \vdots & \vdots & \vdots \\ a_{p1} & \cdots & q_{p,p-1} & 0 \end{vmatrix}}{\begin{vmatrix} a_{11} & \cdots & a_{1,p-1} & 1 \\ \vdots & \vdots & \vdots & \vdots \\ a_{i1} & \cdots & a_{i,p-1} & 1 \\ \vdots & \vdots & \vdots & \vdots \\ a_{p1} & \cdots & a_{p,p-1} & 1 \end{vmatrix}} \tag{10.22}$$

The numerator is the determinant of the state transition matrix, **A**, with the last column replaced with zeros except in the *i*th row, which is set to unity. The denominator is the determinant of the state transition matrix with all entries in the last column set to unity.

The steady-state system availability can then be obtained from

$$A(\infty) = \sum_{i=1}^{k} Q_i(\infty) \tag{10.23}$$

and the steady-state unavailability from

$$Q(\infty) = \sum_{i=k+1}^{p} Q_i(\infty) \tag{10.24}$$

Alternatively, the system unavailability can be obtained from

$$Q(\infty) = \frac{\begin{vmatrix} a_{11} & \cdots & a_{1,p-1} & 0 \\ \vdots & \vdots & \vdots & \vdots \\ a_{k,1} & \cdots & a_{k,p-1} & 0 \\ a_{k+1,1} & \cdots & a_{k+1,p-1} & 1 \\ \vdots & \vdots & \vdots & \vdots \\ a_{p1} & \cdots & a_{p,p-1} & 1 \end{vmatrix}}{\begin{vmatrix} a_{11} & \cdots & a_{1,p-1} & 1 \\ \vdots & \vdots & \vdots & \vdots \\ q_{p1} & \cdots & a_{p,p-1} & 1 \end{vmatrix}} \tag{10.25}$$

Example – repairable component

For the Markov diagram of the repairable component shown in Fig. 10.1 the state equation is

$$\dot{\mathbf{Q}} = \mathbf{Q}[\mathbf{A}]$$

where

$$[\mathbf{A}] = \begin{bmatrix} -\lambda & \lambda \\ \nu & -\nu \end{bmatrix}$$

The steady-state solutions for Q_1 and Q_2 are then:

$$Q_1 = \frac{\begin{bmatrix} -\lambda & 1 \\ \nu & 0 \end{bmatrix}}{\begin{bmatrix} -\lambda & 1 \\ \nu & 1 \end{bmatrix}} = \frac{\nu}{\lambda + \nu}$$

$$Q_2 = \frac{\begin{bmatrix} -\lambda & 0 \\ \nu & 1 \end{bmatrix}}{\begin{bmatrix} -\lambda & 1 \\ \nu & 1 \end{bmatrix}} = \frac{\lambda}{\lambda + \nu}$$

10.7 Standby systems

Standby systems, where there is a component which can take the place of the active component should it fail, are commonly analysed using Markov methods. This is due to the potential dependency of the probability of failure of the standby component on the failure of the primary component. The system will fail when both primary and standby components have failed simultaneously. Components in standby redundancy have three phases: standby, operation, and repair. In standby the component can function if called upon. Depending on the failure characteristics during the standby phase a component is classified as follows.

10.7.1 Hot standby

A component has the same failure rate regardless of whether it is in standby or operation. Since the failure rate of a component in this category does not change when it goes from standby to operation, and its failure probability is not affected by the state of other components, the hot standby system consists of statistically independent components.

10.7.2 Cold standby

Components do not fail when they are in standby. The components have non-zero failure rates only when they are in operation. Thus the failure likelihood of the standby component is affected by the failure of the principal component. Cold standby results in mutually dependent component failures.

10.7.3 Warm standby

A standby component can fail but with a smaller probability than when it is operating. As with cold standby, warm standby also has dependent component failures.

Example – two-component cold standby system

Consider a two-component standby system arranged such that when one component fails the standby is brought into operation while the other is repaired. Neither of the components, which are identical, can fail in standby. Failure and repair rates are λ and ν respectively.

The system starts with one of the components working and the other in standby, denoted A_w and B_s – state 1. If the working component (A) fails then the component will enter to the failed state while its repair is carried out and the standby (B) becomes the active working component (state 3). Failure of the active component prior to the other component repair will leave both components in the failed state (state 5). If in state 3 and the failed component A is repaired it is placed in standby (state 2). There are five possible states in which the system can reside. These, together with the transition rates, are shown on the Markov diagram in Fig. 10.4.

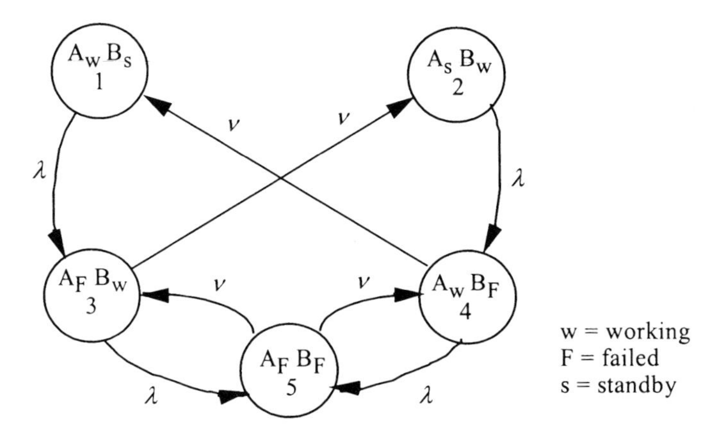

Fig. 10.4 Two-component cold standby Markov diagram

From the diagram we can formulate the following equation:

$$\left(\dot{Q}_1(t), \dot{Q}_2(t), \dot{Q}_3(t), \dot{Q}_4(t), \dot{Q}_5(t)\right) = \left(Q_1(t), Q_2(t), Q_3(t), Q_4(t), Q_5(t)\right) [\mathbf{A}] \tag{10.26}$$

where

$$[\mathbf{A}] = \begin{bmatrix} -\lambda & 0 & \lambda & 0 & 0 \\ 0 & -\lambda & 0 & \lambda & 0 \\ 0 & \nu & -(\lambda+\nu) & 0 & \lambda \\ \nu & 0 & 0 & -(\lambda+\nu) & \lambda \\ 0 & 0 & \nu & \nu & -2\nu \end{bmatrix}$$

For a steady-state solution we get $Q_i(t) = 0, = 1, \ldots, 5$.

Therefore the steady-state equations are

$$\begin{array}{llllll} -\lambda Q_1 & & & +\nu Q_4 & & = 0 \\ & -\lambda Q_2 & +\nu Q_3 & & & = 0 \\ \lambda Q_1 & & -(\lambda+\nu)Q_3 & & +\nu Q_5 & = 0 \\ & \lambda Q_2 & & -(\lambda+\nu)Q_4 & +\nu Q_5 & = 0 \\ & & \lambda Q_3 & +\lambda Q_4 & -2\nu Q_5 & = 0 \end{array} \tag{10.27}$$

These equations are linearly dependent and so additional information is required. This can be provided in the form

$$Q_1 + Q_2 + Q_3 + Q_4 + Q_5 = 1 \tag{10.28}$$

Taking four of the five equations of (10.27) together with equation (10.28) gives the resulting set of linear simultaneous equations which can be solved by Gaussian elimination:

$$\begin{pmatrix} 1 & 1 & 1 & 1 & 1 \\ -\lambda & 0 & 0 & \nu & 0 \\ 0 & -\lambda & \nu & 0 & 0 \\ \lambda & 0 & -(\lambda+\nu) & 0 & \nu \\ 0 & 0 & \lambda & \lambda & -2\nu \end{pmatrix} \begin{pmatrix} Q_1 \\ Q_2 \\ Q_3 \\ Q_4 \\ Q_5 \end{pmatrix} = \begin{pmatrix} 1 \\ 0 \\ 0 \\ 0 \\ 0 \end{pmatrix} \tag{10.29}$$

Carrying out row manipulation ('$R'_i =$ ' denotes new row i is formed from) the stages to get the augmented matrix in triangular form are

$$\begin{matrix} R'_2 = R_2 + \lambda R_1 \\ R'_4 = R_4 - \lambda R_1 \end{matrix} \quad \left(\begin{array}{ccccc|c} 1 & 1 & 1 & 1 & 1 & 1 \\ 0 & \lambda & \lambda & \nu+\lambda & \lambda & \lambda \\ 0 & -\lambda & \nu & 0 & 0 & 0 \\ 0 & -\lambda & -(2\lambda+\nu) & -\lambda & \nu-\lambda & -\lambda \\ 0 & 0 & \lambda & \lambda & -2\nu & 0 \end{array}\right)$$

$$\begin{matrix} R'_3 = R_3 + R_2 \\ R'_4 = R_4 + R_2 \end{matrix} \quad \left(\begin{array}{ccccc|c} 1 & 1 & 1 & 1 & 1 & 1 \\ 0 & \lambda & \lambda & \nu+\lambda & \lambda & \lambda \\ 0 & 0 & \lambda+\nu & \lambda+\nu & \lambda & \lambda \\ 0 & 0 & -(\lambda+\nu) & \nu & \nu & 0 \\ 0 & 0 & \lambda & \lambda & -2\nu & 0 \end{array}\right)$$

$$\begin{matrix} R'_4 = R_4 + R_3 \\ R'_5 = R_5 - \dfrac{\lambda R_3}{\lambda+\nu} \end{matrix} \quad \left(\begin{array}{ccccc|c} 1 & 1 & 1 & 1 & 1 & 1 \\ 0 & \lambda & \lambda & \nu+\lambda & \lambda & \lambda \\ 0 & 0 & \lambda+\nu & \lambda+\nu & \lambda & \lambda \\ 0 & 0 & 0 & \lambda+2\nu & \lambda+\nu & \lambda \\ 0 & 0 & 0 & 0 & -2\nu-\dfrac{\lambda^2}{\lambda+\nu} & \dfrac{-\lambda^2}{\lambda+\nu} \end{array}\right)$$

The matrix is now in the required triangular form; back-substitution gives the solution:

$$Q_1 = \frac{\nu^2 - \lambda^2 + \lambda}{2\nu^2 + 2\nu\lambda + \lambda^2}$$

$$Q_2 = \frac{\nu^2 + \lambda^2 - \lambda}{2\nu^2 + 2\nu\lambda + \lambda^2}$$

$$Q_3 = Q_4 = \frac{\lambda\nu}{2\nu^2 + 2\nu\lambda + \lambda^2}$$

$$Q_5 = \frac{\lambda^2}{2\nu^2 + 2\nu\lambda + \lambda^2} \tag{10.30}$$

The system can be regarded as having failed when both components have failed – state 5 on the Markov transition diagram.

Therefore the steady-state probability of system failure is

$$\frac{\lambda^2}{2\nu^2 + 2\nu\lambda + \lambda^2}$$

Example – two-component warm standby

In the case of a warm standby system the standby component can fail in standby as well as in operation. Failure in standby is assumed to occur with a rate $\bar{\lambda}$, which is less than the rate when active, λ.

The Markov state transition diagram for the warm standby system is similar to that for the cold standby system shown in Fig. 10.4. To model warm standby requires the addition of two transitions to this diagram. These transitions represent the failure of the standby components. If B fails in state 1 an edge linking states 1 and 4 with transition rate $\bar{\lambda}$ is added. An edge with the same rate links state 2 to state 3 to represent the failure of component A in standby.

10.8 Reduced Markov diagrams

One of the major drawbacks of the Markov method is that the models very quickly become large. For a ten-component system where the components can be working or failed there are 2^{10} = 1024 states on the transition diagram. This means 1024 differential equations. It is evident that the models need to be kept as small as possible to achieve the objectives of the analysis. One way that this can be achieved is by defining the states in a slightly different manner. For example rather than defining which components are working and failed we could just specify the number in each state.

Consider the two-component model discussed in Section 10.3. The states in which the system can reside could be specified as:

State 1 – two components work
State 2 – one component works
one component fails
State 3 – two components failed.

This effectively combines states 2 and 3 (A_WB_F and A_FB_W) from the previous model. The Markov state transition diagram is shown in Fig. 10.5. The component failure and repair rates are λ and ν respectively.

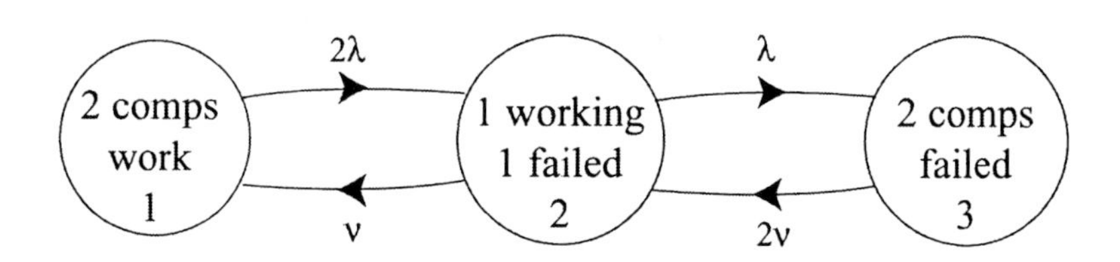

Fig. 10.5 Two-component reduced Markov model

The rates of transition need to be considered carefully. There are two working components in state 1 which can each fail with rate λ. This gives a transition from state 1 to state 2 of 2λ. With only one working component in state 2 the transition to state 3 is λ. Similar arguments apply to the component repair processes.

This model gives the following state equations

$$\left[\dot{Q}_1, \dot{Q}_2, \dot{Q}_3\right] = \left[Q_1, Q_2, Q_3\right] \begin{bmatrix} -2\lambda & 2\lambda & 0 \\ \nu & -(\lambda+\nu) & \lambda \\ 0 & 2\nu & -2\nu \end{bmatrix} \tag{10.31}$$

$$\begin{aligned} \dot{Q}_1 &= -2\lambda Q_1 + \nu Q_2 \\ \dot{Q}_2 &= 2\lambda Q_1 - (\lambda+\nu)Q_2 + 2\nu Q_3 \\ \dot{Q}_3 &= \lambda Q_2 - 2\nu Q_3 \end{aligned} \tag{10.32}$$

When these are solved we have

$$Q_{\text{SYS}} = Q_2 + Q_3 \quad \text{for series systems}$$
$$Q_{\text{SYS}} = Q_3 \quad \text{for parallel system}$$

If we consider the state equations for the four-state, two-component Markov model (Fig. 10.3) [equation (10.17)], where $\lambda_A = \lambda_B = \lambda$ and $\nu_A = \nu_B = \nu$:

$$\begin{aligned}
\dot{Q}_1 &= -2\lambda Q_1 + \nu Q_2 + \nu Q_3 \\
\dot{Q}_2 &= \lambda Q_1 - (\nu + \lambda) Q_2 + \nu Q_4 \\
\dot{Q}_3 &= \lambda Q_1 - (\nu + \lambda) Q_3 + \nu Q_4 \\
\dot{Q}_4 &= \lambda Q_2 + \lambda Q_3 - 2\nu Q_4
\end{aligned} \tag{10.33}$$

If we combine states 2 and 3, i.e. let $Q_{23} = Q_2 + Q_3$ in the equations and add the equations for Q_2 and Q_3 we get the system of equations:

$$\begin{aligned}
\dot{Q}_1 &= -2\lambda Q_1 + \nu Q_{23} \\
\dot{Q}_{23} &= 2\lambda Q_1 - (\nu + \lambda) Q_{23} + 2\nu Q_4 \\
\dot{Q}_4 &= \lambda Q_{23} - 2\nu Q_4
\end{aligned} \tag{10.34}$$

which are equivalent to those of the three-state model [equation (10.32)], but because of their reduced dimension they are easier to solve.

10.8.1 Steady-state solutions

Using equations (10.32) we have the state transition matrix given in equation (10.31) and so the steady-state solution can be obtained from equation (10.22):

$$Q_1 = \frac{\begin{vmatrix} -2\lambda & 2\lambda & 1 \\ \nu & -(\lambda+\nu) & 0 \\ 0 & 2\nu & 0 \end{vmatrix}}{\begin{vmatrix} -2\lambda & 2\lambda & 1 \\ \nu & -(\lambda+\nu) & 1 \\ 0 & 2\nu & 1 \end{vmatrix}} = \frac{\nu^2}{(\lambda+\nu)^2}$$

$$Q_2 = \frac{\begin{vmatrix} -2\lambda & 2\lambda & 0 \\ \nu & -(\lambda+\nu) & 1 \\ 0 & 2\nu & 0 \end{vmatrix}}{\begin{vmatrix} -2\lambda & 2\lambda & 1 \\ \nu & -(\lambda+\nu) & 1 \\ 0 & 2\nu & 1 \end{vmatrix}} = \frac{2\lambda\nu}{(\lambda+\nu)^2}$$

$$Q_3 = \frac{\begin{vmatrix} -2\lambda & 2\lambda & 0 \\ \nu & -(\lambda+\nu) & 0 \\ 0 & 2\nu & 1 \end{vmatrix}}{\begin{vmatrix} -2\lambda & 2\lambda & 1 \\ \nu & -(\lambda+\nu) & 1 \\ 0 & 2\nu & 1 \end{vmatrix}} = \frac{\lambda^2}{(\lambda+\nu)^2}$$

Example – two-component warm standby

The five-state Markov model of the warm standby system can also be represented as a three-state reduced alternative. This is shown in Fig.10.6.

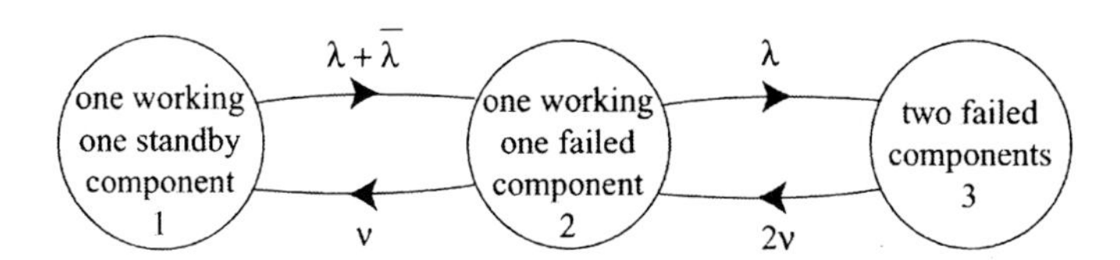

Fig. 10.6 Three-state warm standby model

The rate of exit from state 1 is $\lambda + \bar{\lambda}$ as there is one working and one standby component which can fail to cause a transition to state 2. If we set $\bar{\lambda} = 0$ we get a model for cold standby.

10.9 General three-component system

A system consisting of three components, each of which can work or fail, has $2^3 = 8$ different system states. If it is assumed that each component i has failure and repair rates λ_i and ν_i respectively then the Markov transition diagram is as given in Fig. 10.7.

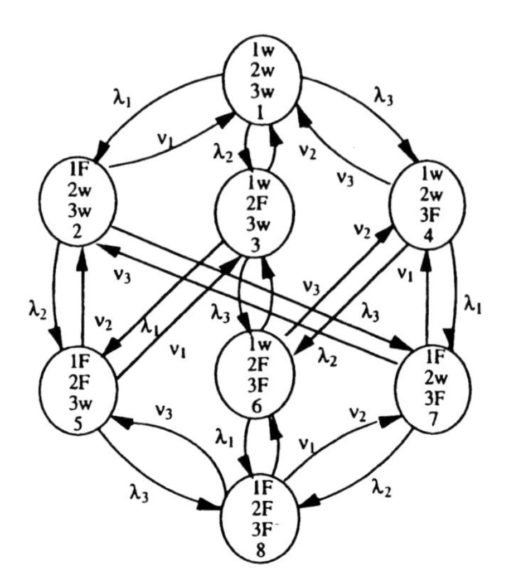

Fig. 10.7 Three-component system

The state equation for the three-component system is

$$\dot{\mathbf{Q}} = \mathbf{QA} \tag{10.35}$$

where state transition matrix **A** is given by:

$$\mathbf{A} = \begin{pmatrix} -(\lambda_1+\lambda_2+\lambda_3) & \lambda_1 & \lambda_2 & \lambda_3 & 0 & 0 & 0 & 0 \\ \nu_1 & -(\nu_1+\lambda_2+\lambda_3) & 0 & 0 & \lambda_2 & 0 & \lambda_3 & 0 \\ \nu_2 & 0 & -(\nu_2+\lambda_1+\lambda_3) & 0 & \lambda_1 & \lambda_3 & 0 & 0 \\ \nu_3 & 0 & 0 & -(\nu_3+\lambda_1+\lambda_2) & 0 & \lambda_2 & \lambda_1 & 0 \\ 0 & \nu_2 & \lambda_1 & 0 & -(\nu_1+\nu_2+\lambda_3) & 0 & 0 & \lambda_3 \\ 0 & 0 & \nu_3 & \nu_2 & 0 & -(\nu_2+\nu_3+\lambda_1) & 0 & \lambda_1 \\ 0 & \nu_3 & 0 & \nu_1 & 0 & 0 & -(\nu_1+\nu_3+\lambda_2) & \lambda_2 \\ 0 & 0 & 0 & 0 & \nu_3 & \nu_1 & \nu_2 & -(\nu_1+\nu_2+\nu_3) \end{pmatrix}$$

with initial conditions $P_1(0) = 1$, $P_i(0) = 0$, $i = 2, ..., 8$.

We have only considered a three-component system and eight equations as shown in equation (10.35). Even for a steady-state solution to this equation the algebra becomes tedious to perform. This highlights one of the difficulties encountered when solving a system using Markov methods. Unless the system is trivial and the number of states on the transition diagram is very small, a computer is required to solve the resulting equations. If steady-state solutions are required then one of the many methods, such as Gaussian elimination, can be implemented to solve the set of linear simultaneous equations.

The model illustrated in Fig. 10.7 is general in the sense that no account has been taken of how each component can affect the performance of the system. Probability of system failure is given below for series, parallel, and two-out-of-three configurations.

Series:

$$P(\text{system works}) = Q_1$$

$$P(\text{system fails}) = Q_2 + Q_3 + Q_4 + Q_5 + Q_6 + Q_7 + Q_8 \tag{10.36}$$

Parallel:

$$P(\text{system works}) = Q_1 + Q_2 + Q_3 + Q_4 + Q_5 + Q_6 + Q_7$$

$$P(\text{system fails}) = Q_8 \tag{10.37}$$

Two-out-of-three:

$$P(\text{system works}) = Q_1 + Q_2 + Q_3 + Q_4$$

$$P(\text{system fails}) = Q_5 + Q_6 + Q_7 + Q_8 \tag{10.38}$$

If a three-component system has minimal cut sets {1} and {2, 3} then in this case

$$P(\text{system works}) = Q_1 + Q_3 + Q_4$$

$$P(\text{system fails}) = Q_2 + Q_5 + Q_6 + Q_7 + Q_8$$

10.10 Time duration in states

To determine the duration that the process spends in any state consider a typical state i in the Markov state transition diagram. Let the random variable T_i be the residence time in state i. This variable will be initialized as soon as state i is entered after a transition from one of the other states on the diagram and will measure the duration until the process exits state i due to a transition to another state. Figure 10.8

shows the typical state i. The transition rate between any two states i and j is expressed as a_{ij}.

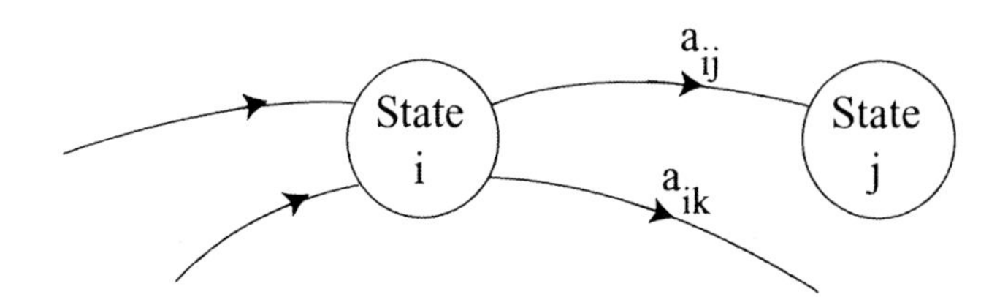

Fig. 10.8 Transitions to/from typical state i

We want to establish the distribution form for the random variable T_i. The duration in state i will end when a transition is made to another state, say state j, with which i communicates. If this transition takes place between t and $t + \mathrm{d}t$ then:

P(entering j between t and $t + \mathrm{d}t$)

= P(remaining in state i from 0 to t and then making a transition to state j between t and $t + \mathrm{d}t$)

= P(no transition from $i \rightarrow k$ ($k \neq j$) occurs prior to t, and transition $i \rightarrow j$ occurs between t and $t + \mathrm{d}t$)

(10.39)

As the transition rates are constant, the time to events are governed by the negative exponential distribution. The probability of no transition (no event) is therefore given by e^{-at} where a is the transition rate and t the time duration.

Therefore

$$P(\text{no transition from } i \rightarrow k \text{ prior to } t) = e^{-a_{ik}t} \quad (10.40)$$

and

P(transition from $i \rightarrow j$ for the first time between t and $t + \mathrm{d}t$)

$= f(t)\,\mathrm{d}t$

$$= a_{ij}\,\mathrm{e}^{-a_{ij}t}\,\mathrm{d}t \tag{10.41}$$

Accounting for no transition to all possible states other than j gives

P(entering j between t and $t + \mathrm{d}t$)

$$= \prod_{\substack{k=1 \\ (k \neq i,j)}}^{p} \mathrm{e}^{-a_{ik}t}\, a_{ij}\mathrm{e}^{-a_{ij}t}$$

$$= a_{ij}\mathrm{e}^{-\sum_{\substack{k=1 \\ (k \neq i)}}^{p} a_{ik}t}\ \mathrm{d}t$$

and since a_{ii} is defined as $-\sum_{\substack{k=1 \\ (k \neq i)}}^{p} a_{ik}$

$$= a_{ij}\mathrm{e}^{a_{ii}t}\ \mathrm{d}t \tag{10.42}$$

Now returning to consider the density function for random variable T_i:

$$\begin{aligned} f(T_i = t)\ \mathrm{d}t &= P(\text{leaving state } i \text{ at time } t) \\ &= \sum_j P(\text{leaving state } i \text{ and entering state } j \text{ between } t \text{ and } t + \mathrm{d}t) \\ &= \sum_{\substack{j=1 \\ j \neq i}}^{p} a_{ij}\mathrm{e}^{a_{ii}t}\ \mathrm{d}t \\ &= -a_{ii}\mathrm{e}^{a_{ii}t}\ \mathrm{d}t \end{aligned} \tag{10.43}$$

Therefore the probability density function for T_i is the form of the negative exponential distribution with rate $-a_{ii}$. The mean value of T_i, the mean duration in state i, $\overline{T}_i$ is:

$$\bar{T}_i = \frac{1}{-a_{ii}} = \frac{1}{\sum_{\substack{j=1 \\ j \neq i}}^{p} a_{ij}} = \frac{1}{\text{rate of departure from state } i} \tag{10.44}$$

10.10.1 Frequency of encountering a state

In the previous section it was established that the duration in each state has a negative exponential distribution. An expression for the mean time in state i is given by equation (10.44). If we again focus our attention on state i we can develop an argument to determine the frequency with which this state will be encountered. The process will either be in state i or in one of the other states represented on the transition diagram.

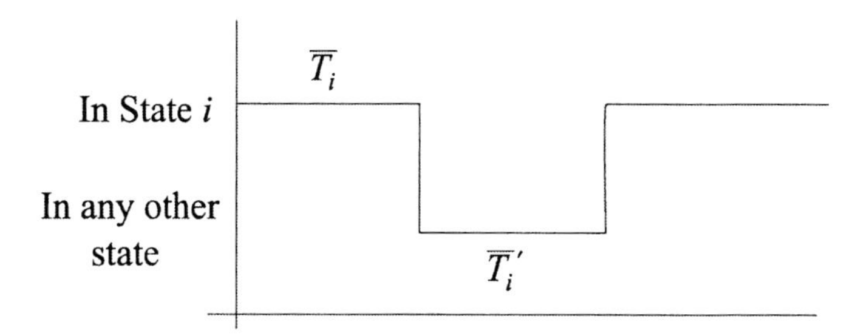

Fig. 10.9 Mean residence time for state i

The mean time in state i, $\bar{T}_i$, is represented on the diagram shown in Fig. 10.9. Let $\bar{T}_i'$ denote the mean residence time in other states before returning back to state i. This is also illustrated.

The steady-state probability of residing in state i is

$$Q_i(\infty) = \frac{\bar{T}_i}{\bar{T}_i + \bar{T}_i'} \tag{10.45}$$

If fr_i is the asymptotic (steady-state) frequency of encountering state i then

$$fr_i = \frac{1}{\text{mean time before returning to state } i} = \frac{1}{\overline{T}_i + \overline{T}_i'} \tag{10.46}$$

From equation (10.45):

$$\overline{T}_i' = \frac{\overline{T}_i\left[1 - Q_i(\infty)\right]}{Q_i(\infty)} \tag{10.47}$$

and

$$fr_i = \frac{1}{\overline{T}_i + \frac{\overline{T}_i\left[1-Q_i(\infty)\right]}{Q_i(\infty)}} = \frac{Q_i(\infty)}{\overline{T}_i} = -a_{ii}Q_i(\infty) \tag{10.48}$$

Therefore the asymptotic rate of encountering a state is its rate of departure multiplied by the asymptotic probability of being in state i.

Example – two-component system (parallel)

The Markov state transition diagram is given in Fig. 10.3. From this we can determine the transition rate matrix **A**:

$$\mathbf{A} = \begin{bmatrix} -(\lambda_1+\lambda_2) & \lambda_1 & \lambda_2 & 0 \\ \nu_1 & -(\nu_1+\lambda_2) & 0 & \lambda_2 \\ \nu_2 & 0 & -(\nu_2+\lambda_1) & \lambda_1 \\ 0 & \nu_2 & \nu_1 & -(\nu_1+\nu_2) \end{bmatrix}$$

Using equation (10.22) we can then determine the steady-state probability of residing in each state $Q_i(\infty)$. For example

$$Q_1(\infty) = \frac{\begin{vmatrix} -(\lambda_1+\lambda_2) & \lambda_1 & \lambda_2 & 1 \\ \nu_1 & -(\nu_1+\lambda_2) & 0 & 0 \\ \nu_2 & 0 & -(\nu_2+\lambda_1) & 0 \\ 0 & \nu_2 & \nu_1 & 0 \end{vmatrix}}{\begin{vmatrix} -(\lambda_1+\lambda_2) & \lambda_1 & \lambda_2 & 1 \\ \nu_1 & -(\nu_1+\lambda_2) & 0 & 1 \\ \nu_2 & 0 & -(\nu_2+\lambda_1) & 1 \\ 0 & \nu_2 & \nu_1 & 1 \end{vmatrix}}$$

The solution for each state gives

$$Q_1(\infty) = \frac{\nu_1\nu_2}{(\nu_1+\lambda_1)(\nu_2+\lambda_2)}, \quad Q_2(\infty) = \frac{\lambda_1\nu_2}{(\nu_1+\lambda_1)(\nu_2+\lambda_2)}$$

$$Q_3(\infty) = \frac{\lambda_2\nu_1}{(\nu_1+\lambda_1)(\nu_2+\lambda_2)}, \quad Q_4(\infty) = \frac{\lambda_1\lambda_2}{(\nu_1+\lambda_1)(\nu_2+\lambda_2)}$$

From equation (10.44) the mean duration in each state is as follows:

$$\overline{T}_i = \frac{1}{\sum_{j \neq i} a_{ij}}$$

$$\overline{T}_1 = \frac{1}{\lambda_1+\lambda_2}, \quad \overline{T}_2 = \frac{1}{\nu_1+\lambda_2}$$

$$\overline{T}_3 = \frac{1}{\nu_2+\lambda_1}, \quad \overline{T}_4 = \frac{1}{\nu_1+\nu_2}$$

Since state 4 is the only failed state for the system we can establish the mean up-time from the relationships shown in Fig. 10.10. The mean time to repair is the mean time in state 4.

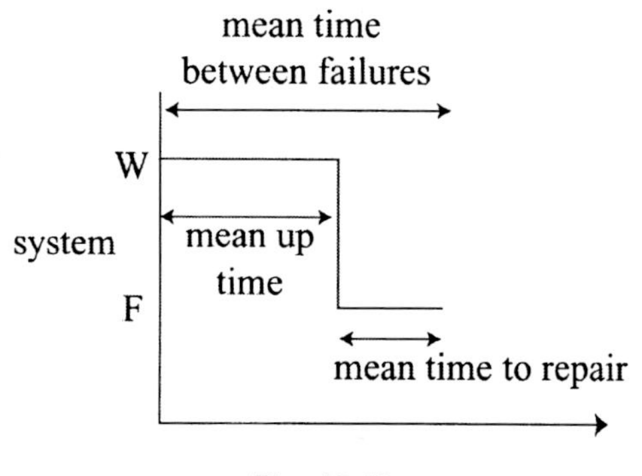

Fig. 10.10

Mean time to repair (MTTR) $\overline{T}_4 = \frac{1}{(\nu_1 + \nu_2)}$

The mean time between failures is the mean time before returning to state 4. So

$$\text{MTBF} = \frac{1}{fr_4}$$

where

$$fr_4 = -a_{44} Q_4(\infty)$$

$$= (\nu_1 + \nu_2) \frac{\lambda_1 \lambda_2}{(\nu_1 + \lambda_1)(\nu_2 + \lambda_2)}$$

Therefore

$$\text{MTBF} = \frac{(\nu_1 + \lambda_1)(\nu_2 + \lambda_2)}{\lambda_1 \lambda_2 (\nu_1 + \nu_2)}$$

Mean up-time (MUT) = MTBF – MTTR

$$= \frac{(\nu_1 + \lambda_1)(\nu_2 + \lambda_2)}{\lambda_1 \lambda_2 (\nu_1 + \nu_2)} - \frac{1}{\nu_1 + \nu_2}$$

$$=\frac{(\nu_1+\lambda_1)(\nu_2+\lambda_2)-\lambda_1\lambda_2}{\lambda_1\lambda_2(\nu_1+\nu_2)}$$

10.11 Transient solutions

Due to the number of simultaneous differential equations whose solution is required to derive an expression for the transient behaviour of complex Markov models, numerical solutions are sought. Numerical solutions provide the probability of residing in any of the states at time t by progressing from the initial situation in very small time steps. One of the most popular of the many methods available to solve first-order differential equations of this type is the Runge–Kutta method.

To illustrate a very simple numerical approach to solving a Markov model, consider the general system of first-order ordinary differential equations which can be developed [see equation (10.15)]. A numerical time-stepping routine can be derived from this using a forward difference approximation to the first derivative:

$$\frac{\mathrm{d}Q_i(t)}{\mathrm{d}t}\approx\frac{Q_i(t+\mathrm{d}t)-Q_i(t)}{\mathrm{d}t} \tag{10.49}$$

Substituting this into equation (10.15) will give a scheme where the state probabilities can be evaluated at discrete time points given the step length $\mathrm{d}t$ and the initial condition $Q_i(0)$, $i = 1, \ldots, p$.

For example, if we return to the two-component system whose state equation is given in equation (10.17):

$$\begin{aligned}
\dot{Q}_1 &= -(\lambda_A+\lambda_B)Q_1+\nu_A Q_2+\nu_B Q_3\\
\dot{Q}_2 &= \lambda_A Q_1-(\nu_A+\lambda_B)Q_2+\nu_B Q_4\\
\dot{Q}_3 &= \lambda_B Q_1-(\nu_B+\lambda_A)Q_3+\nu_A Q_4\\
\dot{Q}_4 &= \lambda_B Q_2+\lambda_A Q_3-(\nu_A+\nu_B)Q_4
\end{aligned}$$

A general form of equation (10.50) is:

$$Q_i(t+\mathrm{d}t)=\sum_{\substack{j=1\\ j\neq i}}^{p} a_{ji}Q_j(t)\,\mathrm{d}t+\left[1+a_{ii}\,\mathrm{d}t\right]Q_i(t)$$

for $i = 1, \ldots, 4$.

Using the forward difference approximation in equation (10.49) gives

$$
\begin{aligned}
Q_1(t+\mathrm{d}t) &= Q_1(t)\left[1-(\lambda_A+\lambda_B)\,\mathrm{d}t\right]+\nu_A Q_2(t)\,\mathrm{d}t+\nu_B Q_3(t)\,\mathrm{d}t \\
Q_2(t+\mathrm{d}t) &= Q_2(t)\left[1-(\nu_A+\lambda_B)\,\mathrm{d}t\right]+\lambda_A Q_1(t)\,\mathrm{d}t+\nu_B Q_4(t)\,\mathrm{d}t \\
Q_3(t+\mathrm{d}t) &= Q_3(t)\left[1-(\nu_B+\lambda_A)\,\mathrm{d}t\right]+\lambda_B Q_1(t)\,\mathrm{d}t+\nu_A Q_4(t)\,\mathrm{d}t \\
Q_4(t+\mathrm{d}t) &= Q_4(t)\left[1-(\nu_A+\nu_B)\,\mathrm{d}t\right]+\lambda_B Q_2(t)\,\mathrm{d}t+\lambda_A Q_3(t)\,\mathrm{d}t
\end{aligned}
\qquad (10.50)
$$

Setting dt and knowing the initial conditions $Q_i(0)$, i=1, 4 these can be substituted into the left-hand side of equation (10.50) to give **Q**(dt). Substituting this in turn into the equations will then yield **Q**(2dt). Stepping through time in increments of dt will provide the numerical solution. Problems can be experienced due to the difference in the numerical values for failure rates and repair rates. This can result in the equations being stiff, in which case a purpose-designed numerical routine will produce more accurate solutions.

Numerical solutions can also be more flexible and can model situations which would not be possible with an analytical solution. Periodic inspection can be modelled using numerical methods. Consider a simple example of a single repairable component where the failure is only revealed at inspection (see Fig. 10.11). The failure rate is λ, and the inspection interval is θ, i.e. the failure is a dormant, unrevealed failure typical of safety systems. This means that if the component experiences failure and passes from state 1 to state 2, the fact that it is at this point non-operational will not be revealed and the potential to undergo repair will not be instigated.

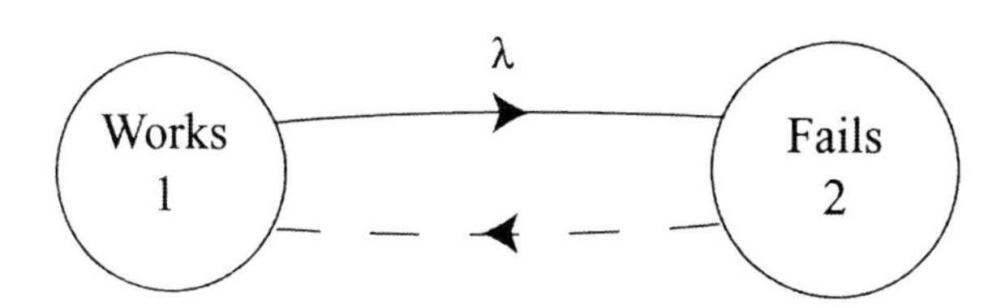

Fig. 10.11 Repairable component where failures are revealed on inspection

This repair process can only commence when inspection reveals the component failure. At this point in time the repair of the failed component will start. This will transfer any probability of being in state 2 (failed) to state 1 (repair), i.e. when $t = \theta$

$$\begin{aligned} Q_1(t+\mathrm{d}t) &= Q_1(t) + Q_2(t) \\ Q_2(t+\mathrm{d}t) &= 0 \end{aligned} \tag{10.51}$$

In a general p-state Markov model with failed state, state p and failure revealed by inspection will return the system to state 1 (all components working). Then this is given by the following.

If $t + \mathrm{d}t \neq \theta,\ 2\theta,\ \ldots,\ n\theta$

$$Q_i(t+\mathrm{d}t) = \sum_{\substack{j=1 \\ j\neq i}}^{p} a_{ji} Q_j(t)\,\mathrm{d}t + [1 + a_{ii}\ \mathrm{d}t] Q_i(t) \quad i = 1,\ \ldots,\ p \tag{10.52}$$

If $t + \mathrm{d}t = \theta,\ 2\theta,\ \ldots,\ n\theta$

$$Q_i(t+\mathrm{d}t) = \sum_{\substack{j=1 \\ j\neq i}}^{p} a_{ji} Q_j(t)\ \mathrm{d}t + [1 + a_{ii}\ \mathrm{d}t] Q_i(t) \quad i = 2,\ \ldots,\ p-1$$

$$Q_1(t+\mathrm{d}t) = Q_1(t) + Q_p(t)$$

$$Q_p(t+\mathrm{d}t) = 0.0$$

Since the numerical methods are looking at discrete time elements then another possible approach to deriving a solution would be to look at discrete time–discrete space Markov models generally referred to as **Markov chains**.

Consider the single, repairable two-state component model shown in Fig. 10.1 with specific values of the transition rates $\lambda = 0.2$ and $\nu =$ 0.9. If we consider the probability of transferring between the discrete states in time interval $\mathrm{d}t = 1$ then the transition diagram is as shown in Fig 10.12.

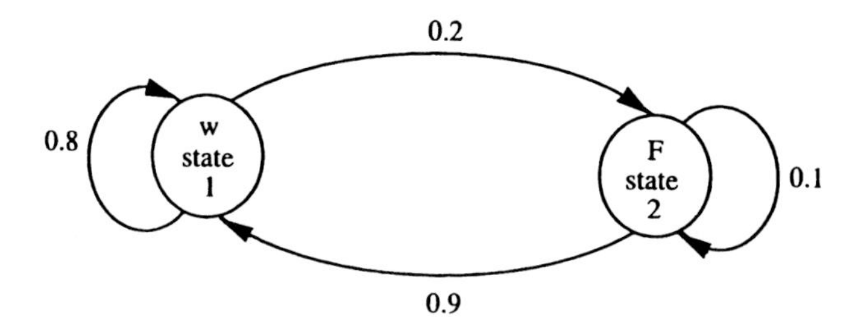

Fig. 10.12 Repairable two-state component

We can now form the **transition probability matrix** represented by **P** which contains the probability of making a transition from one state to another in a single time step. **P** is a square $m \times m$ matrix where m is the total number of states in the diagram. Its elements P_{ij} are given by:

$P_{ij} =$ probability of making a transition from state i to state j in the specified time interval given that the system was in state i at the beginning of the time interval

That is:

$$\begin{array}{c} \text{to states} \\ \begin{array}{cc} 1 & 2 \end{array} \end{array}$$

$$\mathbf{P} = \begin{bmatrix} 0.8 & 0.2 \\ 0.9 & 0.1 \end{bmatrix} = \begin{bmatrix} P_{11} & P_{12} \\ P_{21} & P_{22} \end{bmatrix} \tag{10.53}$$

Since each row of matrix **P** represents the probability of making a transition from a state or remaining in the original state its elements must sum to unity. **P**, the transition probability matrix, tells us the chances of state transitions during one time element. If we multiply **P** by itself we get

$$\mathbf{P}^2 = \begin{bmatrix} (P_{11}P_{11} + P_{12}P_{21}) & (P_{11}P_{12} + P_{12}P_{22}) \\ (P_{21}P_{11} + P_{22}P_{21}) & (P_{21}P_{12} + P_{22}P_{22}) \end{bmatrix}$$

$$= \begin{bmatrix} 0.82 & 0.18 \\ 0.81 & 0.19 \end{bmatrix} \tag{10.54}$$

The first element of the first row of $\mathbf{P}^2$ is 0.82. This is obtained from $P_{11}P_{11} + P_{12}P_{21}$. The first term of this is the probability of being in state 1 initially and remaining in this state for both time intervals. The second term is the probability of making a transition from state 1 to state 2 during the first time interval and returning back to state 1 during the second. Together these two terms provide the only ways which the system can start and end in state 1 after two time intervals.

Similarly the term in the first row of the second column of $\mathbf{P}^2$ gives the probability of being in state 2 after two time intervals, having started in state 1. Elements of the second row of $\mathbf{P}^2$ give the probabilities of being in states 1 and 2 having started in state 2.

Extending this argument we can see that the probability of being in any state after n time intervals can be obtained from multiplying $\mathbf{P}$ n times to give $\mathbf{P}^n$. If the initial system state is known then

$$\mathbf{P}(n) = \mathbf{P}(0)\mathbf{P}^n \tag{10.55}$$

where $\mathbf{P}(0)$ is a row vector containing the probabilities of starting in each state.

Considering the third and fourth time intervals having started in the working state we get

$$\mathbf{P}(3) = \mathbf{P}(0)\mathbf{P}^3$$

$$= [1 \quad 0]\begin{bmatrix} 0.818 & 0.182 \\ 0.819 & 0.181 \end{bmatrix} = [0.818 \quad 0.182]$$

$$\mathbf{P}(4) = \mathbf{P}(0)\mathbf{P}^4$$

$$= [1 \quad 0]\begin{bmatrix} 0.8182 & 0.1818 \\ 0.8181 & 0.1819 \end{bmatrix} = [0.8182 \quad 0.1818]$$

After only four time steps these probabilities are close to their steady-state values (0.818 181, 0.181 818) given by equation (10.20). However, this example only provides an illustration of the method. In practice much smaller time-step lengths would be required to model the transient system behaviour accurately.

10.12 Reliability modelling

Reliability assessment for systems which feature repairable or non-repairable components can also be performed using Markov methods. To achieve this the states on the Markov transition diagram which represent system failure are made **absorbing**. An absorbing state is one which, having been entered, cannot be left. So any transitions from a state corresponding to a system failure to a system success state are deleted in order to calculate this reliability. This is illustrated for a two-component (repairable) parallel system in Fig. 10.13. State 4, both components failed, represents the system failure condition and is made absorbing.

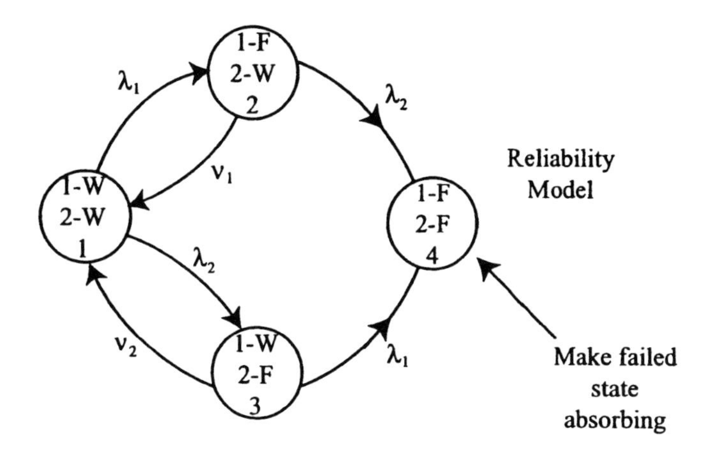

Fig. 10.13 Reliability modelling for a two-component parallel system

To derive the equations for this model with a total of p states, again number all states which represent the system in a functional condition first; therefore states 1 through to k are working system states. States $k + 1$ through to p are failed system states. Since the failed states are absorbing, any transitions from these states back to those which represent the system functioning are removed:

$$a_{ij} = 0 \quad \text{for} \quad i > k, j \leq k$$

Let $P_i(t)$ represent the probability of residing in state i at time t. Then, as with the availability modelling, a set of first-order differential equations can be formulated:

$$\frac{dP_i}{dt}(t) = \sum_{j=1}^{p} P_j(t)\, a_{ji} \qquad \text{for} \quad i = 1, 2, \ldots, p$$

That is

$$\left[\frac{dP_1(t)}{dt}, \frac{dP_2}{dt}(t), \ldots, \frac{dP_p}{dt}(t)\right] = \left[P_1(t), P_2(t), \ldots, P_p(t)\right]\left[\mathbf{A}'\right] \tag{10.56}$$

where $[\mathbf{A}']$ is the $p \times p$ transition matrix with elements:

a_{ij} : transition rate from state i to state j

$a_{ij} = 0$: if $i > k$, $j \leq k$

a_{ii} : $-\sum_{\substack{j=1 \\ j \neq i}}^{p} a_{ij}$

From $[\mathbf{A}']$ a **reduced transition rate matrix** $[\mathbf{A}'_k]$ can be formed from the first k rows and columns, i.e.

$$\left[\frac{dP_1(t)}{dt}, \frac{dP_2(t)}{dt}, \ldots, \frac{dP_k(t)}{dt}\right] = [P_1(t), P_2(t), \ldots, P_k(t)][\mathbf{A}'_k] \tag{10.57}$$

Following the solution of these equations with initial conditions $P_i(0)$, i=1, …, k, the system reliability can be evaluated by

$$R(t) = \sum_{i=1}^{k} P_i(t) \tag{10.58}$$

and the mean time to first failure (MTFF):

$$\text{MTFF} = \int_0^{\infty} R(t)\, dt \tag{10.59}$$

Note that $P_i(t) \to 0$ as $t \to \infty$.

Solving equations (10.57), (10.58), and (10.59) gives

$$\text{MTFF} = \frac{1}{|\mathbf{A}'_k|} \begin{vmatrix} 0 & P_1(0) & P_2(0) & \cdots & P_k(0) \\ 1 & a_{11} & a_{12} & \cdots & a_{1k} \\ 1 & a_{21} & a_{22} & \cdots & a_{2k} \\ \vdots & \vdots & \vdots & \vdots & \vdots \\ 1 & a_{k1} & a_{k2} & \cdots & a_{kk} \end{vmatrix} \tag{10.60}$$

Thus unlike the system availability, the system reliability depends on the initial state:

$$\text{Asymptotic failure rate} = \frac{1}{\text{MTFF}} \tag{10.61}$$

Example – two-component parallel system

Consider the Markov model illustrated in Fig. 10.13 with initial condition $P_1(0) = 1$, $P_2(0) = P_3(0) = P_4(0) = 0$. The transition rate matrix is

$$\mathbf{A}' = \begin{bmatrix} -(\lambda_1 + \lambda_2) & \lambda_1 & \lambda_2 & 0 \\ \nu_1 & -(\nu_1 + \lambda_2) & 0 & \lambda_2 \\ \nu_2 & 0 & -(\nu_2 + \lambda_1) & \lambda_1 \\ 0 & 0 & 0 & 0 \end{bmatrix}$$

From this the reduced transition rate matrix is

$$\mathbf{A}'_k = \begin{bmatrix} -(\lambda_1 + \lambda_2) & \lambda_1 & \lambda_2 \\ \nu_1 & -(\nu_1 + \lambda_2) & 0 \\ \nu_2 & 0 & -(\nu_2 + \lambda_1) \end{bmatrix}$$

giving

$$\text{MTFF} = \frac{1}{|\mathbf{A}'_k|}\begin{vmatrix} 0 & 1 & 0 & 0 \\ 1 & -(\lambda_1+\lambda_2) & \lambda_1 & \lambda_2 \\ 1 & \nu_1 & -(\nu_1+\lambda_2) & 0 \\ 1 & \nu_2 & 0 & -(\nu_2+\lambda_1) \end{vmatrix}$$

$$= \frac{\lambda_1(\lambda_1+\nu_1)+\lambda_2(\lambda_2+\nu_2)+(\lambda_1+\nu_1)(\lambda_2+\nu_2)}{\lambda_1\lambda_2(\lambda_1+\nu_1+\lambda_2+\nu_2)}$$

10.13 Summary

For a system which is characterized by a lack of memory so that its future state depends only on its present state, and in addition the chances of changing between states does not vary with time, then the system can be assessed by the Markov techniques presented in this chapter.

The method is particularly useful for representing situations where component failures are not independent. As such, standby redundancy systems are a common application area for this method. Markov analysis can become difficult to apply when the number of discrete system states gets large. This can be resolved by applying Markov methods to the parts of the system which are dependent and then using these results as sub-system failure probabilities in other techniques such as fault tree analysis or reliability networks.

10.14 Bibliography

Billington, R. and **Allan, R.** (1983) *Reliability Evaluation of Engineering Systems*. Pitman.

Davidson, J. (Ed.). (1988) *The Reliability of Mechanical Systems*. IMechE Publishers Limited.

Foster, J. W., Phillips, D. R., and **Rogers, R. T.** (1981) *Reliability, Availability and Maintainability*. M/A Press.

Frankel, E. G. (1988) *Systems Reliability and Risk Analysis*. Kluwer Academic Publishers.

Henley, E. J. and **Kumamoto, H.** (1981) *Reliability Engineering and Risk Assessment*. Prentice-Hall.

Villemeur, A. (1991) *Reliability, Availability, Maintainability and Safety Assessment*. John Wiley.

Chapter 11

Simulation

11.1 Introduction

In some instances fault trees or reliability networks will not be an appropriate means to model a system. Some characteristics of the system will mean that the preferred deterministic methods are not appropriate for the analysis. For example the model may be too large or complex to perform a deterministic analysis; the reliability of components or sub-systems may not be independent, and prevent the use of a general fault tree analysis code; component failure or repair distributions may not have a constant failure or repair rate. This means that the transition times from one component state to another is not governed by the negative exponential distribution which therefore prevents the use of Markov methods and the majority of fault tree analysis codes. In these situations the system performance can be simulated using Monte Carlo methods.

Simulation seeks to 'duplicate' the behaviour of the system under investigation by studying interactions among its components. The output is normally presented in terms of selected measures of system performance. Simulation must be treated as a statistical experiment with each run of the model an observation. In this case the experiment is conducted totally on the computer.

To perform a Monte Carlo analysis the probabilistic model of the system is constructed and usually implemented on a computer. A number of trial runs are then made. After each trial run or simulation the outcome is recorded. When a significant number of trial runs have been performed the probability of a particular outcome can be estimated by the number of times that the outcome was achieved divided by the total number of simulations.

Simulation methods are used to investigate many different engineering systems – particularly manufacturing systems. However, the nature of reliability problems, where the occurrence of failure events is rare, means that in order to reduce the computer time expended on the analysis, efficient sampling routines must be used.

11.2 Uniform random numbers

The success of a Monte Carlo simulation of a system is dependent upon the generation of random numbers which form a uniform distribution. Many methods exist for generating random numbers, for example Von Neumann suggested that the mid-square method could be used. His procedure was to square the preceding random number and extract the middle digits. For example, for a four digit number: if 5232 was one random number ($5232 \times 5232 = 27\,373\,824$), the next random number is 3738. This method is not ideal since when zero is encountered the sequence terminates. Since these sequences are defined in a deterministic way, while they are uniform they are not really random – they are pseudo-random. Pseudo-random numbers are those which are independent and have a uniform distribution.

The simplest way to generate random numbers is by the outcomes of random experiments such as tossing coins or rolling dice. Regular 20-sided solid dice with the digits 0–9 appearing twice each on opposite sides of the dice can be used. These icosahedral dice, while generating random number sequences, are too slow to use for any purposes other than to demonstrate a simulation procedure since actual simulations require large volumes of random numbers.

Random number tables such as that shown in Fig. 11.1 provide an alternative to the dice but again this method is also too slow to use in reliability simulation. The tables are a sequence of random digits; to use them any entry can be selected as a starting point and subsequent digits can be obtained by reading across or down the table. For example, if it were required to generate a sequence of random numbers to four decimal places then the first three would be (starting at the top of the table and reading across):

0.0347
0.4373
0.8636

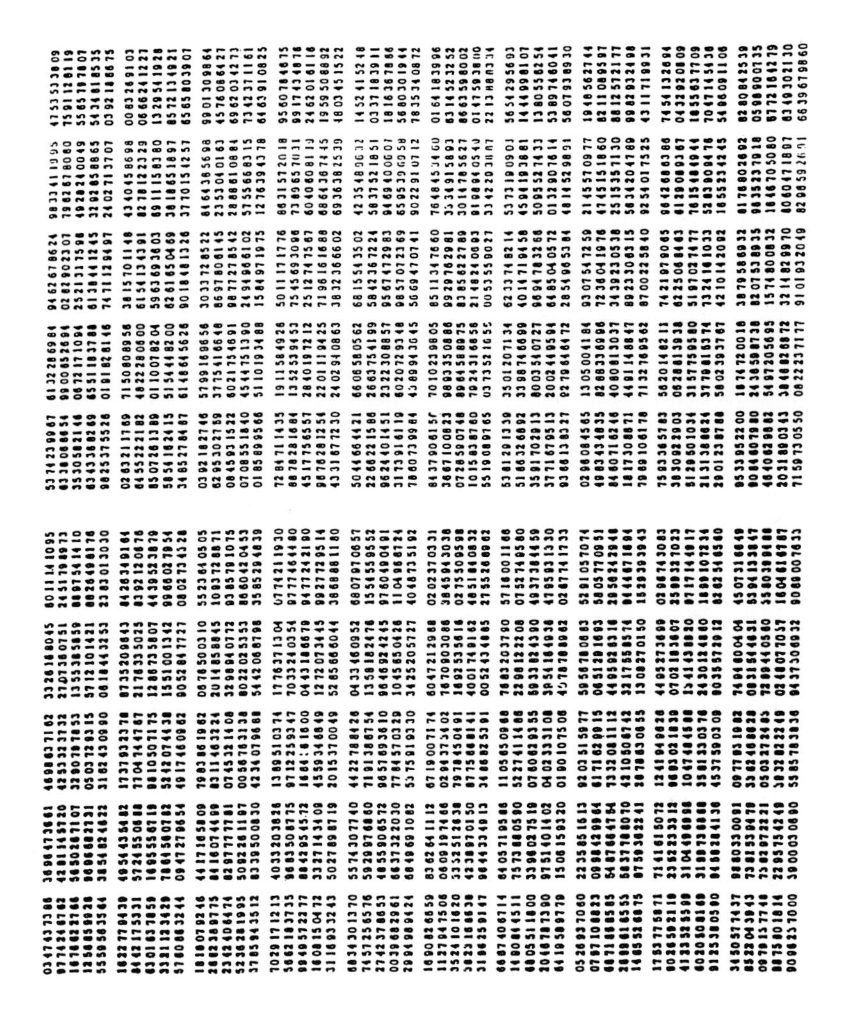

Fig. 11.1 Random number table

To generate random numbers in the quantities required to simulate real engineering systems a digital computer must be used. Since the rules used on the computer to generate the number sequences are deterministic then the numbers can be pseudo-random or random. We require random numbers so that they form a uniform distribution over [0,1], they are independent, and are obtained quickly. Pseudo-random numbers can have these properties but are more efficiently generated; they are also repeatable. Recursion formulae are the most suitable means of generating pseudo-random numbers and have the advantage that the properties of the numbers they produce can be investigated mathematically. Recursion formulae need some specified number to start the sequence. This initial number is called the 'seed' and the same number sequence will be produced if the 'seed' specified is the same. When developing and de-bugging computer simulation software this property can be very useful to test the reproducibility of results.

The recursion formulae most commonly used are linear congruential generators of the type

$$x_{n+1} = ax_n + b \quad (\text{mod } m) \tag{11.1}$$

$$x_0 - \text{seed} \qquad R_i = \frac{x_i}{m}$$

where x_i is the sequence of numbers produced, R_i is the random number produced in the range [0, 1], (mod m) means modulus m, where the remainder is given after the expression has been divided by m, and a, b, and m are integer constants suitably chosen to give the desired properties of pseudo-random numbers which can be calculated efficiently and produce a sequence of numbers with a large cycle length. This means that a large sample of numbers are produced before the sequence returns to its original 'seed' value and repeats itself.

For example:

$$X_{i+1} = (aX_i + c) \quad (\text{mod } m)$$

Let $a = 5$, $c = 3$, and $m = 16$:

i	X_i	R_i	i	X_i	R_i
0	7	—	10	9	0.563
1	6	0.375	11	0	0.000
2	1	0.063	12	3	0.188
3	8	0.500	13	2	0.125
4	11	0.688	14	13	0.813
5	10	0.625	15	4	0.250
6	5	0.313	16	7	0.438
7	12	0.750	17	6	0.375
8	15	0.938	18	1	0.063
9	14	0.875	19	8	0.500

11.3 Direct simulation method

Reliability modelling achieved by direct statistical simulation is illustrated in Fig. 11.2. Two types of input describe the simulation problem. The first type, labelled 'input 1', lists the types of statistical distributions for times to failure and times to repair for each component together with the distribution parameters. 'Input 2' is the system logic. This set of data indicates how all the components are linked together in the system and the effects which each component failure will have on the system performance.

The system operation is then simulated over the time period of interest by taking random samples from the statistical distributions and logging how the changing component states affect the system functionality.

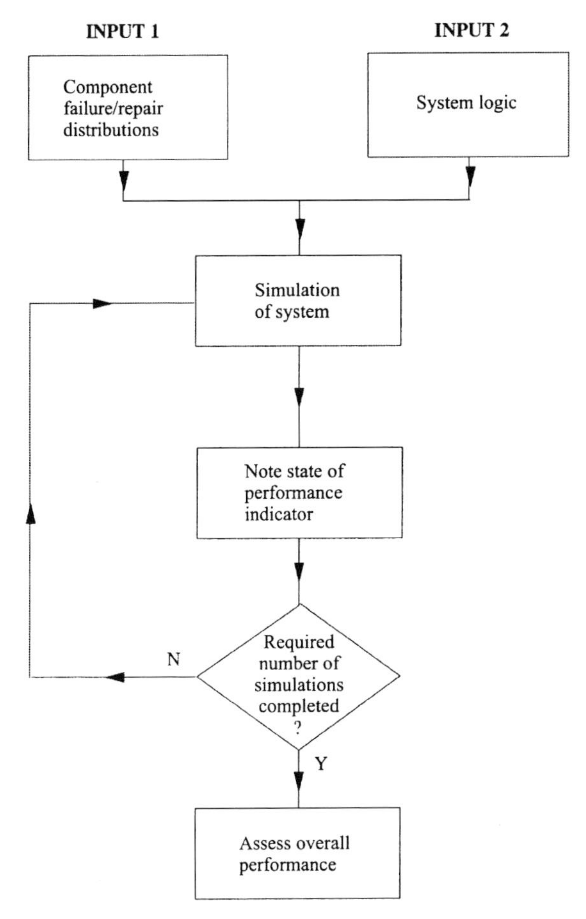

Fig. 11.2 Direct statistical simulation

Example
Consider the two-component parallel system shown in Fig. 11.3.

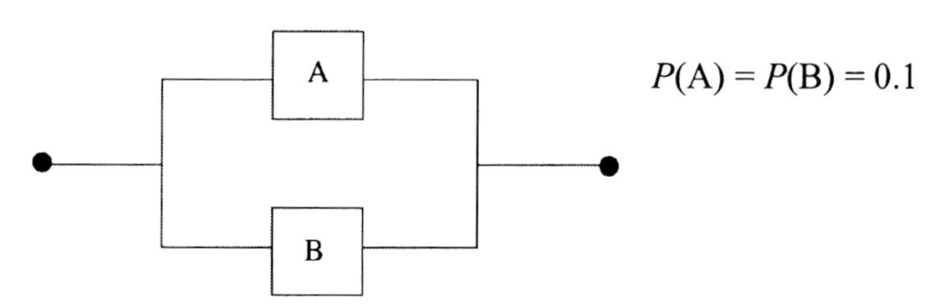

Fig. 11.3 Two-component redundant system

Components A and B each have probability of failure of 0.1. Both inputs 1 and 2 are defined since the probability of component failures is constant and the system functional logic requires both A and B to fail for the system to fail.

The first simulation is carried out by generating a random number for each component in the system. If the random number is less than the failure probability then the component is assumed to fail, or if the random number lies between the failure probability and unity then the component remains working (Fig. 11.4).

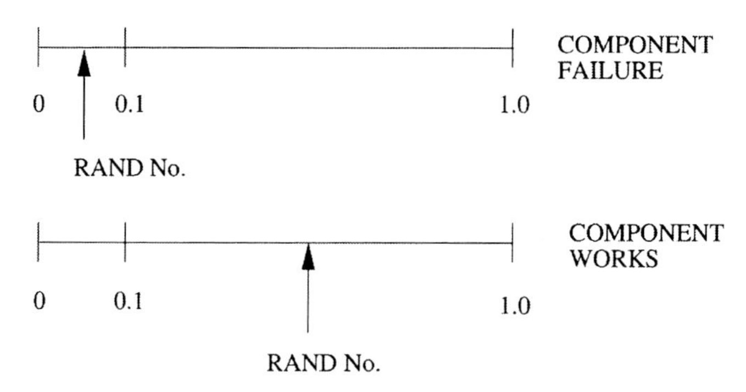

Fig. 11.4 Direct sampling

The two-component system requires two random numbers for the first simulation. Therefore:

For A:

Random no. = 0.002 since this is less than 0.1 component A is failed

For B:

Random no. = 0.432 since this is greater than 0.1 component B works

Since at least one of the components is working the first simulation results in a working system. Many such simulations would need to be performed to measure the system reliability.

Using this direct method N simulations of a k-component system need kN random numbers to be generated. If the probability of system failure was 10^{-4} then we could expect one system failure to occur every 10 000 k random numbers generated. Statistically significant results therefore need many simulations to be performed, and more efficient routines are required for all but the simplest of systems. One such improved method of sampling, 'dagger sampling', is described below.

11.4 Dagger sampling

Dagger sampling provides a means of generating several samples using only a single random number. Consider a component whose failure probability is 0.1. Using direct sampling, each time a component state is generated a random number is required. If ten states for the component were generated then on average it would be expected that one would be failed and nine working. Ten sample states for such a component can also be generated with the same expected failure occurrence using a single random number as shown in Fig. 11.5.

Since the probability of component failure is 0.1, the unit interval is divided up into ten equal lengths. A random number is thus equally likely to fall into each of these sub-divisions. From the lines with ten equal intervals ten samples are made with a different portion of the line considered active in each sample. Sample number 1 has active region $0 \leq x < 0.1$, sample number 2 has active region $0.1 \leq x < 0.2$, and so on. On generating a random number (say 0.681) a slice (hence the name dagger) is taken through all these ten samples noting on which sample the active region is encountered. For all samples where the slice did not cut through an active region the component is considered to be working. The active region for the random number 0.681 was encountered on sample number 7 and so the component is failed for this sample.

Great savings in efficiency can be made using this method as a single random number has produced ten samples. For components with more realistic failure probabilities such as 1×10^{-4} the efficiency improvement is even greater. The trade-off for the fact that many samples are generated from a single random number is the loss of independence of samples. Even so the convergence rate is superior with this method.

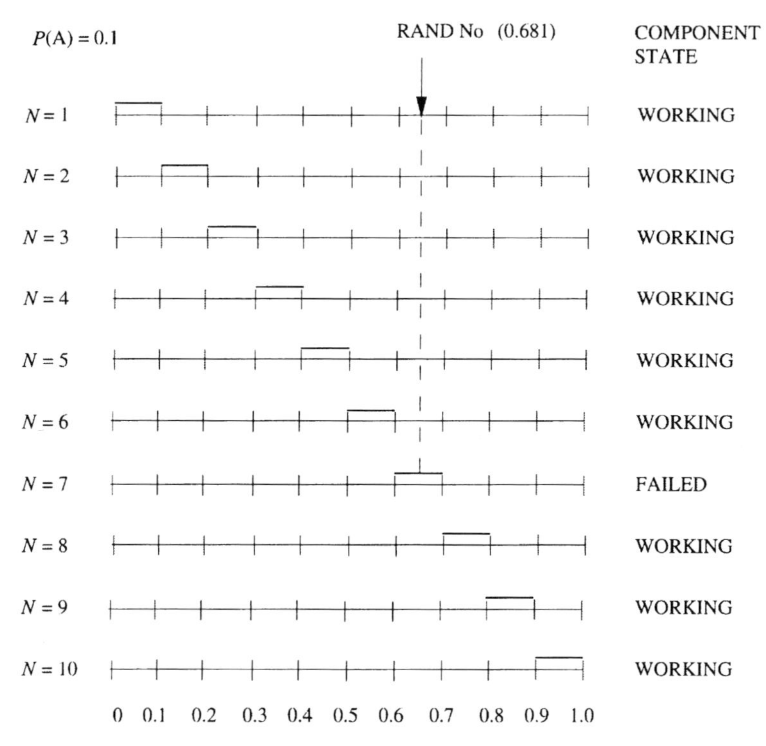

Fig. 11.5 Dagger sampling

Direct sampling is not always appropriate. Commonly component downtime characteristics are described in terms of their failure and repair distributions. In these cases the simulation needs to advance through time dealing with each failure or repair event chronologically. The times at which components fail or are repaired is provided by a random sample from the relevant distribution.

11.5 Generation of event times from distributions

11.5.1 *Exponential distribution*

The density function for the exponential distribution with mean μ is

$$f(t) = \frac{1}{\mu} \mathrm{e}^{-t/\mu} \tag{11.2}$$

If this distribution represents the variability in times to failure or repair for a component with mean μ then random samples from this distribution can be obtained by first integrating to get the cumulative failure distribution $F(t)$:

$$F(t) = \int_0^t f(u)\ \mathrm{d}u = 1 - \mathrm{e}^{-t/\mu} \tag{11.3}$$

The cumulative failure distribution has the same range and properties as the distribution of random numbers. Therefore to take a random sample, generate some random number X and equate to $F(t)$ (with $0 \leq F(t) \leq 1$). So

$$X = 1 - \mathrm{e}^{-t/\mu} \tag{11.4}$$

Rearranging gives the time to failure

$$t = -\mu \ln(1 - X) \tag{11.5}$$

If X is uniform over [0,1] then so is $1 - X$ and this can be simplified:

$$t = -\mu \ln X \tag{11.6}$$

This approach is the same as randomly generating the failure probability and reading back on the cumulative distribution graph to obtain a time to failure as illustrated in Fig. 11.6.

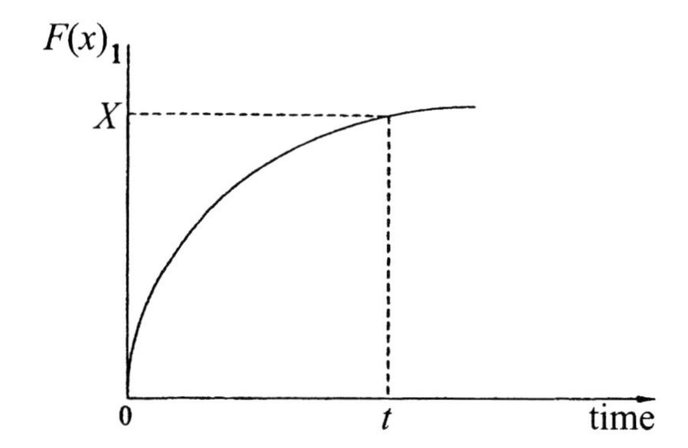

Fig. 11.6 Exponential distribution

11.5.2 Weibull distribution

Random samples can be obtained directly from the Weibull distribution in the same manner as for the exponential distribution. The two-parameter Weibull distribution with parameters β and η has a density function of the form

$$f(t) = \beta \frac{t^{\beta-1}}{\eta^{\beta}} \mathrm{e}^{-(t/\eta)^{\beta}} \tag{11.7}$$

Integrating gives the cumulative distribution

$$F(t) = 1 - \mathrm{e}^{-(t/\eta)^{\beta}} \tag{11.8}$$

Random times can then be obtained from utilizing random number X by

$$\begin{aligned} X &= 1 - \mathrm{e}^{-(t/\eta)^{\beta}} \\ \mathrm{e}^{-(t/\eta)^{\beta}} &= 1 - X \\ (t/\eta)^{\beta} &= -\ln(1-X) \\ t &= \eta\left[-\ln(1-X)\right]^{1/\beta} \quad \text{or} \quad t = \eta\left[-\ln(X)\right]^{1/\beta} \end{aligned} \tag{11.9}$$

11.5.3 Normal distribution

Sampling from the normal distribution with mean μ and standard deviation σ cannot be achieved by simple transposition of its formulae since its density function

$$f(t) = \frac{1}{\sigma\surd(2\pi)} e^{-\frac{1}{2}[(t-\mu)/\sigma]^2} \tag{11.10}$$

cannot be integrated to yield a formula for $F(t)$ which can be inverted to give t in terms of $F(t)$.

One approach to obtain normal variates from uniformly distributed random variables is to use the central limit theorem.

Central limit theorem

Let $X_1, X_2, \ldots, X_n$ be independent random variables which are identically distributed and have mean μ and variance σ^2. Then if $S_n = X_1 + X_2 + \cdots + X_n$, the random variable $(S_n - n\mu)/(\sigma\surd n)$ is asymptotically normally distributed with mean 0 and standard deviation 1.

The random numbers $U(0, 1)$ are identically distributed and can be used to form S_n. In practice we have to settle for some finite value of n so that the resulting S_n will only be approximately normal. If n is selected as 2 then the resulting triangular distribution is unsuitable. For $n = 3$ a better bell-shaped distribution is obtained so $n \geq 3$ is suitable. A convenient number to select from the mathematical point of view is $n =$ 12 [since X_i has $\mu = 0.5$ and $\sigma^2 = 1/12$, therefore S_n is N(6,1)].

To obtain a random sample from the normal distribution then twelve $U(0, 1)$ random numbers $X_1, X_2, \ldots, X_{12}$ are generated:

$$X = \sum_{i=1}^{12} X_i \tag{11.11}$$

By the central limit theorem X is normally distributed with mean 6 and standard deviation 1. So

$$t = (X - 6)\sigma + \mu \tag{11.12}$$

is normally distributed with mean μ and standard deviation σ.

Example – single-component simulation

A component fails with an exponential distribution with mean time to failure 500 h. The repair process is normally distributed with mean time to repair 25 h and standard deviation 5 h. Calculate the percentage downtime in 2500 h.

The flowchart to simulate the performance of the component transferring between working and failed states is shown in Fig. 11.7. One hundred independent simulations are required to provide an estimate of the percentage component downtimes.

An example calculation obtained by proceeding through one 2500-h period of the flowchart is given below.

Initialize variables:

NSIM = 1	(number of the simulation being performed)
SIMTIM = 0.0	(total simulation time)
DOWNT = 0.0	(total downtime recorded)

Generate component failure EXP(500):

RAND No. = 0.039

$$t_f = -500 \ln(0.039)$$
$$= 1622.0 \text{ h}$$

SIMTIM = 1622.0

Generate component repair $N(25, 5)$:

12 RAND Nos:	0.664	0.979	0.949
	0.734	0.742	0.056
	0.176	0.816	0.326
	0.333	0.290	0.951

$$X = \sum_{i=1}^{12} X_i = 7.016$$

$$t_R = (X - 6)\sigma + \mu$$
$$= (7.016 - 6)5 + 25$$
$$= 30.1 \text{ h}$$

SIMTIM = 1652.1
DOWNT = 30.1

Second component failure EXP(500):

$$\text{RAND No.} = 0.216$$
$$t_f = -500\ \ln(0.216)$$
$$= 766.2\text{ h}$$

SIMTIM = 2418.3

Second component repair $N(25, 5)$:

RAND Nos:	0.280	0.092	0.407
	0.474	0.927	0.346
	0.969	0.259	0.900
	0.006	0.296	0.882

$$X = \sum_{i=1}^{12} X_i = 5.838$$
$$t_R = (X - 6)\sigma + \mu$$
$$= (5.838 - 6)5 + 25$$
$$= 24.2\ \text{h}$$

SIMTIM = 2442.5
DOWNT = 54.3

Third component failure EXP(500):

$$\text{RAND No.} = 0.699$$
$$t_f = -500\ \ln(0.699)$$
$$= 179.0\text{ h}$$

SIMTIM = 2621.5

This goes beyond the 2500 h operating time. Therefore in the 2500 h, the component is down for a total of 54.3 h.

$$\text{Percentage downtime} = \frac{54.3}{2500} \times 100 = 2.172 \text{ per cent}$$

This procedure provides just one sample and should be repeated many times to obtain an accurate expectation of system performance.

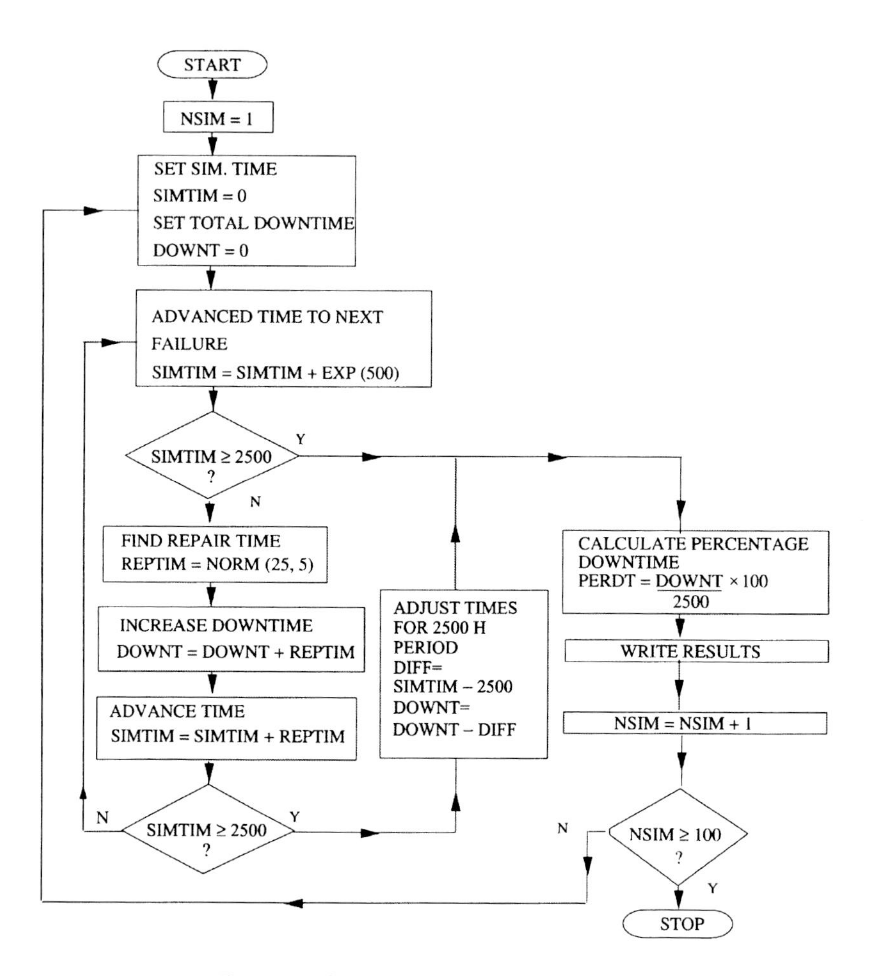

Fig. 11.7 Single-component simulation

11.6 System logic

For systems the order in which their components fail needs to be considered.

If A, B, and C are non-repairable components with times to fail t_A, t_B, and t_C then the time to failure of the system is as follows.

Series system

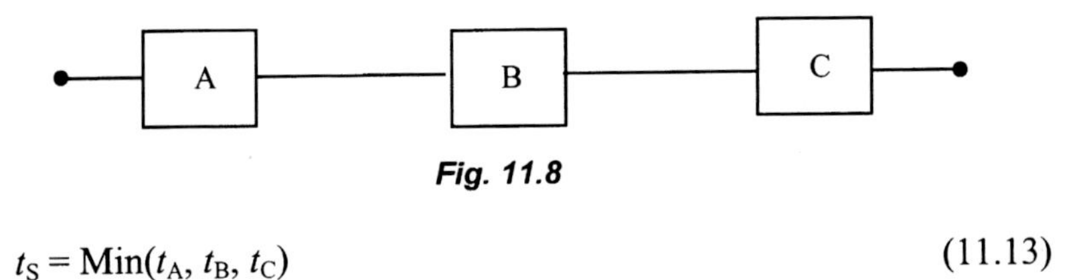

Fig. 11.8

$$t_S = \text{Min}(t_A, t_B, t_C) \tag{11.13}$$

Parallel system

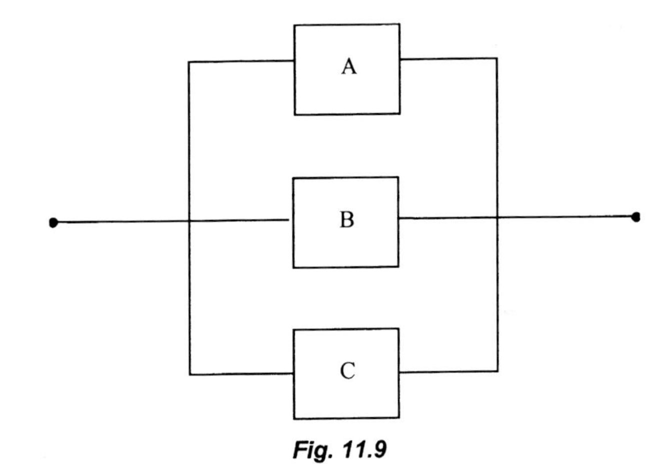

Fig. 11.9

$$t_S = \text{Max}(t_A, t_B, t_C) \tag{11.14}$$

Series–parallel system

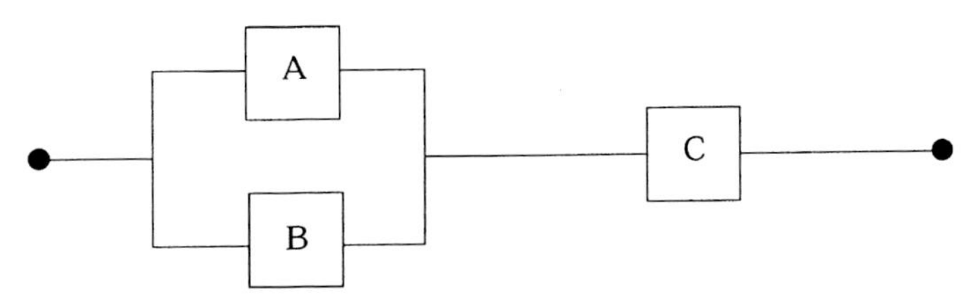

Fig. 11.10

$$t_S = \text{Min}[t_C, \text{Max}(t_A, t_B)] \tag{11.15}$$

11.7 System example

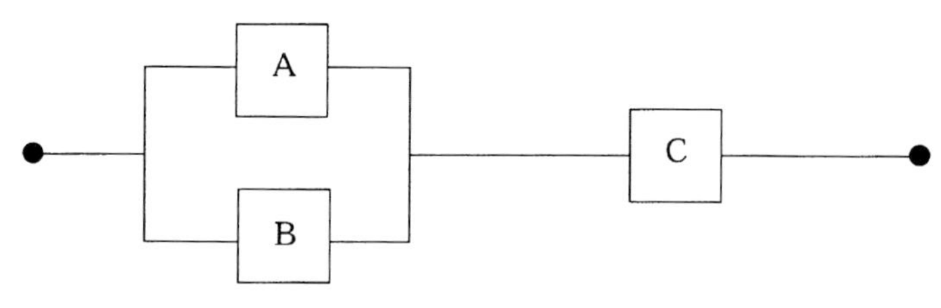

Fig. 11.11

The system shown in Fig. 11.11 consists of three components A, B, and C. Components have an exponential life distribution with mean times to failure of 500, 200, and 600 h for A, B, and C respectively. Component B is in standby mode and comes into operation on the failure of A. B cannot fail in standby mode. Each component is non-repairable and it is required to derive the time to failure distribution of the system.

Efficiency in the logical algorithm is required so that the system can be simulated generating as few random numbers as possible. The algorithm to solve this problem is shown in Fig. 11.12. It should be noted that the method does not generate all component failure times and then check on the status of the system. Failure times are only produced for those components which are relevant. The additional 'housekeeping' procedures which organize the methodology of a resulting code may appear to be too complex and time consuming and therefore not worth the effort. It may be difficult to see the advantages for such a small system with only three components such as this but significant reductions in computer time are achieved for more complex systems. Indeed, an efficient 'housekeeping' system can be the difference between a problem being soluble or not using simulation.

Example calculations for this system are shown below.

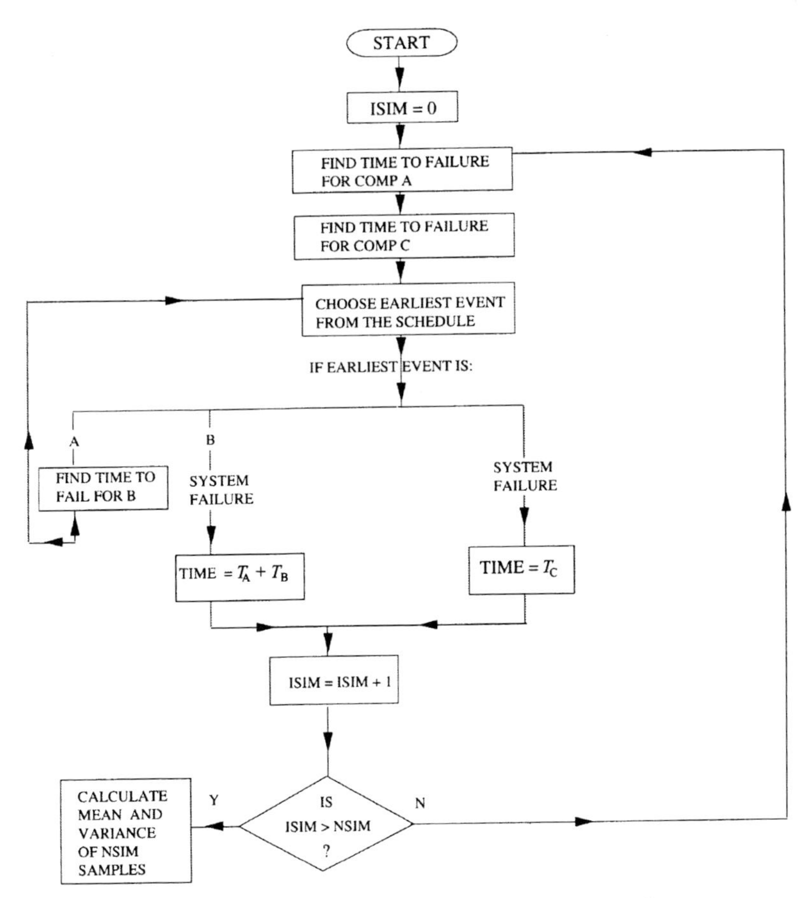

Fig. 11.12 Three-component system example

Simulation 1
As shown in Fig. 11.13:

Time to failure for component A:

$$X = 0.674$$
$$t_A = -500\ln(0.674)$$
$$= 197.26$$

Time to failure for component C:

$$X = 0.221$$
$$t_C = -600\ln(0.221)$$
$$= 905.76$$

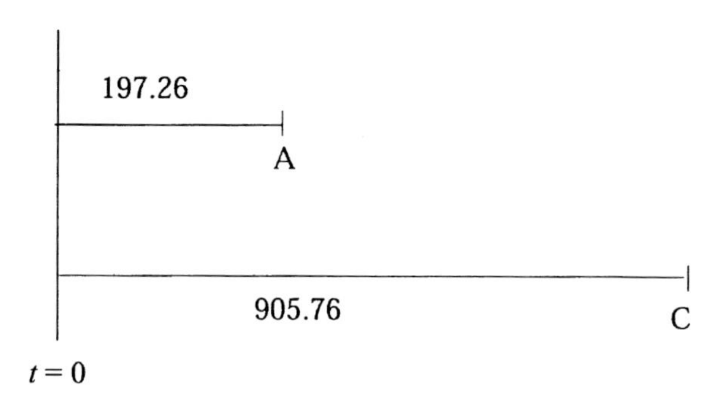

Fig. 11.13

In Fig. 11.14:

Time to failure for component B:

$$X = 0.334$$
$$t_{\mathrm{B}} = -200\ln(0.334)$$
$$= 213.42$$

Time to system failure = 197.26 + 213.42 = 410.68 h.

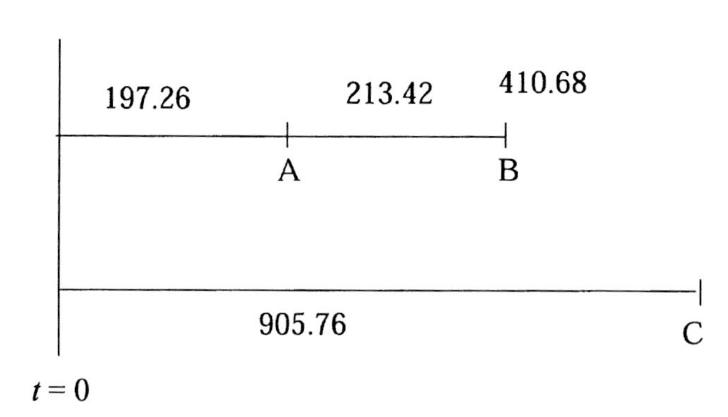

Fig. 11.14

Simulation 2
As shown in Fig. 11.15:

Time to failure for component A:

$$X = 0.783$$
$$t_A = -500 \ln(0.783) = 122.31$$

Time to failure for component C:

$$X = 0.913$$
$$t_C = -600 \ln(0.913) = 54.61$$

Time to system failure = 54.61 h.

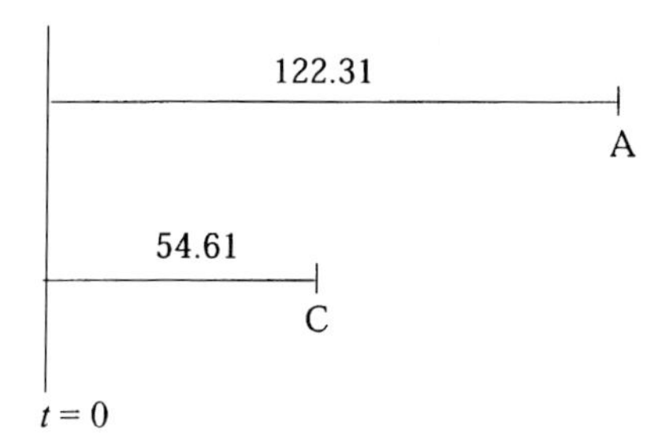

Fig. 11.15

Prior to the development of kinetic tree theory, Monte Carlo simulation was the method most commonly used to calculate top event probabilities in fault trees. Today the analytical approaches are the most commonly used though simulation may be the only option available to solve some of the more complex fault trees. The primary disadvantage of using simulation, particularly with the nature of reliability problems, is the time and expense involved in first developing the model and then executing the simulation program so that enough simulations are carried out to provide some confidence in the result. This usually means a large number of iterations and a vast amount of computer time.

11.8 Terminating the simulation

Several methods can be used to terminate the simulation process when 'enough' simulations have been performed to yield accurate estimates of the system characteristics. The simplest is to conduct a set number of simulations or experiments. However, it is not possible to know prior to the analysis of a complex system how many experiments need to be conducted to produce a set level of confidence in the predictions. If several assessments are started with unique random number seeds and the same initial conditions, then the output measures of system performance computed across the runs will be independent and identically distributed. Standard statistical methods can then be used to make inferences on experimental results.

The results required are those that we can regard as 'typical' or 'representative' of the real system performance. This can be influenced heavily by the initial conditions. These will have an obvious effect on the output variable value, as illustrated in Fig. 11.16. In a reliability problem our initial conditions may be that all components are in the working state, whereas the system may experience several failed components in the typical state. To gain representative results for the steady-state period when this situation exists it may be necessary to remove the results for the 'warm-up' or transient period from those used to assess system performance.

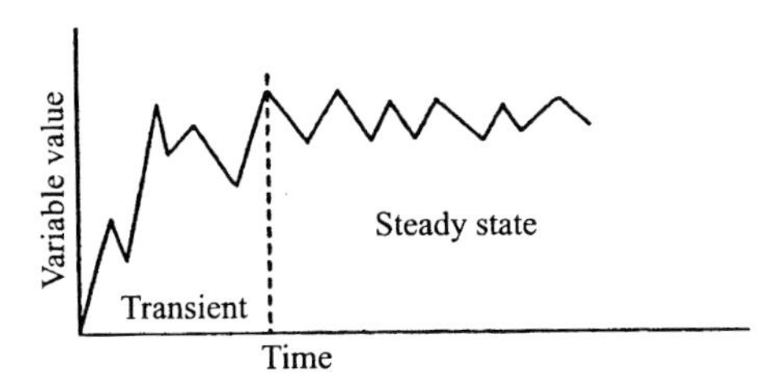

Fig. 11.16 Transient versus steady state

11.9 Summary

Monte Carlo simulation offers the most versatile of all the system analysis methods available. Independence of component failure and constant failure rates, assumptions required for most analytic methods, are not required for simulation. Systems can be modelled in whatever level of detail is required. It is particularly useful in modelling different maintenance strategies.

The trade-off for the versatility of the technique is the demand that this method makes in terms of computer power. Random samples from failure and repair distributions are made which require the generation of very large quantities of random numbers.

11.10 Bibliography

Henley, E. J. and **Kumamoto, H.** (1981) *Reliability Engineering and Risk Assessment*. Prentice-Hall.

Kleijnen, J. P. C. (1974) *Statistical Techniques in Simulation*. Marcel Dekker.

Lewis, P. A. W. and **Orav, E. J.** (1989) *Simulation Methodology for Statisticians, Operations Analysts and Engineers*, Vol. 1. Wadsworth and Brooks/Cale.

Taha, M. A. (1988) *Simulation Modelling and SIMNET.* Prentice-Hall.

Chapter 12

Reliability Data Collection and Analysis

12.1 Introduction

The collection and analysis of equipment failure data plays an important role in reliability studies. It is required in two main areas, namely for the reliability assessment of new designs and for the analysis of in-service reliability performance.

In the first case generic data are required as input to the system reliability models to establish the probability of failure or unavailability of the system. Although not always applied in reliability assessments quantification provides a useful framework to establish the relative importance of particular systems or equipments to the plant's expected reliability performance. In these cases, good quality generic reliability data are vital.

When a plant is built and running there is also a need to confirm that the required reliability performance is being achieved. The collection and analysis of data from operating plant can be used to demonstrate the conformance of specific equipment with reliability requirements and in many cases will identify areas where minor modifications or improved maintenance will significantly improve system reliability.

In both cases it is important that data are assessed against clear definitions and guidelines. Vague descriptions of the equipment's engineering and functional attributes, system boundaries, and failure descriptors can lead to uncertainties in the results which in the extreme case can make the reliability study virtually worthless. The objective of this chapter is therefore to establish guidelines for the collection and analysis of reliability data so that uncertainties can be reduced to a minimum.

12.2 Generic data

In the reliability assessment of new designs a variety of different reliability data are required. The objective will be to identify data from broadly similar equipment operating under similar functional and environmental conditions. Different types of reliability data will be required for different types of study. These can be classified under two main headings as follows:

Safety studies:	Overall failure rates
	Principal failure modes and proportions
	Common cause failure rates
	Human error rates
Availability studies:	Overall failure rates
	Principal failure modes and proportions
	Common cause failure rates
	Spurious trip rates
	Active repair and waiting times

Although most of these data classes are identical for both safety and availability studies the emphasis will be different. In safety studies the objective will be to identify those failure modes which are potentially dangerous whereas in availability studies the proportion of failures which cause a forced-outage or reduction in performance will be the main focus of interest.

Data will be obtained from a variety of sources. These may include in-house failure rate data banks, reliability data handbooks, maintenance records, and private or published data tables. As one progresses from published data tables (where a simple description of the item, e.g. CONTROL VALVE, and an overall failure rate, e.g. 30 f/Mh, will generally be the extent of the information available) to an in-house failure rate data bank the quality of the data will increase.

RELIABILITY DATA SHEET

SHEET NUMBER:	R297	**CLASS:**	Valve
DATA SOURCE:	J1199	**TYPE:**	Globe
ABSTRACT:	A47	**ACTUATION:**	Pneumatic
FAILURE RATE:	44.8 f/Mh	**FUNCTION:**	Control
GENERIC DATA TYPE:	Valves and valve actuators		

FAILURE MODES:		**EFFECT ON SYSTEM:**
Leakage	50%	10% Critical
Actuation	30%	10% Critical
Seize/stiff	10%	5% Critical
Others	10%	5% Critical

SAMPLE STATISTICS:

POPULATION:	16
NUMBER OF FAILURES:	22
CALENDAR TIME INTERVAL:	3.5 years
AV. OPERATING TIME/ITEM:	26 100 h
TEST INTERVAL:	1 year

DESIGN KEYWORDS

DISCIPLINE	**SIZE**	**MOTION**	**SPEED**	**COMPLEXITY**
Instrument	Small	STATIC	Slow	SIMPLE
Electrical	MEDIUM	Rotating	High	Complex
MECHANICAL	Large	Reciprocating	V. high	V. complex

OPERATIONAL KEYWORDS

ENVIRONMENT	**FUNCTION**	**STRESS**	**DISTRIBUTION**	**F. MODES**
Nuclear	Passive	Low	RANDOM	Unimodal
INDUSTRIAL	PROCESS	NORMAL	Wearout	Bimodal
Offshore	Safety	High	Burn-in	MULTI-MODAL

REMARKS

2.5- and 4.5-in pneumatic diaphragm operated control valves in boiler feedwater system 580 lbf/in^2 80 °C. Fail-open and fail-closed failures both critical.

Fig. 12.1 Reliability data sheet example

Compare the above data with the reliability data sheet shown in Fig. 12.1. In addition to an overall failure rate the data sheet identifies:

The type of valve:	Globe
The method of actuation:	Pneumatic
Its function:	Control
The operating medium:	Water

and a variety of other information on its size, operating environment, the failure modes, number of observed failures, sample population size, operating time, plus supporting descriptive information. Clearly a high-quality failure rate data bank built up carefully over the years is a valuable asset to the reliability engineer which will reduce uncertainties in an assessment.

The evaluation and weighting of generic failure rate data for use in specific reliability studies is discussed later in this chapter.

12.3 In-service reliability data

Data collected from operating plant will be in the form of individual failure reports. In most cases these will be records, such as work orders, detailing unscheduled maintenance actions where the emphasis will be on the repair/replacement of maintainable items. For the purposes of reliability analysis this information will need to be interpreted to establish the failure mode and the effect of the failure on the system operating state. The transfer of maintenance records into a suitable reliability data format when such unrecorded information is still fresh in the minds of operations and maintenance staff is clearly desirable.

Data of this type are referred to as failure-event data. These data can be analysed for a sample of several similar pieces of equipment or for a single piece of equipment. The assumption of restoration to 'as good as new' is generally made, that is, that each repair is perfect. Failure-event data will clearly build up more rapidly as the population size is increased; however, it is important that all members of the sample population are of the same type, operating in broadly similar conditions. In a statistical sense the pieces of equipment can then be assumed to have lifetimes which are independent and identically distributed – an important concept in the analysis of lifetime data.

Theoretically, lifetime distribution analysis should only be carried out at component level where the assumption is made that the item can only fail once. From an engineering standpoint, however, the situation is quite different and many analyses are successfully carried out on equipment to reveal problems due to one or more dominant failure modes which exhibit increasing or decreasing failure rate. It is worth noting that all pieces of equipment are assembled from a number of component parts, each of which will have a different lifetime distribution. This mixture of lifetime distributions will generate an apparent exponential distribution of times to failure for the equipment unless a dominant failure mode exists for one of the components.

For a newly commissioned plant, failure event data will gradually accumulate as the operational life increases. In the early stages only relatively coarse data analysis will be worthwhile; the objectives will be to rank the plant and equipment areas with respect to their contributions to downtime, to check that equipment is operating within its reliability specification (frequently based on equipment MTBFs and MTTRs), and to monitor for any adverse trends in reliability. More detailed analysis will become possible as the operating hours increase. However, it is best to establish, by simple statistical analysis, that performance is unsatisfactory before proceeding to more detailed studies.

Data collection and analysis should be progressive and should increase as failure experience increases. For example, for an offshore platform the progression could be:

Stage 1: Analysis at system level – wellhead, gas processing, gas export systems, etc., involving analysis to establish importance ranking of systems and major equipment by downtime and failure frequency.

Stage 2: Analysis at equipment level – compressors, pumps, instruments, valves, etc., involving analysis to establish MTBF and MTTR of equipment for comparison with best practice elsewhere.

Stage 3: Analysis at equipment/component level – bearings, seals, etc., involving lifetime distribution analysis to identify wearout or other characteristics.

These data should be compared with past experience and fed back into the system reliability models to help with the optimization of operational procedures and maintenance and test strategies.

It is clearly important that the reliability data collection system is designed to be compatible with the system reliability models, with system and equipment boundaries well defined. Preferably it should be integrated with the maintenance record system, but in all cases it should be possible to extract all the relevant reliability information from the record system as an exclusive data set. The definition of failure modes, system operating states, and many other attributes also needs to be clearly specified at an early stage. Attention to these matters will ensure that the various analyses can be carried out quickly at the appropriate times without introducing additional uncertainties.

12.4 Data collection

12.4.1 General

For reliability studies three different types of data need to be collected.

1. Inventory data. Information on the design and functional characteristics of the item.
2. Failure-event data. Details of each failure event within the surveillance period including the effect on the system operational state.
3. Operating time data. The dates and times when surveillance commenced and ended, dates and times of transitions between different system operating states, time on standby, active repair times, waiting times, etc.

In many cases operating time data are the most difficult to obtain.

12.4.2 Inventory data

Inventory data are the information which defines the type of equipment in terms of its design, functional, operational, and environmental characteristics. The data are required for each piece of equipment and will require information under four main headings.

1. Equipment identification data
2. Manufacturing and design data
3. Maintenance and test data
4. Engineering and process data.

The first three groups are required for most types of process equipment; the fourth group will be specific to each particular class of equipment. Additional information may also be recorded on the inventory data form to identify any features not covered by the specific categories.

Equipment identification data will comprise:

(a) A unique identification number, so that the failure history of each specific piece of equipment can be followed through the period under surveillance.

(b) A hierarchical plant location code, so that all failure histories for a particular system can be linked to identify the areas contributing most to plant outages.
(c) A generic class code, so that the failure event reports for similar classes of equipment can be grouped together for statistical analysis.

Manufacturing and design data should include:

(a) The manufacturer's name
(b) The equipment model number
(c) Details of the driver, if any
(d) Size and design parameters
(e) Date of manufacture
(f) Date of installation.

Maintenance and test data will include:

(a) Major scheduled maintenance tasks and frequencies
(b) Test schedules and frequencies.

As noted previously the engineering and process data will depend on the class of equipment. Typically for centrifugal pumps this should contain information on the following features:

- Body type
- Number of stages
- Rotational speed
- Seal type
- Bearing type
- Lubrication type
- Flow rate
- Suction pressure
- Discharge pressure
- Nominal power rating
- Operating temperature
- Operating medium.

Recommended codes for inventory descriptors are given for a number of equipment classes in Euredata Report No. 1 (**1**). Not all of these data are necessarily available on plant that has been operating for some time; nevertheless in a recent exercise on mechanical valves (**2**) it was shown that about 70 per cent of the inventory data were readily available. A higher proportion should clearly be available on plant in the design phase or recently commissioned.

Adequate inventory data are essential to support data analysis studies since they ensure that the sample populations chosen for detailed statistical analysis are broadly identical. The information required is relatively large; however, since inventory data collection is a one-off exercise, it can generally be completed within a relatively short time. In most cases it will not be possible to complete all the data fields on the inventory form, but the adoption of standard inventory descriptors and formats will ensure that a consistent approach is adopted throughout the plant and will facilitate comparison with other sources of data. For complex equipment or sub-systems a block diagram is a useful aid to identify which components are within the equipment boundary. An example of a boundary block diagram for a centrifugal pump from the *OREDA 92 Handbook* (**3**) is shown in Fig. 12.2. Figure 12.3 shows a completed inventory data form for a centrifugal pump.

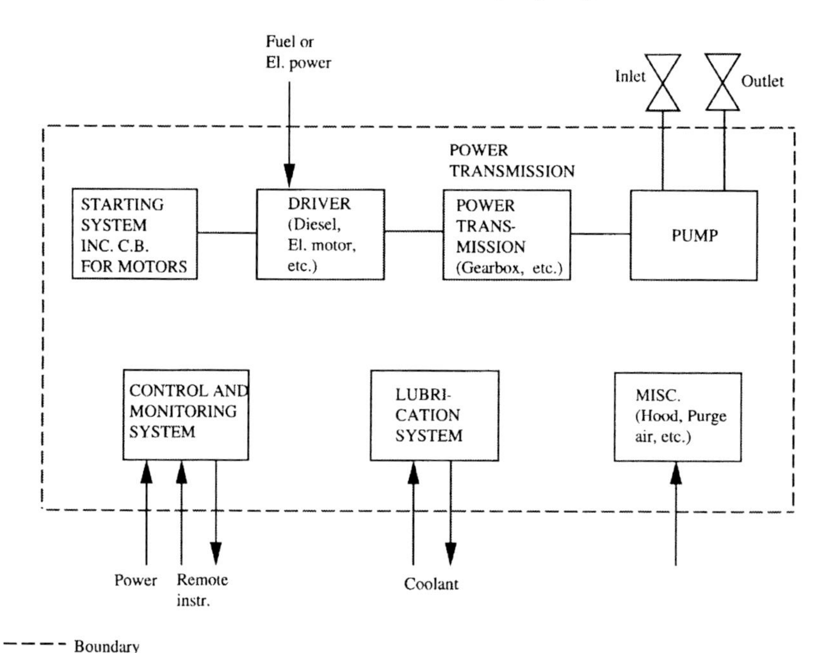

Fig. 12.2 Pumps, boundary definition

INVENTORY DATA SHEET			
SHEET NUMBER:	146	**CLASS:**	Pump
ABSTRACT:	A345	**TYPE:**	Centrifugal
UNIQUE NUMBER:	XY3726/1	**PLANT LOCATION:**	Tag. AP2671
GENERIC DATA TYPE:	Rotating machinery		

MANUFACTURER/DESIGN PARAMETERS

MANUFACTURER:	XYZ
MODEL NUMBER:	AB321
DRIVER:	Electric motor
BODY TYPE:	Horizontal split casing
NO. OF STAGES:	2
SPEED:	1450 r/min
SEAL TYPE:	Packed gland
BEARING TYPE:	Roller and ball
LUBRICATION:	Grease and water
FLOW RATE:	8500 gal/min
SUCTION PRESSURE:	5 lbf/in^2
DISCHARGE PRESSURE:	80 lbf/in^2
NOMINAL POWER:	15 kW
OP. TEMPERATURE:	20 °C
OP. MEDIUM	Water
MAINTENANCE DATA	Maintenance Manual MAP2671 refers

DESIGN KEYWORDS

DISCIPLINE	SIZE	MOTION	SPEED	COMPLEXITY
Instrument	Small	Static	Slow	Simple
Electrical	MEDIUM	ROTATING	HIGH	COMPLEX
MECHANICAL	Large	Reciprocating	V. high	V. complex

OPERATIONAL KEYWORDS

ENVIRONMENT	AMB. TEMP.	VIBRATION	OP. PRESSURE	MEDIUM
Nuclear	Low	Low	Low	Good
INDUSTRIAL	NORMAL	NORMAL	NORMAL	NORMAL
Offshore	High	High	High	Poor

REMARKS

Two-stage centrifugal pump in cooling water system. 4-in suction 3-in discharge. Direct drive by electric motor. Installed 30/4/89.

Fig. 12.3 Inventory data sheet example

12.4.3 Failure-event data

Failure-event data are records of individual failures associated with particular equipment. Each failure will be the subject of a separate record and will be a combination of codes, numerical data, and supporting descriptive information. These data, including some cross-reference to the inventory data form, should be entered into fields on the record sheet with appropriate codes defining the failure mode and the effect of the failure on system operation. The codes will generally be qualified with textual descriptions of the repair action and, in some cases, with details of the failure cause.

Typically, the information entered on the failure event record will include:

The plant reference:	For example, platform X
Equipment reference number:	Tag or unique number
System and sub-system reference:	For example, oil processing – 1st separator
Report number:	A sequential number for each failure record
Completed by:	The name of the data collector
Checked by:	The name of the project supervisor
Event type:	Generally 'Failure' but dummy events may be inserted to label the start and end of surveillance, outages for planned maintenance, changeover from operating to standby duty, etc.
Temporal data:	Dates and times of failure detection, start and end of repair action and return to service
Failure data:	Failure mode and description, severity code and effect on system
Repair data:	System outage time comprising active repair and waiting times, craft hours
Plant operational state:	Operating, standby, etc.

These data will generally be supported by some textual description of the failure and other details considered relevant.

Codes should be used for system and equipment failure modes and for failure severity classification since they lessen the chance of error. Particularly where a computerized database management system is employed the use of codes facilitates automatic checking for wrong coding and typographical errors. Despite the use of codes there will still be uncertainties in some failure-event records because of the communication chain (sometimes involving five or six people between the operator detecting the failure and the reliability analyst) and the elapsed time between the reporting of the failure and the data analysis. It is important therefore that clear guidelines with unambiguous definitions of the different descriptors and codes are produced and explained to everyone in the communication chain.

A typical failure-event data sheet is shown in Fig. 12.4.

EVENT DATA SHEET

SHEET NUMBER:	146	**CLASS:**	Pump
ABSTRACT:	A345	**TYPE:**	Centrifugal
UNIQUE NUMBER:	XY3726/1	**PLANT LOCATION:**	Tag. AP2671
GENERIC DATA TYPE:	Rotating machinery		
PLANT	Plant A		
SYSTEM	Cooling water AP2671		
SUB-SYSTEM	Pump AP2672/1		

FAILURE MODE	**EFFECT ON SYSTEM**	**RESTORATION MODE**
1 Bearing seized	Critical	Replace bearing
2		
3		

ACTION TAKEN	**DATE**	**TIME**
Failure detection	1/2/**	14.30
Start maintenance	1/2/**	15.30
Complete maintenance	1/2/**	18.30
Ready for operation	1/2/**	19.00

Active repair time 3 h

Waiting time 1 h 30 min

FAILURE MODE/SEVERITY

Failed while running, critical, loss of cw supply

FAILURE CAUSE

Bearing overheated

REMARKS

Drive side thrust bearing seized. Pump dismantled and new bearing fitted.

Completed by: J Bloggs

Checked by: E Sykes

Fig. 12.4 Failure-event data sheet example

12.4.4 Operating time data

Operating and repair time data frequently feature in the inventory or failure-event records. There is, however, a case for establishing a separate file of information which records the status of plant operation at particular times in terms of production rate, start-up and shutdown of particular processing trains, systems, and equipment, with the reasons for the transitions. On offshore platforms this is generally recorded in a Daily Production Report or telex – a valuable source of information for the reliability engineer concerned with offshore systems.

Obtaining operating hours for rotating machinery can be difficult, particularly where standby redundancy is involved. For larger machines a logbook recording information such as cumulative running hours, number of starts, running and start-up failures, etc. is frequently maintained by operations staff. For other equipment such as cooling water pumps, heat exchangers, valves, etc. no such logbooks will be kept, so the ratio of operating hours to calendar time will need to be estimated.

The way that the total calendar time is broken down for typical production equipment is illustrated in Fig. 12.5. Unless the residence times in these different operational states is recorded, estimating operating hours, repair and waiting time hours, hours on standby, etc. can be difficult. This lack of information will lead to uncertainties when analysing generic and in-service failure data.

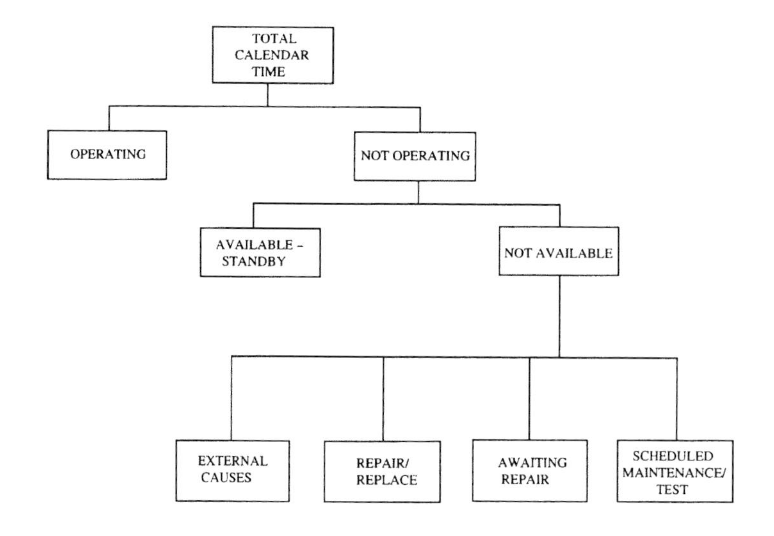

Fig. 12.5 Equipment operational states

12.5 Data quality assurance

Many reliability data exercises fail because of lack of attention to quality assurance. Before starting data collection it is necessary to audit the source of the data so that realistic objectives can be established within the available budget. A quality plan should then be developed which clearly defines the objectives of the exercise and details the procedures to be followed in data collection and analysis. Finally a quality review should be carried out on completion of the exercise to assess its success and to feed back processed data and recommendations to the host system management. Feedback of information is essential as a data validation exercise and to ensure co-operation in future studies.

12.5.1 Quality plan

The quality plan should cover five main areas, namely:

(a) Organization
(b) Documentation
(c) Data collection
(d) Data processing
(e) Reporting.

The requirements in these areas are itemized as follows.

Organization

- Agree the objectives.
- Define the responsibilities of staff by a project family tree and job specifications.
- Provide suitable staff training. (Where trained staff are involved, training may take the form of a project seminar.)

Documentation

- List objectives and amplify the requirements in an outline description. List the items on which the data are to be collected. Describe the different stages in the programme plan.
- Design data collection forms and procedures for the appropriate level of reliability/maintainability data processing. As a guide, the following areas need to be covered:

Data type	Information requirements
Inventory data:	Item identification
	Boundary specifications
	Design details
	Operational/environmental parameters
	Descriptions and definitions
Failure data:	Time interval (calendar time)
	Operational hours
	No. of failures
	No. of demands (starts/stops, tests)
	System operating states
	State/time transitions
	Time on standby
	Failure modes
	Special operational/environment conditions
Maintainability data:	Manhours/craft/disciplines
	Active repair times
	Waiting times
	Special tools/equipment
	Types of maintenance

- Institute appropriate document control procedures.

Data collection

- Identify the overall reliability/maintainability data requirements.
- Define equipment boundaries.
- Provide definitions for each reliability/maintainability parameter.
- Identify input data needed for parameter calculation and data routing by information flow diagrams.
- Establish the route for feedback of data to maintenance staff for validation and information.

Data processing

- Establish data input procedure for manual or computer interface.
- Introduce suitable data validation procedures for monitoring the quality of the information collected. These should include:
  (a) internal coherence checks
  (b) cross-checks between files
  (c) analysis for trends

(d) cross-checks within samples
(e) cross-checks between samples
(f) cross-checks with other databases.

- Define statistical methods/programs to be employed at the appropriate analysis level. Areas which may need to be covered are:
  (a) Tests for exponentiality
  (b) Methods for non-exponential distribution fitting
  (c) Calculation of distribution parameters
  (d) Estimating confidence/uncertainty limits
  (e) Methods for combining samples.

Reporting

- Identify milestones in the programming and arrange meetings to coincide. A meeting brief and agenda should be distributed prior to the meeting.
- Minutes of the meeting should be distributed within 2 weeks.
- Issue regular progress reports. A suitable interval is monthly. The report should cover financial and technical aspects. Deviation from the programme and problem areas must be highlighted.
- The format of the final report should be agreed. The draft final report should generally be distributed within 8 weeks of completion of data processing. In large or long exercises one or more interim reports may be useful to circulate results and tentative conclusions and recommendations.
- Data sheets summarizing the reliability/maintainability data produced should be incorporated in the final report and may be worth issuing separately as a reliability/maintainability data handbook.

12.6 Reliability data analysis

12.6.1 General

Data collected during plant operation provides an on-going indication of the effectiveness of maintenance and the critical areas and equipment in the plant. It can also be a valuable source of in-house generic reliability data since it reflects the operation and maintenance strategies of the organization. Analysis of plant operations data is likely to be considered at three levels, namely: system level, equipment level, and equipment/component level. Although there is some common ground it is important to differentiate between components, equipment, and systems since different methods are required for their analysis. The emphasis here is on the analysis of data from operating systems for use in maintenance and availability studies.

The analysis of global generic data for use in conjunction with in-house information to provide relevant input to reliability models is considered later in the chapter.

12.6.2 Component reliability

Components, by definition, are not repairable; they have a finite life and the lifetime characteristics can generally be represented by the Weibull distribution for a sample of identical components. If the components conform to the typical characteristic of the bath-tub curve with decreasing, constant, and increasing failure rate phases the differences will become evident from a Weibull plot. These are the characteristics of a population (of components) the members of which fail at different ages. Theoretical lifetime characteristics may be determined by a laboratory test programme. In service the situation can vary due to the local conditions which apply during operation. The electrochemical reaction of materials to process and atmospheric conditions and the way in which the host equipment is operated and maintained will clearly have a significant impact on component in-service lifetimes. This can be expected to be particularly evident during the constant failure rate period – shortening or lengthening it as a function of the deterioration mechanisms induced by the interaction of load and strength distributions. The standard bath-tub relating failure rate – sometimes called hazard rate – $h(t)$ and time to component failure may deteriorate to a hip-bath or extend to a swimming-bath as represented in Fig. 12.6.

As noted in Chapter 5 the basic Weibull equation defines the expected reliability performance of a component:

$$F(t) = 1 - \exp(-t/\eta)^{\beta} \tag{12.1}$$

where

$F(t)$ = probability of failure (end of life) in the period (0–t)
η = characteristic life
β = shape factor
$\beta < 1$ indicates a decreasing failure rate
$\beta > 1$ indicates an increasing failure rate
$\beta = 1$ indicates a constant failure rate

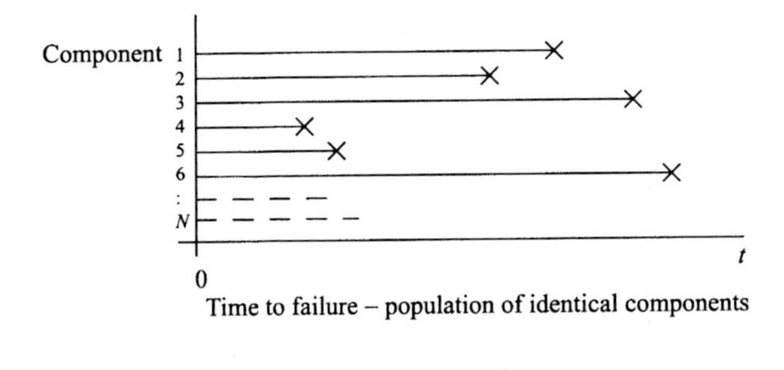

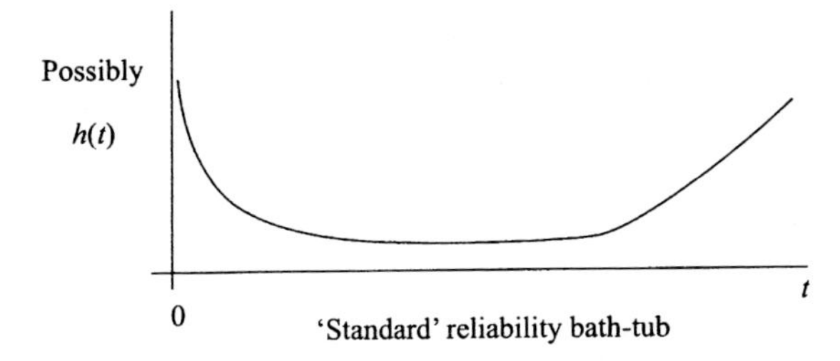

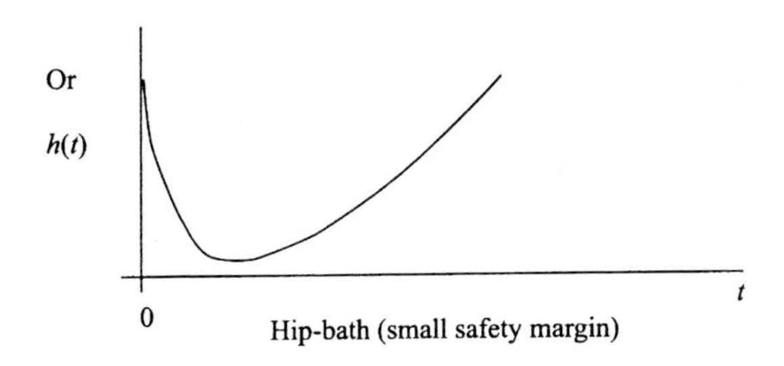

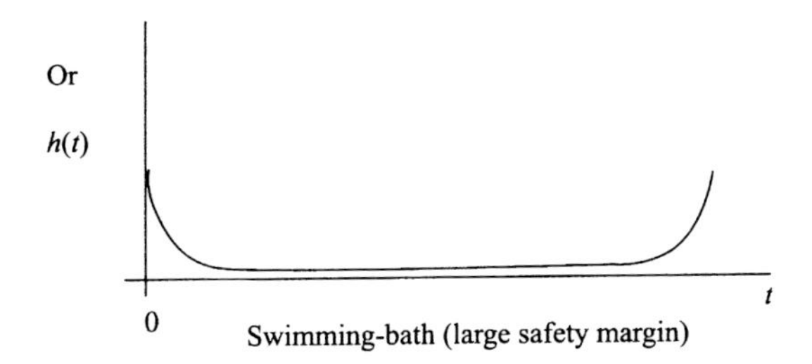

Fig. 12.6 Component failure characteristics

When $\beta = 1$ $F(t)$ reduces to the well-known negative exponential distribution function:

$$F(t) = 1 - \exp(-\lambda t)$$

and its complement

$$R(t) = \exp(-\lambda t)$$

then (and only then) $1/\eta$ = the (constant) failure rate λ.

These are the characteristics of a population of components operated to failure and then removed from test. Examples of mechanical components are bearings, seals, pipes, flanges, etc. A 'component' may be an assembly of several component parts but is defined as a 'component' if it is non-repairable when it fails.

12.6.3 Equipment reliability

An equipment is an assembly of components which operates to provide a specific function. It will fail as a result of component failure and (assuming no catastrophic effects) can be restored to operation by replacement of the failed component. The mapping of component failures to represent the failure history of one item of equipment is shown in Fig. 12.7.

Since an item of equipment is not replaced every time a component fails the representative failure statistic is mean time between failures (MTBF) rather than the mean time to failure (MTTF). Here MTBF is defined as the mean **operating** time between failures – its reciprocal is the equipment failure rate λ_E.

Assuming components with constant hazard rates and equipment failures which occur randomly with time, the process is a homogeneous Poisson process (HPP) giving interfailure times that are exponentially distributed.

The probability of x failures in a time interval t is

$$P(x,t) = \frac{e^{-\lambda_E t} \cdot (\lambda_E t)^x}{x!}$$

Equipment comprising components A, B, ..., N which are immediately replaced by new components when they fail.

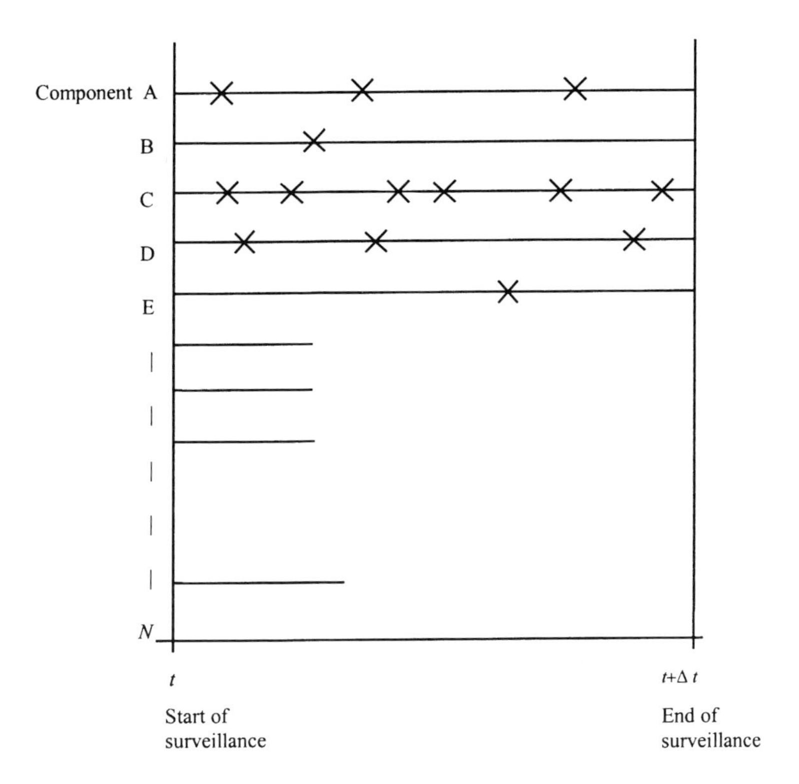

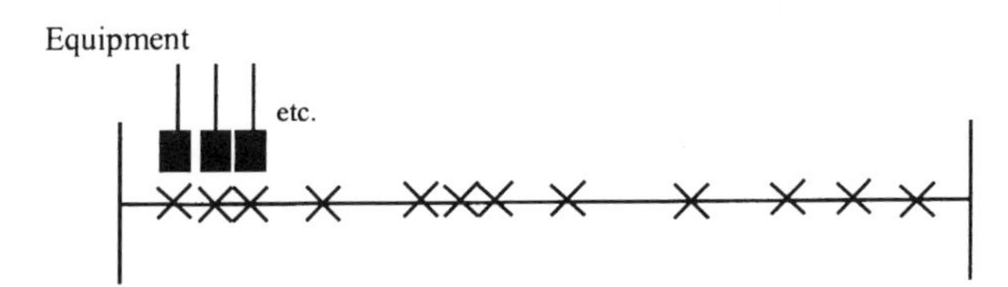

Fig. 12.7 Representation of equipment failures

where

x = number of failures
λ_E = component failure rate
t = time interval

If $x = 0$, i.e. (no failure)

$$P(0,t) = e^{-\lambda_E t}$$

since $(\lambda_E t)^0$ and $0! = 1$. This is the same as the expression for the reliability $R(t)$ of a non-repairable system.

The probability of one failure in a time interval t is $e^{-\lambda_E t} \cdot (\lambda_E t)/1!$; the probability of two failures is $e^{-\lambda_E t} \cdot (n\lambda_E t)^2/2!$; and so on.

Mechanical components in equipment seldom exhibit constant hazard rates but the equipment failure rate will often appear to be constant because of the mixture of components with different lifetime characteristics. The process is a non-homogeneous Poisson process (NHPP) so the basic Poisson equation still applies.

Weibull analysis can be useful for analysing in-service equipment failures where the assumption of 'good-as-new' after repair is considered valid or where there is a dominant failure mode. Otherwise the Duane model, based on exponential learning curves, can be applied for equipment exhibiting increasing or decreasing failure rates.

Examples of equipment are compressors, pumps, valves, etc. The level at which data should be collected and analysed is of a working unit – say, a centrifugal pump with motor, motor starter, coupling, and step-up/down gearbox if fitted, or a population of identical equipment operated and maintained in identical conditions.

12.6.4 System reliability

Systems are more complex. They generally comprise a combination of equipment (in series and/or parallel) and components. During the system lifetime, components may be replaced, equipment repaired in situ, or replaced by new or completely overhauled equipment. Modification may also be carried out to improve performance or to meet new operational requirements. The reliability characteristics (MTTF/MTBF, MTTR, and availability) will need to be synthesized by network reduction or some other system reliability assessment technique.

The example of a simple system shown in Fig. 12.8 has three items of equipment all rated at 100 per cent output. Total system failure results when equipment Z fails OR equipment X AND equipment Y fail simultaneously. If only X or Y fails it constitutes a partial system failure, and if equipment X and equipment Y are rated at 50 per cent then partial failures put the system into a 50 per cent output state. Total loss of output is, as before, when Z fails OR X AND Y fail simultaneously. For process systems multiple operating states are quite common. Analysis of system data must, therefore, be approached with caution.

Examples of systems are gas compression systems, cooling water systems, power generation systems, etc.

12.6.5 In-service data analysis

In-service reliability data analysis will generally start at system level to identify the systems contributing most to plant downtime. These systems will then be partitioned to enable the more detailed analysis of single or multiple items of equipment.

The first task is to establish a set of base statistics for comparison with subsequent plant and equipment performance and published 'best' results from the industry. The reliability, availability, and maintainability (RAM) performance indicators proposed by UNIPEDE (**4**) are probably sufficient in the first instance; subsequently more detailed analysis to determine the underlying lifetime characteristics may be worthwhile. Typically for continuously operated plant the UNIPEDE performance indicators are:

Reliability: characterized by mean time between failures (MTBF)

$$\text{MTBF} = \frac{\text{theoretically available production hours}}{\text{number of corrective maintenance actions}}$$

(or its reciprocal: mean failure rate)

Availability: characterized by percentage downtime (DT%)

$$\text{DT\%} = \frac{\text{number of non-productive hours} \times 100}{\text{theoretically available production hours}}$$

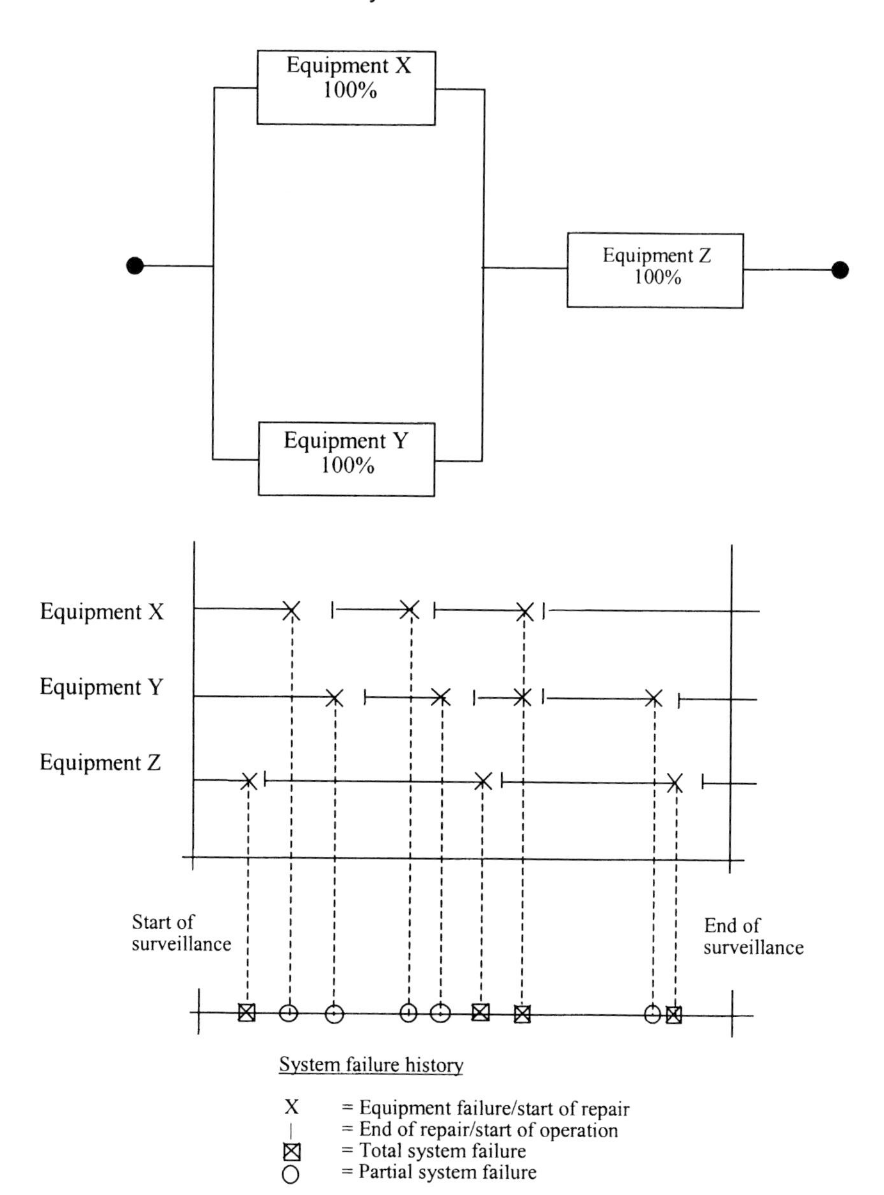

Fig. 12.8 Representation of system failures

Maintainability: characterized by mean time to repair (MTTR)

$$\text{MTTR} = \frac{\text{number of non-productive hours}}{\text{number of corrective maintenance actions}}$$

These are robust, easily computed statistics which can be applied at plant, system, or equipment level. Typical ranges for equipment reliability, availability, and maintainability are given in a number of source documents, for example, Bloch and Geitner (**5**).

For plant that can operate at several different output levels slightly more complex definitions may be required. At some stage it may also be worthwhile adding other indicators such as start-reliability (for machines which are critical to production) and cost-based statistics so that the impact of equipment unreliability can be more finely gauged. For the initial analyses the UNIPEDE RAM performance indicators are generally considered sufficient.

When analysing maintenance (or any other data) it is important to be clear about the objectives. Defining the required outputs defines the necessary inputs. The boundaries assumed for the systems and equipment must be unambiguous, as must be the assumptions on the availability of interacting services (electricity, water, air, etc.) on plant, systems, and equipment reliability and availability performance. Writing down the basic objectives at the beginning of a project is a necessary discipline when identifying and reviewing likely problems and data requirements. Attention to these aspects at an early stage should ensure that minimum effort is required later in the project.

Each of the important systems may need to be sub-divided into sub-system, equipment, or component level – where component in this case means an item which will be replaced rather than repaired. The golden rule is to keep everything as simple as possible within the constraints of obtaining the required output. Generally the analyst should concentrate on corrective maintenance reports. Although useful data can be embedded in preventive maintenance records the sheer volume of reports will make the exercise significantly more time consuming without equivalent returns. At some stage it may be worthwhile introducing a customized reliability analysis module into the maintenance management system to allow the impact of incipient failures, which are generally addressed during preventative maintenance, to be included in the on-going analysis.

Procedures need to be developed for carrying out the analysis with periodic quality assurance checks; these need to be tested in a pilot exercise. In many cases this will highlight the problems involved in handling the data. Information flow diagrams (IFDs) showing the paths from the inventory data, operating information, and corrective maintenance reports can be useful for identifying the basic information required for calculating the output statistic. The results of a well-specified pilot exercise can provide a template for more wide-ranging studies.

Think about the resources required – particularly for the more advanced analytical methods: these often require significant additional input from the analyst. Many projects founder after a relatively short time because the instigator was too ambitious, so make sure that time and costs are realistically estimated. Transferring the basic data into one of the standard spreadsheet programs such as LOTUS 123 or EXCEL will make life much easier because of the simple database facilities they provide for ordering data, carrying out calculations, and creating graphs. For more complex analyses standard routines can provide input to more sophisticated database management systems such as ACCESS or one of the statistical analysis modules that are now available.

12.6.6 System level analysis

One of the simplest and most effective techniques for identifying critical areas in plant, systems, or equipment is Pareto analysis. Pareto was a 19th-century Italian economist who observed that many real-life problems were caused by a relatively small proportion of causes. Pareto analysis aims to segregate the 'vital few' from the 'trivial many'. The results are frequently presented as a histogram relating frequencies of occurrence to output causes. A Pareto plot of the 'top-ten' systems contributing to plant downtime in a 1000 tonne/day ammonia plant, averaged over the first 3 years of operation, and a comparison with the downtime predicted during the conceptual design stage is shown in Fig. 12.9. It can be seen that three systems are responsible for over 50 per cent of plant downtime. Concentrating on the 'vital few' systems will obviously have the greatest impact on the availability performance of the plant.

The procedure for carrying out a Pareto analysis is quite simple. For the equipment represented in Fig. 12.9 it requires the equipment populations to be identified and the boundaries of each system defined so that only failures within this boundary are included in the analysis.

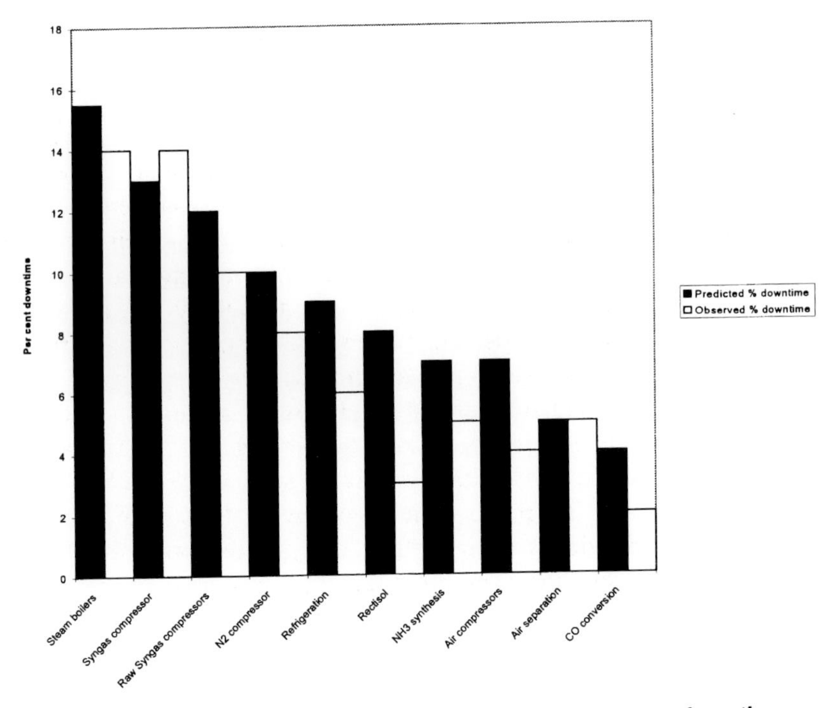

Fig. 12.9 Ammonia plant Pareto plot – 'top-ten' system downtime analysis

The number of downtime outages are counted for each system for the specified period and presented as a histogram, generally using one of the sophisticated spreadsheet programs such as EXCEL. More detailed analysis (possibly just by reading through some corrective maintenance reports) may show there is more than one failure mechanism responsible for system downtime. As with all studies the conclusions must be supported by sound engineering judgement.

'Top-ten' analysis, based on the Pareto model, is frequently employed by major companies to identify areas where the greatest reliability improvements can be achieved. It is equally applicable for use with the other performance indicators, i.e. times between failure and repair times etc. The analysis of the different performance indicators can help to define maintenance strategy as well as areas that require more detailed analysis.

12.6.7 Equipment level analysis

Initially analysts will probably make the assumption that equipment failures are independent identically distributed (IID) events with approximately constant rates of failure. The MTBF is then the simple ratio

$$\mathrm{MTBF} = \frac{T}{N_\mathrm{f}} \tag{12.2}$$

where N_f is the number of failures and T is the total operating time. The equipment failure rate (λ_E) is the reciprocal of the MTBF.

For items that are repaired or replaced during the surveillance period to 'as-good-as-new', the data will comprise a mixture of failure and survival times which may be randomly censored. The mean time between failures (MTBF) for these multiple randomly censored samples can be calculated from the expression

$$\lambda_E = \frac{N_\mathrm{f}}{\Sigma t_i + \Sigma s_i} \tag{12.3}$$

where the t_i values are the times between failures, the s_i values are the survival times, and N_f is the number of failures. The pattern of failures for a population of eight items operated for a surveillance period of 1 year is shown in Fig. 12.10. The calculations of failure and survival times are shown in Table 12.1.

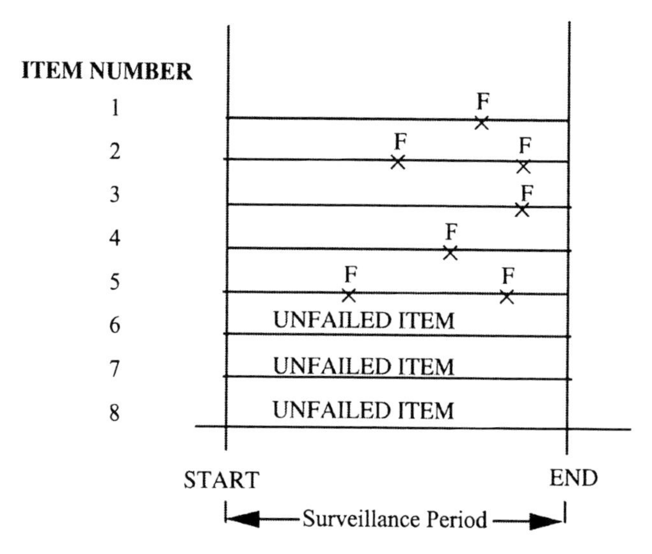

Fig. 12.10 Multiple censored example

Table 12.1

Item no.	Time to first failure		Time between first and second failure	Survival time
1	6250			8760 – 6250 = 2510
2	4320		7490 – 4320 = 3170	8760 – 7490 = 1270
3	7620			8760 – 7620 = 1140
4	6030			8760 – 6030 = 2730
5	2840		7150 – 2840 = 4310	8760 – 7150 = 1610
6				8760
7				8760
8				8760
	27 060	+	7480	
			$\sum t_i = 34\,540$	$\sum s_i = 35\,540$

For the times to failure and survival times shown in Table 12.1 the mean time to failure is therefore

$$\text{MTBF} = \frac{27\,060 + 7480 + 35\,540}{7} = 10\,011\,\text{h}$$

$$\lambda_{\text{E}} = 1/\text{MTBF} = 0.875 \text{ failures/year}$$

This assumes the repair/replacement times are instantaneous, so the total operating time $8 \times 1 = 8$ item years; hence

$$\lambda_{\text{E}} = \frac{N_{\text{f}}}{T} = \frac{7 \text{ failures}}{8 \text{ years}} = 0.875 \text{ failures/year}$$

The assumption in both of these calculations is that all failure modes were the same. This may be acceptable where the failure mode is defined at a high level (e.g. forced-outage failure) but not when the interest is in, say, failure of the motors in a population of pump sets. In this case failures of the pump would be discounted but the censored times would still be included because the motor would have operated (without failure) for these times.

Combining the data from similar equipment to analyse lifetime characteristics can be difficult. For complex equipment, such as gas turbines, which have relatively high failure rates and where there may be dependent failure modes it is often politic to consider each item of equipment separately. In these cases some indication of the consistency of reliability performance is required.

12.6.8 Trend analysis

Retrospective analysis of failure data can show the trends over past time, production output, number of starts, miles travelled, or some other parameter. These data, useful in themselves, also provide the basis for on-going trend analysis. Plotting failure frequencies at say monthly intervals can often show an apparently random pattern from which the underlying trend may be difficult to see. This is evident from the histogram of gas turbine outages in Fig. 12.11.

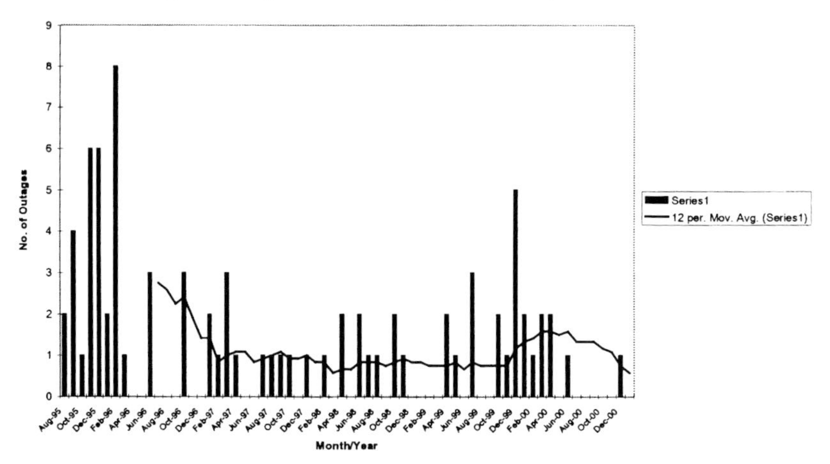

Fig. 12.11 Gas turbine – histogram of monthly outages and 12-month moving average overlay

Analysts often use moving averages to smooth out short-term variation in data. Three-month and 12-month moving averages are quite typical. The method involves computing the average of the last 3 (or 12) observations before plotting against month number or date of observation. A combination of 3-month and 12-month moving averages for the 'top-ten' problem areas can often be very informative. The 12-month moving average overlaid on the histogram in Fig. 12.11 shows the changes in the failure pattern for an offshore gas turbine over 5 years of surveillance; note the deviations from a constant failure rate evident at the beginning and end of the 12-month moving average.

The deviation from constant failure rate can sometimes be seen more readily from a linear plot of cumulative failures against cumulative time. This is shown in Fig. 12.12 with the line of constant failure rate in the central period of about 4 years inserted.

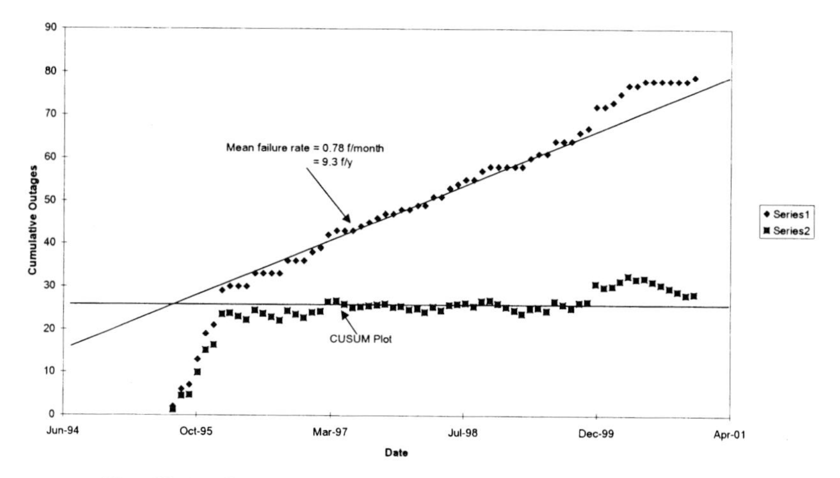

Fig. 12.12 Cumulative outages versus cumulative time – gas turbine data

Other methods of trend analysis such as cumulative sum (CUSUM) techniques are well documented – for example by Newton (**6**). A CUSUM plot is also shown in Fig. 12.12, based on the constant failure rate of 0.78 failures/month derived from the slope of the cumulative plot. The horizontal section of the CUSUM plot shows where the assumption of a constant failure rate is true and the deviation in the early and late periods. In circumstances where the trend is not clear, use can be made of the Laplace test for trend. For the case of data analysed at some random aggregate time t_0 where N_f sequenced failure times are denoted by t_i, the test statistic

$$u_L = \sqrt{(12N_f)}\left[(\Sigma t_i / N_f t_0) - 0.5\right] \quad (12.4)$$

is distributed approximately as a standardized normal variate, so the probability of a trend can be established by comparison of the test statistic with tabulated data from the cumulative normal frequency distribution. If the value of u_L is less than zero the process is improving so that times between failures are increasing on average; if greater than zero the process is deteriorating.

Clearly the gas turbine failure rate computed from the period of steady-state operation (constant failure rate) is the most relevant for maintenance engineers since this provides the basis for maintenance planning. It is also a necessary reliability statistic for availability studies. Basing the estimate on the 68 failures over the 5-year period of surveillance gives a much higher failure rate, 13.6 f/year versus 9.3 f/year, leading to pessimistic predictions for future conceptual design studies and estimates of maintenance staff and spares requirements for normal operation.

More detailed analysis can sometimes be worthwhile to identify the equipment's underlying failure characteristics, particularly the expected period of steady-state operation and the extent of likely deviations. Weibull analysis has often been offered as the technique for determining these characteristics. However, analysing times between failure presents problems, particularly for equipment which has operated for some time before the start of surveillance. Weibull analysis of ordered interfailure times will often show shape factors of unity or close to unity. This is frequently what is observed and generally indicates that more than one failure mode is involved. Partitioning a set of interfailure times so that a specific failure mode can be identified may show higher shape factors when some form of deterioration mechanism is evident or can be assumed.

Weibull analysis can be useful for analysing in-service equipment failures when there is a dominant failure mode. If there is no obviously dominant failure mode then models such as Duane, based on exponential learning curves, can be applied to equipment failure data. The US Army Materials System Analysis Activity (AMSAA) refined the basic Duane model several years ago, particularly for application to mechanical equipment. This model (known as the AMSAA–Duane or DA model) is used for statistical analysis of equipment reliability growth (or deterioration) processes. The analytical process is basically graphical.

The natural logarithm of cumulative failure events $N(T)$ versus log cumulative time is a linear plot if the model applies:

$$N(T) = AT^{B} \tag{12.5}$$

Taking logarithms gives

$$\ln N(T) = \ln (A) + B \ln (T) \tag{12.6}$$

which is the equation of a straight line.

The scale parameter A is determined from the intercept on the y axis of ln $N(T)$ and B, the slope. It should be noted that B is not the same as the β in the Weibull equation although both slopes indicate the same characteristics, i.e.:

$B < 1$ failure rate decreasing
$B = 1$ failure rate constant
$B > 1$ failure rate increasing

The model intensity function $w(t)$ measures the instantaneous failure rate at each cumulative failure time and is the first derivative of $N(T)$ hence

$$W(T) = ABT^{B-1} \tag{12.7}$$

The mean failure rate (MFR) for the surveillance period $(0,T)$ is

$$\text{MFR}(T) = AT^{B-1} \tag{12.8}$$

Hence, if the points are approximately on a straight line with $B = 1$ the cumulative mean failure rate is identical to the intensity function. For $B > 1$ the MFR is increasing with time and decreasing for $B < 1$.

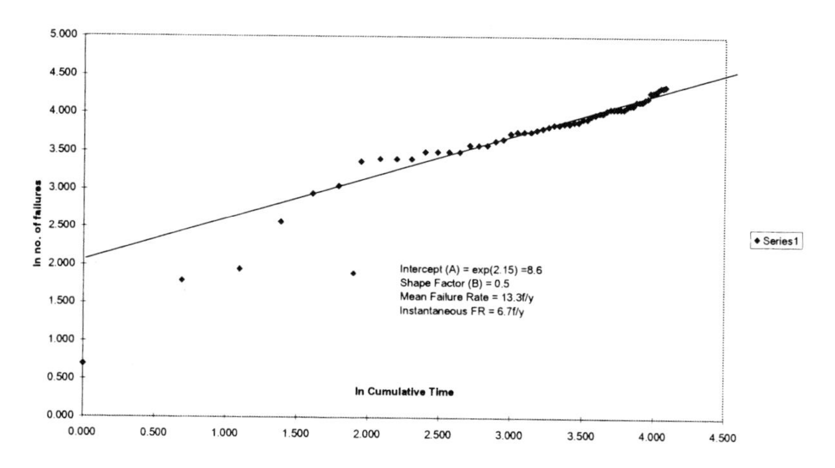

Fig. 12.13 Duane/AMSAA plot of gas turbine data

Figure 12.13 shows a DA plot for the gas turbine data. The intercept on the y axis is ln 2.15 so $A = e^{2.15} = 8.6$. The slope (B) is 0.5. (The dx

and dy values need to be adjusted for the different scales on Fig. 12.13 x and y axes.)

The decreasing failure rate, indicated by the slope $B < 1$, is confirmed by comparison of the instantaneous failure rate at the end of the surveillance period of 60 months with the mean failure rate for the same period. For the parameters computed from the DA plot:

$A = 8.6$
$B = 0.5$
$T = 60$ months

The mean failure rate is

$$\text{MFR}(T) = AT^{(B-1)} = 8.6 \times 60^{(0.5-1)} = 1.11/\text{month} = 13.3 \text{ f/year}$$

which is in close agreement with the previous estimate computed from the 68 failures in 5 years (cf. 13.3 versus 13.6 f/year). The instantaneous failure rate at the end of the 5-year period:

$$W(T) = ABT^{(B-1)} = 8.6 \times 0.5 \times 60^{(0.5-1)} = 0.56/\text{month} = 6.7 \text{ f/year}$$

It can be seen that the instantaneous failure is significantly less than the estimate from a constant failure rate model with the indication that (in this case) gas turbine reliability is improving with time in service. It is also evident that together the different methods of analysis provide engineers with a better overall picture of the reliability characteristics of the gas turbine during operation.

Basically the difference between the Weibull and DA models is the timescale. With the Weibull model individual times to failure are plotted as order statistics whereas cumulative failures versus cumulative time in operation are employed in the DA model. Weibull handles one failure mode at a time with the assumption that equipment is replaced or restored to 'as-good-as-new' after each failure. The DA model can cope with either single failure modes or a mixture of modes as a time-series analysis where the equipment failure rate may be changing with time.

An example of Weibull analysis of equipment failure data for a specific failure mode is shown in Fig. 12.14. This analysis features the special probability plotting paper by Chartwell (Graph Data Ref. 6572) to represent the failure characteristics of a population of electric motor bearings. The scales are arranged so that the ordered failure times can be plotted against the median rank of the cumulative percentage of the population failed. A perpendicular drawn to the estimation point from the best straight line through the points intersects two nomograms from which the shape factor (β) and mean life (μ) can be obtained for this distribution. The dotted line across the page at 63 per cent failed provides the estimator for the characteristic life (η). The shape factor $\beta = 3.3$ indicates a clear wear-out characteristic with a mean life of 27 000 h. To limit failures in service to less than 1 in 10 the B10 life (10 per cent probability of failure) is the recommended replacement interval for these components. This can be read from the Weibull plot as 15 500 h so replacement every 2 years would probably be acceptable, with about 90 per cent of bearings expected to operate without failure for this period.

It can be seen that the basic reliability statistics can be obtained from relatively simple calculations although it must be stressed that consistent data must be used in the calculations. As shown here this can best be confirmed by histograms with a moving average or by plotting cumulative failures against cumulative time. Within the different time periods of different IID data (particularly for the steady-state period) a further breakdown in order to rank sub-systems, failure modes, and failure causes by importance will clearly be valuable for maintenance planning and availability prediction. Tables 12.2 to 12.5 show typical results from a small population of aero-derivative gas turbines. The data in the tables are from 1 year's operation of three gas turbines driving electric generators.

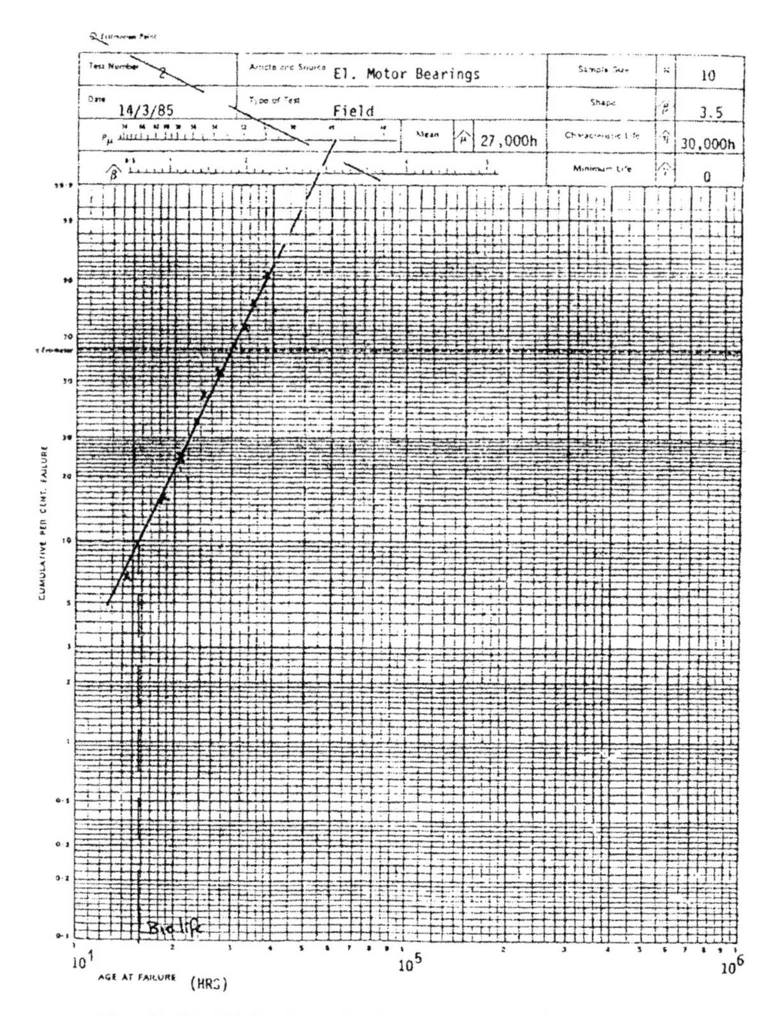

Fig. 12.14 Weibull analysis – motor bearings example

Table 12.2 Gas turbine failure – breakdown by sub-system

Sub-system	Outages (%)
Control/monitoring	43
Gas generator	33
Lubrication	6
Starting system	4
Power turbine	3
Others	11

Table 12.3 Gas turbine failures – breakdown by failure mode

Failure mode	Outages (%)
External leaks	16
Fail while running	11
Reduced output	11
Fail to start	9
Overheating	3
Vibration	2
Others	48

Table 12.4 Gas turbine failures – breakdown by failure cause

Failure cause	Outages (%)
Control and instrumentation	36
Material damage	13
Leakage	8
External events	7
Mechanical defect	7
Blocked/plugged	3
Others	26

Table 12.5 Gas turbine failures – breakdown by trade

Trade	No. of outages	Outage (h)
Mechanical		
Fuel system	8(4)	38
Ignition	2(2)	Negligible
Combustion	1(1)	Negligible
Material damage	1(0)	1464
Air intake	1(1)	Negligible
Other mechanical	4(3)	25
Electrical	2(1)	77
Instrumentation	11(5)	126
Inspection	6(4)	375

Note: numbers in parenthesis indicate zero time outages.

The scope for Duane or Weibull analysis can be limited during normal operation because of difficulties in identifying a representative population of equipment of identical design, operating in identical plant and environmental conditions. Monitoring trends in equipment failure frequencies will sometimes indicate where more detailed analysis using the Weibull or Duane models is likely to be worthwhile; however, for maintenance planning and availability assessment, recourse will also need to take account of the information in generic source documents. Combining generic data and the limited in-house data will generally provide the most robust estimates for use in reliability assessments.

12.7 Generic reliability data analysis

For new plant designs there is clearly no experience of actual reliability performance that can be used in the assessment. In these circumstances it is necessary to employ generic data with adjustments for the difference in the operating duty and environmental conditions.

The amount of data available in the public domain is relatively small, and significant effort and engineering judgement must be applied to ensure that the input data are valid for the application. Failure data will generally be obtained in the first instance from generic sources published in textbooks, reliability data handbooks, or generic reliability data banks; however, for items which are safety or production critical these data may not provide sufficient information for confidence in the validity of estimated failure characteristics.

Best estimates from generic data sources need to be adjusted for the specific conditions under which the item is expected to operate. Recommended failure rate prediction models are currently of the form

$$\lambda_{XA} = \lambda_{\mathrm{b}} \alpha_A \prod k_i \tag{12.9}$$

where λ_{XA} is the predicted failure rate for equipment X in failure mode A, λ_{b} is the base failure rate for equipment types similar to X, α_A is the proportion of failures in mode A, and k_i is the stress factor for stress i. This is the principal model featured in several textbooks, for example Green and Bourne (7), and is based on experience with non-repairable electronic systems where the assumption is that deterioration is unlikely to become evident within the time frame of the mission.

Since electronic components are small and relatively cheap, significant programmes of testing have been carried out to establish the effect of different stresses (for example temperature effects) so that a strong body of experience exists to allow the calculation of robust k factors and failure mode proportions for a specific set of expected operating conditions. With electronic systems, factors such as the effect of different conditions, for example voltage, current, operating temperature, etc., can be clearly specified.

Similar experience is not available for mechanical systems. The modification of the basic model for mechanical systems generally assumes only two *k* factors – one for environment effects and the other for the expected operating duty; *k* factors for these two types of stress are shown in Chapter 4 and are included here in Tables 12.6 and 12.7 for convenience.

Table 12.6 Environmental stress factors (k_1)

General environmental conditions	k_1
Ideal static conditions	0.1
Vibration-free, controlled environment	0.5
Average industrial conditions	1
Chemical plant	1.5
Sea-going ships	2
Road transport	3
Rail transport	4

Table 12.7 Duty stress factors (k_2)

Component nominal rating (%)	k_2
140	4
120	2
100	1
80	0.6
60	0.3
40	0.2
20	0.1

Note: for equipment designed against a specific design code (e.g. vessels) nominal rating = 100 per cent. For other equipment (e.g. valves, pumps, etc.) a nominal rating of 80 per cent or less may be appropriate.

In practice the environmental factors have been shown to be adequate for process system applications. Duty factors on the other hand are very open to interpretation. Except in the case of mechanical equipment designed to a specific code (e.g. pressure vessels designed to the British Standard) the operating duty parameters associated with generic data available in the open literature, reliability data handbooks, or reliability data banks are, at best, rather vague. Clearly a number of factors are likely to influence operating duties and therefore system reliability performance.

The factors which are considered to need particular consideration are associated with three areas, namely:

(a) Design
(b) Operation
(c) Environment (industry and local).

These, to a greater or lesser extent, can be expected to influence system reliability performance.

Design features

Clearly the safety margin will affect a component's resistance to failure. For example, stiffness will dictate pipeline wall thickness for situations where pressure, volume, and temperature conditions are low, hence the safety margin is likely to be very large. At the other end of the spectrum, with high-pressure/temperature applications, wall thickness will be determined by the stresses generated in the vessel by the operating conditions. Safety margins can be assumed to be correspondingly smaller. For rotating assemblies high speeds will attract lower safety margins than those for low-speed applications since stresses in the disc are a function of the diameter and the rotational speed. In both of these cases it is likely that safety margins will be lower for more demanding conditions with the expectation of higher failure rates during operation. Other features, such as power levels and complexity, can also be expected to have an effect.

Operational effects

Operational conditions vary from industry to industry; hence a variety of external influences such as the quality of maintenance, mode of operation, operational constraints, and interactions with other systems are likely to have an effect on operational reliability performance – these factors are mentioned from time to time in the literature but there are seldom any attempts to quantify the effects.

Environmental conditions

In addition to the general environmental conditions associated with different industries, local environments in terms of dust, humidity, vibration, etc. may have a significant impact on component, equipment, and system reliability performance, leading to greater uncertainty in the validity of the generic data available. Internal environments may also have an impact, for example the aggressiveness of the process medium, its chemical compatibility with materials of construction, etc. It is very difficult to quantify these effects from the generic data generally available.

12.7.1 Estimating equipment failure rates

There are a number of different methods for predicting failure rates for a specified item of equipment depending on the availability of representative data and the criticality of the failure mode. It is worth looking at a typical method for an item on which significant published reliability data exists. As an example, assume a forced-outage failure rate is required to evaluate the contribution to production unavailability due to a pump set with the following specification:

Single-stage centrifugal pump, electric motor driven	
Duty:	Cooling water supply to condenser
Power rating:	75 kW
Operating duty:	2000 gal/min at 150 ft head
Operational mode:	Continuous operation
Environment:	Industrial plant, sheltered area
Medium:	River water

At the simplest level overall failure rates for centrifugal pumps might be obtained from textbooks as follows:

Andrews and Moss (**8**):	Centrifugal pump, boiler feed 220 f/Mh
Davidson and Hunsley (**9**):	Centrifugal pump, small, 100 f/Mh
	Centrifugal pump, large, 300 f/Mh
Du Pont (**10**):	Centrifugal pump, (fail to run) 57 f/Mh

Taking the median of the three samples with unspecified failure mode (therefore assumed to include all failures) an estimate of a base failure rate for this pump set would be

$$\lambda_b = 220 \text{ f/Mh}$$

For average industrial conditions and normal duty the k factors assumed are unity, i.e.

$$k_1 = 1.0$$
$$k_2 = 1.0$$

and – for want of other information – an engineering judgement assumption of the proportion of failures which would be production critical for the fail-while-running (FWR) failure mode is, say, 50 per cent. Hence $\alpha_A = 0.5$. An estimate of the FWR failure rate is therefore

$$\lambda_{XA} = \lambda_b \alpha_A \prod k_i$$
$$= 220 \times 0.5 \times 1.0 \times 1.0 = 110 \text{ f/Mh}$$

cf. Du Pont estimate of 57 f/Mh.

An alternative model was proposed by Andrews and Moss (**11**) to determine expected reliability performance of rotating machinery which took into account the specific design and the operational and environmental features of the equipment application. The model was subsequently developed for more general application to generic reliability data. This model (the DOE model) is based on research by Moss (**11**) which showed that the failure rates of sample populations of identical control equipment operating in different conditions cover a wide range, typically with a factor of 10 between the highest and lowest ranked failure rates for the different equipment classes. For example, the failure rate range of 24 samples of air-operated control valves was 5–68 failures/10^6 h. About 60 per cent of equipment failure rates were within a factor of 2 and 90 per cent within a factor of 4 of the median of the distribution. The development of the DOE model was based on this research.

The DOE model employs simple linguistic variables such as high, moderate, and low (H, M, and L) which have weighting coefficients (–1), (0), and (+1) respectively. These coefficients are combined for the expected design, operational, and environmental attributes to estimate a mean weight (X), which is applied as an exponent to estimate a modifying k factor. The categories and proposed weights are listed in Table 12.8.

Table 12.8 DOE model equipment attributes and weights

	Low (L)	Moderate (M)	High (H)
Weight	(–1)	(0)	(+1)
Design attributes			
Size/power	< 100 kW	< 1000 kW	> 1000 kW
Speed	Static	< 5000 r/min	> 5000 r/min
Complexity	Simple	Average	Complex
Operation attributes			
Mode	Continuous	Intermittent	Standby
Medium	Benign	Average	Aggressive
Pressure/temp.	Ambient	< 100 lbf/in^2/30 °C	< 500 lbf/in^2/100 °C
Environment attributes			
Plant	Nuclear	Av. industrial	Offshore
Local	Controlled	Sheltered	Exposed
Temp./humidity	Controlled	Av. industrial	Extreme

The modifying k factor is computed from the expression

$$k = 2^X$$

where X is the mean weight of the design, operational, and environmental attributes. For example, for the single-stage, electric motor driven centrifugal pump considered previously the weights ranked by engineering judgement are as follows.

Design attributes:	Size/power:	< 100 kW = L = –1
	Speed:	< 5000 r/min = M = 0
	Complexity:	average = M = 0
Mean design weight:	$(-1+0+0)/3 = -0.33$	
Operational attributes:	Mode:	continuous = L = –1
	Medium:	river water = M = 0
	Press./temp.:	65 lbf/in^2/20 °C = L = –1
Mean operation weight:	$(-1+0-1)/3 = -0.66$	
Environmental attributes:	Plant:	industrial = M = 0
	Local:	sheltered = M = 0
	Amb. temp.:	moderate = M = 0
Mean environment weight:	$(0+0+0)/3 = 0$	

Combined weight (X) = $(-0.33) + (-0.66) + 0 = (-1.0)$

Modifying stress factor (k) = $2^X = 2^{-1} = 0.5$

Assuming the same base failure rate (λ_b = 220 f/Mh) and a similar proportion of FWR failures as previously, i.e. 50 per cent, then

$$\lambda_{XA} = 220 \times 0.5 \times 0.5 = 55 \text{ f/Mh}$$

This is significantly lower than the original estimate and much closer to the Du Pont recommended failure mode rate of 57 f/Mh for the fail-to-run failure mode.

Checking the result against the larger database shown in Fig. 12.15, 20 sample failure rates are assumed to be from centrifugal pumps of various types, operating in different industrial and plant environments. The failure rate range is 8–2785 f/Mh with the median failure rate between the tenth and eleventh samples, that is:

$$\lambda_b = \frac{150 + 187}{2} = 167 \text{ f/Mh}$$

hence:

$$\lambda_{XA} = 167 \times 0.5 \times 0.5 = 42 \text{ f/Mh} \qquad \text{(approximately 3 years MTBF)}$$

This estimated failure mode rate is lower again but consistent with the experience of pumping engineers and in line with the generally held conclusion that failure rate estimates from generic data sources tend to be pessimistic.

This is a relatively trivial example for an item of equipment on which many data are recorded in the open literature. An example of a more extreme case, where there is less generic data, is a gas turbine system used offshore for electricity generation of 25 MW output and continuously operated with intermittent periods on standby [see p. 350 of the 1992 *OREDA Handbook* (**3**)]. In this case the weightings for the design, operation, and environment attributes are estimated as follows.

Design attributes:	Size/power:	> 1000 KW = (+1)
	Speed:	< 5000 r/min = (0)
	Complexity:	complex = (+1)
Mean design weight:	(+1 + 0 + 1)/3 = 0.67	
Operational attributes:	Mode:	continuous/intermittent = (0)
	Medium:	average = (0)
	Press./temp.:	< 100 lbf/in^2/30 °C = (0)
Mean operation weight:	(0 + 0 + 0)/3 = 0	
Environmental attributes:	Plant:	offshore = (+1)
	Local:	sheltered = (0)
	Temp./humidity:	extreme = (+1)
Mean environment weight:	(+1 + 0 + 1)/3 = 0.67	
Combined weight:	+ 0.67 + 0 + 0.67 = 1.34	

The rotating machinery database in Fig. 12.15 lists ten samples for gas turbines in the range 35–4212 f/Mh. The median is approximately the arithmetic mean of the failure rates for samples 5 and 6, i.e. (320 + 491)/2 = 406 f/Mh. A best-estimate of the base failure rate for this application is, therefore

$\lambda_b = 406$ f/Mh

Again assuming a FWR proportion of 50 per cent:

$\lambda_{XA} = 406 \times 0.5 \times 2^{1.34} = 514$ f/Mh

This compares reasonably well with the mean FWR failure rate in the *OREDA Handbook* of 693 f/Mh (range 446– 951 f/Mh).

These examples suggest that the DOE model is reasonably robust and likely to be more representative for mechanical equipment than employing simple environment and duty *k* factors, since it is based on a structured evaluation of the attributes of the equipment design and the expected operational and environmental conditions.

12.7.2 Generic reliability database

A generic reliability database for many types of industrial equipment is included here in addition to the rotating machinery data used in the previous example. The complete list is:

Group 1 Rotating machinery (Fig. 12.15)
Group 2 Heat exchangers, vessels, and piping (Fig. 12.16)
Group 3 Valves and valve actuators (Fig. 12.17)
Group 4 Ancillary mechanical (Fig. 12.18)
Group 5 Ancillary electrical (Fig. 12.19)

A number of reliability data handbooks have also been published which feature mechanical items. The most notable examples are the *Non-Electronic Component Reliability Data Handbook* (**12**), covering mechanical and electrical components in military and commercial applications, and the IEEE Standard 500 (**13**), featuring mechanical and electrical equipment and instruments in nuclear power stations. The *OREDA Handbook* (**3**), mentioned previously, provides a range of detailed reliability and maintainability information for offshore mechanical, electrical, and instrument systems and equipment. Failure mode information, failure rates, and confidence limits are given in the *OREDA Handbook* and in certain other publications.

Class						Type		
Pumps	***Compressors***	***Turbines***	***Motors***	***Generators***	***Others***		***Failure rate failures/ Mhours***	***Failure rate range (f/year)***
	Compressor					System	5582	
		Turbine				Gas	4212	
		Turbine				Gas	3550	32
Pump						Export	2795	
	Compressor					Centrifugal	2694	
Pump						Dosing	2500	
	Compressor					Set	1810	16
	Compressor					Centrifugal	1700	
		Turbine				Gas	1667	
				Generator		Gas turbine	1542	
		Turbine				Gas	1484	
			Engine			Diesel	1278	
		Turbine				Gas	865	8
Pump						Air turbine	684	
	Compressor					Centrifugal	640	
Pump						Auxiliary feed	615	
		Turbine				Gas	525	
		Turbine				Gas	491	4
Pump						Fire/diesel	457	
Pump						Diaphragm	439	
Pump						Water	438	
	Compressor					Reciprocating	413	
		Turbine				Gas	320	
Pump						Cement	320	
	Compressor					Package	251	
Pump						Centrifugal	250	
		Turbine				Gas	240	
Pump						Turco	228	2
Pump						Boiler-feed	225	
Pump						Motor	208	
		Turbine				Gas	205	
Pump						Centrifugal	194	
Pump						Fire/electric	183	
				Generator		Diesel	177	
Pump						Reagent	171	
Pump						Seawater	150	1
	Compressor					Air	126	
Pump						Fire	120	
Pump						Reciprocating	95	
Pump						Centrifugal	88	
Pump						Mud	69.3	
					Fan	Ventilation	59	
					Blower	Axial	57.1	
					Fan	Centrifugal	57.1	
Pump						Diesel drive	57	
				Generator		Electric	55	
Pump						Motor driven	48.6	
Pump						Canned rotor	45.6	0.5
				Generator		Motor	45	
			Motor			AC	45	
				Generator		DC	37	
		Turbine				Gas	35	
Pump						Canned	34.2	
Pump						Glandless	34.2	
Pump						Centrifugal	32	
Pump						Ecc	30	
Pump						Lubrication	30	
				Generator		AC	30	
			Engine			I C	29	
		Turbine				Steam	29	0.25
			Motor			Starter	25	
Pump						Vane	22.8	
Pump						Centrifugal	22	
Pump						Vacuum	20	
Pump						Fuel	17	
					Exhauster		11.4	
Pump						Rotary	11.2	0.125
				Generator		All	10.1	
Pump						Centrifugal	8	
					Fan	Axial	5.7	
			Motor			Induction	5.7	
			Motor			DC	5	
			Motor			Stepper	3.4	
					Blower	All	3.13	
				Generator		AC	3	
					Fan	General	2.8	
			Motor			Electric	2	0.02
Pump						Jet	0.93	
					Blower	Circulator	0.018	

Fig. 12.15 Group 1, rotating machinery

Class						Type		
Boilers	***Vessels***	***Piping and pipeline***	***Heat exchangers***	***Heaters/ coolers***	***Others***		***Failure rate failures/ Mhours***	***Failure rate range (f/year)***
					Dryer	Chemical	2225	20
Boiler						Vessel	1940	
				Heater		Boiler	1800	
				Cooling		Tower	342	3
		Pipeline				Kill	310	
Boiler						Gas fired	264	
Boiler						Steam	217	2
Boiler						Steam	195	
	Vessel					Air receiver	172	
		Pipeline				Choke line	160	
	Vessel					Contactor	138	
					Dryer	Desiccant	111	1
	Vessel					Separator	97	
					Column	Stripping	91.3	
				Heater		Electric	73	
					Column	De-aerator	57.1	0.5
			H. ex.			Plate	39	
		Pipeline				Booster	38	
	Vessel					Condenser	34.2	
			H. ex.			Tubular	32.7	
					Condenser	Refrigeration	28.5	0.25
					Condenser	Vapour	26	
	Vessel					Scrubber	16	
				Calorifier		Storage	11.4	
			H.ex.			General	11.4	
					Evaporator	Refrigeration	11.4	
	Vessel					Reactor	3.3	
	Vessel					Dissolver	2.3	
	Vessel					Evaporator	2.3	
				Cooler		Air	2.3	
	Vessel					Dissolver	2.3	
					Evaporator	Vessel	2.3	0.02
	Vessel					Polisher	2.28	
	Vessel					Suppression	1.9	
				Cooler		General	0.67	
				Cooler		Coil	0.57	
				Heater		Coil	0.57	
	Vessel					Pressurizer	0.44	
		Pipeline				Submarine	0.26	
		Pipeline				Gas	0.14	
				Cooler		Jacket	0.11	
		Pipeline				Product	0.0802	
	Vessel					Pressure	0.08	
	Vessel					Cladding	0.046	
		Pipeline				per metre	0.017	
		Pipe				< 3" per metre	0.001	0.001
		Pipeline				Land	0.000 21	
		Pipe				> 3" per metre	0.0001	

Fig. 12.16 Group 2, heat exchangers, vessels, and piping

Class						Type		
Control valves	***Manual valves***	***Relief valves***	***Dampers***	***Valve actuators***	***Others***		***Failure rate failures/ Mhours***	***Failure rate range (f/year)***
	Valve					Choke	189	
Valve						Ball Hov	108	1
	Valve					Ball > 24 inch	83	
	Valve					Ball < 24 inch	80	
Valve						Flowstream	79.9	
Valve						XMV < 24 inch	59	
				Actuator		Mechanical	57.1	0.5
				Actuator		General	56.3	
Valve						Air operated	54.2	
Valve						Master	51.4	
	Valve					Choke/kill	45	
Valve						DHSV	35.1	
			Damper			Automatic	34.2	
Valve						ESD	30	
		Valve				Relief	30	
	Valve					Vacuum	30	
		Valve				Safety/relief	23	0.2
Valve						Control	22.8	
Valve						Injector	22.8	
					Valve	Deluge	21	
					Valve	Vent	18.3	
			Damper			Multi-vane	18	
					Valve	Slam shut	17	
				Actuator		Linear	14.8	
			Damper			Fire	13.7	
	Valve					Ball	12.3	0.1
	Valve					Cylinder	12	
			Damper			Manual	11.4	
					Valve	Sprinkler	11.4	
	Valve					Steam	11.4	
		Valve				PRV	11	
Valve						Deluge AOV	10.2	
Valve						Automatic	10	
Valve						Blowdown	10	
	Valve					Diaphragm	7.5	
					Manifold	Valve	6.8	
	Valve					Butterfly	6.1	0.05
	Valve					Bellows	5.7	
					Valve	Solenoid	5.7	
	Valve					Manual	5.5	
	Valve					Breathing	4.76	
	Valve					Instrument	4.56	
					Valve	Non-return	4.56	
				Actuator		Valve	3.42	
	Valve					Isolation/stop	3.4	
	Valve					Plug	3	
	Valve					Angle	2.7	
				Actuator		Electric	2.3	0.02
				Actuator		Pneumatic	2.3	
					Valve	Check	2	
	Valve					Gate	1.66	
					Valve	Shear	1.14	
					Ejector	Steam	0.22	0.002

Fig. 12.17 Group 3, valves and valve actuators

Class						Type		
Material processing	***Gears and drives***	***Cranes and hoists***	***Hydraulic and pneumatic***	***Fire fighting***	***Others***		***Failure rate failures/ Mhours***	***Failure rate range (f/year)***
		Crane				Boom	7900	
		Crane				35 ton	6279	
Dryer						Chemical	6225	
		Crane				Electric drive	6010	
		Crane				50 ton	4680	
		Crane				10 ton EOT	3500	30
Mill						Grinding	2950	
		Crane				60 ton	1598	
Conveyor						Bucket	1320	
	Engine					Diesel	1278	
Humidrier						Electric	1256	10
		Crane				12 ton	913	
Compensator						Heave	560	
Conveyor						Screw	526	
Conveyor						Band/belt	375	
Centrifuge						Oil	205	
Dryer						Desiccant	111	1
Regulator						Charge	98	
Bellows						General	65	
	Gearbox						65	
Filter						Fluid	64	
	Actuator					Mechanical	57.1	
	Actuator					General	56.3	0.5
Dryer						All	52	
			Bursting disc			Metal	46.7	
Regulator						Feed	45.6	
				System		Fire fighting	45	
	Belt					Drive	40	
				System		Firewater	36	
	Engine					I C	29	
Filter						Inlet air	28	
		Hoist/crane				Hook	23	0.2
Stirrer						Unit	22.8	
				Valve		Deluge	21	
			Relay			Pneumatic	20	
	Transmission					Gearbox	16	
	Actuator					Linear	14.8	
			Fittings			Hose	13	
				Valve		Sprinkler	11.4	0.1
				Valve		Deluge AOV	10.2	
	Coupling					Drive	9.8	
	Transmission					Gear	8.6	
			Coupling			Mechanical	5.3	
			Coupling			Hose	4.6	
Filter						Air	4.6	0.01
Bellows						Expansion	4	
Filter						Mechanical	4	
	Actuator					Pneumatic	2.3	
Filter						HEPA	1.7	
Strainer						Sieve	1.6	
			Connector			Pneumatic	1.5	
Filter						Gas	1.2	
Recombiner							1.1	
	Coupling					Flexible	0.96	
			Accumulator			Pressurized	0.53	
			Fittings			Hydraulic	0.24	
Ejector						Steam	0.22	
	Transmission					Gear	0.2	
Drum						Filler	0.14	
			Accumulator			Hydraulic	0.1	
Filter						Wire	0.1	

Fig. 12.18 Group 4, ancillary mechanical

Class						Type		
Transformers and converters	***Generators and inverters***	***Switchgear and dis.boards***	***Batteries and battery chargers***	***Controllers and annunciators***	***Others***		***Failure rate failures/ Mhours***	***Failure rate range (f/year)***
				Computer		Mainframe	5500	
	Generator					Gas turbine	1542	
				Controller		Microprocessor	265	
	Generator					Diesel	177	
					Line	Communication	100	1
					Starter	Motor	96	
	Generator					Electric	55	
	Generator					Motor	45	
					Timer		42.5	
	Generator					Direct current	37	
	Generator					AC	30	
			Charger			Battery	26	
	Inverter					Electric	26	
			Charger			Battery	26	
					Switch	Diaphragm	20.4	
			Battery			System	17	
Converter						Electronic	10	
					Line	Communication	10	
			Battery			Cell	10	
		Breaker				Circuit-HV	8.3	0.1
		Breaker				Circuit-fixed	8	
Transformer						Power	8	
					Relay	Protective	6.84	
		Relay				Switchgear	6.84	0.05
					Switch	Micro	4.56	
					Rectifier	Electric	4.36	
					Rectifier	Power	4	
	Inverter					DC/AC	3.7	
					Rectifier	AC/DC	3.65	
Converter						Press/current	3	
					Relay	Electrical	2.9	
Transformer						Electric	2.9	
					Line	Communication	2.4	
			Battery			Fuel-cell	2	
Transformer						415V	2	
					Starter	Electrical	1.75	
				Computer		Module	1.3	0.01
		Breaker				Circuit	1.03	
Transformer						Control	1	
Converter						MV-current	0.74	
					Switch	Disconnect	0.7	
				Annunciator		Alarm	0.6	
		Bus				Switchboard	0.57	
Transformer						Oil filled	0.47	
				Annunciator		Module	0.4	
					Relay	Overload	0.4	
		Relay				Power	0.3	
		Board				Distribution	0.11	
		Bus				Switchgear	0.07	
		Busbar				11kV	0.06	
		Board				Printed circuit	0.05	
		Board				Printed wiring	0.03	
					Connector	Electrical	0.01	

Fig. 12.19 Group 5, ancillary electrical

Laboratory testing of mechanical equipment to obtain reliability data is seldom economic, so that the majority of in-house data will be generated from field experience. Most organizations collect data that can be analysed to provide reliability information. These data exist in maintenance and test records, operational logbooks, and other technical information systems. Particularly in the present climate, where a wide range of computer systems are available for maintenance planning, the basis for comprehensive in-house reliability databases exists in most companies.

The problem with in-house data, however, is that population sizes for important items are frequently quite small so that reasonable confidence in the data can only be obtained after relatively long periods of operation. Generic reliability data from one or more of the commercially available data banks can be useful to supplement or reinforce the failure rate estimates obtained from in-house sources.

Maintainability data are generally considered to come within the orbit of reliability data banks but until recently have received little attention. With the increased interest in availability of process plant it is clearly an area needing attention. In a study of the Brent Gas Disposal System Performance **(14)**, for example, it was found necessary to define four specific modes of failure – each associated with a significantly different MTTR – in order to generate a realistic representation of system availability.

Information on hazardous incidents has also been shown to be necessary to support risk assessment studies. The tendency now is therefore towards the extension of the reliability data bank concept to a comprehensive information system which can be used to ensure consistency in data used in reliability and risk assessments.

12.8 Summary

The collection and analysis of reliability data requires a systematic approach with clear definitions of the reliability parameters and comprehensive collection and analysis procedures.

The analysis of plant failure data is an important part of the maintenance function and will help to identify trends in reliability performance and the systems and components causing problems to plant availability.

Generic data are required by reliability engineers to support the reliability and risk assessment of new designs. This generally requires the combination of in-house data with reliability information from the public domain.

A quality assurance plan is a necessary part of any reliability study. It should be written to ensure that the sources of information and the assumptions made in the data analysis are traceable and fully documented.

12.9 References

(1) Reference classification concerning component reliability. Euredata Report No. 1, CEC Joint Research Centre, Ispra, Italy.

(2) Mechanical valve reliability. Euredata Report No. WP/15EI/5ER/SER/2088/91, CEC Joint Research Centre, Ispra, Italy.

(3) *OREDA 92 Offshore Reliability Data*, 2nd edition. Distributed by DNC Technica.

(4) UNIPEDE guidelines on dependability management for the power industries, Part 2 IEC300-1:1993.

(5) **Bloch, R. P.** and **Geitner, F. K.** (1999) Use equipment failure statistics properly. *Hydrocarbon Processing*, January.

(6) **Newton, D.** In (Ed. J. Davidson and C. Hunsley), *The Reliability of Mechanical Systems*, 2nd edition, pp. 20–42. MEP.

(7) **Green, A. E.** and **Bourne, A. J.** (1972) *Reliability Technology*, Wiley Interscience.

(8) **Andrews, J. D.** and **Moss, T. R.** (1993*) Reliability and Risk Assessment*, 1st edition. Longman.

(9) **Davidson, J.** and **Hunsley, C.** (1998) *The Reliability of Mechanical Systems*. MEP.

(10) Some published and estimated failure rates for use in fault tree analysis (1981). Du Pont.

(11) **Andrews, J. D.** and **Moss, T. R.** (1996) Factors influencing rotating machinery reliability. In 10th ESReDA Seminar on *Rotating Machinery Performance*, Chamonix, Mont-Blanc, April.

(12) *Non-Electronic Component Reliability Data Handbook,* NPRD-91 (1991) Reliability Analysis Center, RADC, New York.

(13) IEEE 500 (1984) *Reliability data for Nuclear Power Generating Stations*. IEEE, New York.

(14) **Spanninga, A.** and **Westwell, F.** (1981) Brent gas disposal system performance forecasts by Monte Carlo techniques. In 3rd National Reliability Conference, IQA, London.

Chapter 13

Risk Assessment

13.1 Introduction

Risk is described in the Oxford English Dictionary as 'a chance or possibility of danger, loss, injury, or other adverse consequences'. Traditionally risk assessment has its roots in banking and insurance and has been applied effectively to corporate finance, re-insurance, and a wide variety of other areas involving financial risk. More recently risk assessment employing probabilistic methods has been applied particularly for the control of major accident hazards.

Probabilistic risk assessment concepts are now applied more widely to industrial plant. The focus has broadened to include optimization of planned maintenance, inspections, and renewals of equipment with the aim of increasing the availability and productivity of the plant without compromising safety. This chapter therefore considers risk-based inspection and maintenance optimization in addition to risk assessment for safety management.

13.2 Background

Since the early 1950s the size and complexity of industrial plant has increased significantly to reap the financial benefits of scale. Prior to 1950 a 100 tonne/day ammonia plant was considered large; now the norm is in excess of 1000 tonne/day. These developments, however, have also shown a marked increase in the frequency and severity of major industrial accidents. Many examples of disastrous industrial accidents can be cited, such as: the cyclo-hexane explosion at

Flixborough in 1974 which killed 28 and injured 89 people; the release of methyl-isocyanate at Bhopal in 1984 which killed more than 2000; and the liquid petroleum gas (LPG) explosion and fire in Mexico City soon after which resulted in over 600 deaths and several thousand injuries.

Public concern at multiple fatalities, injuries, and pollution from major accidents such as large explosions, fires, and radioactive or toxic releases invariably result in calls for increased controls at national and international levels. Governments and international bodies have reacted to this concern by the introduction of more stringent legislation, sometimes with significant impact on the industry. For example, the Flixborough enquiry triggered the extension of major accident hazard legislation in the UK from nuclear to industrial installations. In the nuclear and offshore industries one can also cite the Three-Mile Island and Piper Alpha incidents. The former resulted in a fundamental shift in public acceptability of nuclear installations which virtually decimated the industry in the USA and elsewhere although there were no known fatalities. The Piper Alpha blow-out and fire killed over 100 people and resulted in much stricter regulation of offshore safety, including the requirement for a risk assessment to support the safety case.

Clearly projects involving the processing and storage of potentially hazardous materials need to address both on-site and off-site safety. The rapid pace of progress in modern technology, however, allows less opportunity for learning from trial and error, making it increasingly necessary to get design and operating procedures right first time; hence the focus has turned to the use of risk assessment for major accident hazards.

Probabilistic risk assessment first surfaced in the UK in major accident safety studies in the 1950s. The Fleck Committee report of the Windscale incident in 1957 is a particular milestone. Lord Fleck recommended, inter alia, that alternative approaches should be developed for the safety assurance of nuclear installations by an independent branch of the UKAEA. F. R. Farmer, the first director of the UKAEA Health and Safety Branch (now SRD – the UKAEA's Safety and Reliability Directorate) was a pioneer in the use of probabilistic methods for the assessment and control of safety-related risks. The basis of Farmer's proposed risk acceptance criterion was that no plant or structure could be considered entirely risk-free and that so-called 'incredible' accidents were often made up of combinations of very ordinary events (**1**). Hence, as the consequences of a potential hazard increased, so the occurrence probability should decrease. Figure 13.1 shows the original Farmer *F–N* curve, where probabilities above the line reflect high-risk situations and those below the line 'acceptable' or low-risk situations.

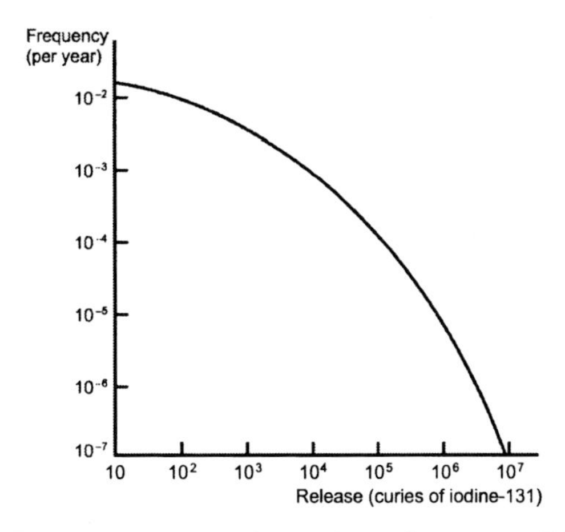

Fig. 13.1 Farmer curve: accident release frequency limit line

The Farmer approach was subsequently applied in the USA. For example, in 1975 a comprehensive assessment of accidental risks in a US commercial nuclear power plant was published (**2**). This probabilistic safety study also contributed substantially to the use of quantitative risk assessment for large-scale hazardous installations.

Probabilistic risk assessment (PRA), usually called quantitative risk assessment (QRA) in the UK, is now a recommended input to safety cases required by regulatory authorities, for example in the Control Of Major Accident Hazards (COMAH) Regulations 1999 (**3**) which implements the Seveso II European Directive. The need for risk assessment has become more apparent over the years and it is now a requirement in other areas, for instance in railway safety.

13.3 Major accident hazards

The term 'Major accident hazard' has been defined as:

> 'An occurrence such as a major emission, fire or explosion resulting from uncontrolled development, in the course of an industrial activity, leading to a serious danger to man,

immediate or delayed, inside or outside the establishment and to the environment from one or more dangerous substances.'

An industrial activity is seen as the aggregate of all installations within a distance of 500 m of each other and belonging to the same factory/plant. When the quantity of hazardous substances exceeds given threshold limits the activity is referred to as a major hazard installation. The list of substances tabled in the Seveso directive consists of 180 chemicals with threshold limits varying from 1 kg for extremely toxic substances to tonne quantities for flammable liquids. For isolated storage a separate list of a few substances is given. In addition to flammable gases, liquids, and explosives the list contains chemicals such as chlorine and ammonia.

Due to the diversity and complexity of industry in general it is not possible to define major hazard installations within certain sectors of industrial activity. Experience has shown, however, that industrial major hazard installations are most common in the following areas:

(a) Offshore and onshore oil and gas installations and associated plant;
(b) Petrochemical plant and refineries;
(c) Chemical works and production units;
(d) LPG/LNG storage and terminals;
(e) Stores and distribution centres for chemicals;
(f) Large fertilizer storage;
(g) Works where chlorine is used in bulk.

The hazard potential obviously varies with the characteristics of the material properties and the quantities involved.

Industrial hazards are generally associated with the potential for fire, explosion, or dispersion of toxic chemicals. Assessments therefore generally focus on the consequences of loss of containment. Flammable material releases have the potential for very significant casualties and property damage on- and off-site. Nevertheless, even for severe accidents the effect is generally limited to a few hundred metres from the site. The sudden release of toxic or radioactive material, however, has the potential to cause deaths and severe injuries to people at considerable distances from the site. The actual number of casualties will depend on the population density in the path of the toxic/radioactive cloud, the weather conditions, and other factors such as the emergency arrangements put in hand subsequent to the release. The release of flammable or toxic materials to the atmosphere may therefore lead to an explosion, a fire, or the formation of a toxic cloud.

13.3.1 Explosions

Explosions are characterized by a shock wave which can cause damage to buildings and eject missiles over distances of several hundred metres. The injuries and damage are initially caused by the shock wave. However, although the effects of overpressure can result in deaths, in most cases it will probably only involve those working in the direct vicinity of the explosion. Experience shows that indirect effects, such as collapsing buildings, flying debris, and the fire which frequently follows, cause far more loss of life and severe injuries.

The effects of the shock wave will vary depending on the characteristics of the material, the quantities involved, and the degree of confinement. Direct injury to people occurs at pressure of 5–10 kPa with loss of life occurring at greater overpressures. The pressure of the shock wave decreases rapidly with distance from the source. As an example, the explosion of a tank containing 50 tonnes of propane results in pressures of 14 kPa at 250 m and 5 kPa at 500 m from the tank.

Explosions can occur either in the form of a deflagration or a detonation depending on the burning velocity during the explosion. Deflagration occurs when the burning velocity is relatively low. In a detonation, the flame front travels as a shock wave with a typical velocity of 2000–3000 m/s. A detonation generates greater pressures and is far more destructive than a deflagration. Whether a deflagration or a detonation takes place depends on the material involved as well as the conditions under which the explosion occurs. It is generally accepted that a vapour-phase explosion requires some degree of confinement for a detonation to take place.

13.3.2 Gas and dust explosions

Generally, catastrophic gas explosions occur when considerable quantities of flammable material are released and mixed with air to form an explosive vapour cloud before ignition takes place. Dust explosions occur when flammable materials are intensively mixed with air. The dispersed solid material is in the form of powder with very small particle sizes. The explosion occurs following an initiating event such as a fire or small explosion that causes powder to become airborne. Due to the fact that grain, milk powder, and flour are flammable, dust explosions have been most common in the agricultural industry with the damaging effects generally confined to the workplace rather than outside the plant.

13.3.3 Confined and unconfined vapour cloud explosions

Confined explosions are those which occur within some form of containment such as a vessel or pipeline. Explosions in buildings also come within this category. Explosions which occur in the open air are referred to as unconfined explosions and produce peak overpressures of only a few kiloPascals. The peak pressures of confined explosions are much higher, of the order of several hundreds of kiloPascals. Probably the most well-known example of a vapour cloud explosion in the UK is Flixborough.

13.3.4 Fires

The effect of fire on people takes the form of skin burns due to exposure to thermal radiation. The severity of the burns depends on the intensity of the heat and the exposure time. Heat radiation is inversely proportional to the square of the distance from the source. The skin can withstand a heat energy of 30 KW/m^2 for only 0.4 s before pain is felt. Fires in industry occur more frequently than explosions and toxic releases although the consequences in terms of loss of life are generally less severe. However, if the ignition of escaping flammable material is delayed an unconfined vapour cloud may be formed.

Fires can take several different forms, including jet fires, pool fires, flash fires, and boiling liquid expanding vapour explosions (BLEVEs). A jet fire appears as a long narrow flame produced, for example, from an ignited gas pipeline leak. A pool fire might be produced, say, if diesel oil leaked from a storage tank on to the floor of an offshore, engine-driven, standby-generator module. A flash fire could occur if an escape of gas reached a source of ignition and rapidly burnt back to the source of the release. The BLEVE is generally far more serious than other fires.

The BLEVE, often referred to as a fireball, is a combination of fire and explosion with an intense heat emission within a relatively short time. If a pressure vessel, in which a liquefied gas is kept above its atmospheric boiling point, fails catastrophically the contents are instantaneously released as a turbulent mixture of liquid and gas expanding rapidly and dispersing in air as a cloud. When this cloud is ignited a fireball results, causing an enormous heat radiation intensity within a few seconds. This heat intensity is sufficient to cause severe skin burns and deaths at several hundred metres from the vessel depending on the quantity of the gas involved. A BLEVE involving a

50 tonne propane tank car, for example, can cause third-degree burns at approximately 200 m and blisters at 400 m. The largest recorded BLEVE was the LPG accident in Mexico City in 1985 which killed 650 people and injured over 2500.

It is sometimes difficult to make a distinction between a fire and an explosion. Frequently an explosion is followed by a fire which may, in fact, be more destructive in terms of casualties and property damage.

13.3.5 Toxic releases

Many chemicals can cause harm to workers even in small quantities if exposure occurs over long periods. The effects of toxic chemicals when considering major hazards, however, are quite different since they are concerned with acute exposure during and soon after a major accident. Thus the storage and use of toxic chemicals in large quantities – which, if released, could disperse with the wind – pose major hazards since they have the potential to kill and injure people living many hundreds of metres away from the plant. Although toxicity data are not particularly abundant the toxicity of certain chemicals is well known. Chlorine, for example, is known to be dangerous to human health at concentrations of 10–20 parts per million (p.p.m.) for 30-min exposure. Chlorine and ammonia are the toxic chemicals most commonly used in quantities that constitute a major hazard and both have a history of major accidents. Nevertheless, other chemicals such as methyl-isocyanate and dioxin must be used with particular care in view of their higher toxicity. For this reason a number of these very toxic chemicals have been included in the Seveso Directive. Radioactive releases pose special problems which are considered to be outside the scope of this chapter.

13.4 Major accident hazard risk assessments

The hazards associated with industrial plant can involve significant risks to the safety of people, the environment, and the plant itself. These risks can be eliminated or significantly reduced by spending money, for example, by including scrubbers at coal-fired power stations and adding additional levels of containment or protection to flammable or toxic chemical process plants; however, it is not possible to avoid risks entirely. In the end society has to judge how much money it is prepared to spend given the law of diminishing returns. The difficulty often faced by safety assessors lies in convincing regulators and decision makers that at some point a process system is ‘safe enough’. The decision process is frequently helped by quantifying the risks.

For potentially hazardous industrial systems, risk or ‘expected loss’ can be defined quantitatively in one of two ways. That is, either as the probability of a specified ‘unwanted event’, for example, a chlorine release greater than 25 tonnes or, alternatively, as the product of the consequences of a specified outcome (range of outcomes) and the probability of its (their) occurrence:

1. $R = P(E > X)$ probability of an event greater than specified threshold
2. $R = C \times P$ consequence × probability of occurrence

The former is most frequently used for a specified safety hazard, say, a ‘loss of coolant accident’ (LOCA) in a nuclear power station, and the latter for a wide range of industrial hazards including the risk of financial loss. In either case, quantified risk assessment (QRA) involves four basic stages:

(a) identification of the potential hazards,
(b) estimation of the consequences of each hazard,
(c) estimation of the probability of occurrence of each hazard,
(d) evaluation of the quantified risk and comparison with acceptability criteria.

13.4.1 Hazard identification

Hazard identification is a requirement of regulatory safety cases. The hazards associated with a particular plant need to be identified to determine their potential for harm to personnel within the plant, people living or working in close proximity to the plant, and to the general public and environment. A range of qualitative methods are employed for hazard identification, including checklists, rapid-ranking, preliminary hazard analysis, and hazard and operability studies (HAZOP). These techniques have been described previously in Chapter 3. The choice of hazard identification method will depend on a variety of factors including perception of the severity of the consequence and perceived public concern. All of the methods are geared (in different degrees) to the identification of potential accidents and their causes. HAZOP is the most rigorous and most frequently used method for industrial plant when major hazards are concerned.

A list of failure cases is generated from the hazard identification phase. Similar events can often be grouped together for analysis of consequences. The list can usually be derived reliably by considering the form in which the dangerous substance is stored or processed, the nature of the potential hazard, and the size of the different equipment containing the material. This is particularly so where loss of containment is the principal concern and the plant is relatively simple (say, a storage vessel with associated pipework and valves). In such cases the range of possible outcomes is limited.

The different paths to loss of containment will include leaks and rupture of equipment but should also include other possible failures such as inadvertent venting via the relief system. A typical list of failure cases for a liquefied gas storage installation could be:

(a) releases from pipework containing liquid for the different pipe diameters on the plant,
(b) gas releases from pipework,
(c) releases from vents and relief valves,
(d) major releases from vessel or pipeline rupture.

In the subsequent consequence analyses the cause of failure is only of concern if it affects the 'source term', i.e. the quantity released, or rate of release, or any other factor that could affect the consequences of the release, for example, whether the release is in the liquid or vapour phase. The cause of the release might also result in the fluid being at a higher or lower temperature or pressure than normal, for example, the possibility of material 'slumping' (i.e. the formation of a pool of refrigerated boiling liquid on the ground or water) due to the Joule–Thomson effect.

The list of basic failure cases will need to be more complex if there is the potential for accidents which have different characteristics. For example, the list of cases for an installation containing liquefied flammable gas should take account of the different possible modes of combustion discussed previously. Such a list would probably include the following:

(a) vapour clouds formed due to vessel pipework failure resulting in vapour cloud explosions or fires,
(b) jet fires from pipework or connection failure, relief valves, or vents,
(c) vessel failure leading to a BLEVE with a fireball and ejection of projectiles.

There may also be instances where loss of containment is not the only consideration. Other events may follow closely, for instance, ejected projectiles or shock waves may also have significant accident potential. The objective of listing failure cases is to ensure that, as far as possible, all the various discrete types of accident have been identified. The different accident scenarios can then be prioritized for consequence analysis in terms of the likely severity of each accident, the different characteristics of the hazards, and the systems and equipment involved. Checklists are particularly useful in drawing up the list of failure cases. The list can also be used constructively when deciding which failure cases might be grouped together for consequence analysis and in assessing how relatively trivial events in combination might escalate into a major accident event. This can be particularly important when considering the relatively few high-consequence/low-frequency events.

13.4.2 Consequence analysis

The list of failure cases from the hazard identification phase will define a set of unwanted events with the potential to cause fatalities, environmental pollution, or major structural damage. The build-up of each accident scenario from the initiating cause to the accident event need to be developed with estimates of the energy potential, the release mechanisms, and the timescales over which the energy may be released. For loss of containment events it is usually necessary to make some simplifying assumptions to estimate the quantity released and the release rate. In some cases it may be reasonable to assume that the whole of a vessel's inventory is released immediately; in others, such as failure of a pipeline, the release may take place over a significant period of time until the pipeline is emptied or the failure isolated. If, as is sometimes the case, the release calculation serves as an input to dispersion calculations, the detailed analysis of release rates will seldom be worthwhile since dispersion models can generally only accommodate instantaneous or continuous constant rates of release. Detailed study of how the release rate changes with time is, therefore, usually of little value if the release is so rapid that it can be treated as effectively instantaneous, otherwise a few representative points from the release profile may be selected for study with a constant rate model.

The precise failure point in a system will be a matter for speculation as will be the severity of the failure. Important parameters such as the internal pressure or liquid level will have a range of possible values. It is, therefore, valuable to consider the sensitivity of the calculations to these uncertainties and to adopt a number of representative cases which will neither seriously under- nor overestimate the virtually infinite range of possibilities. At every stage it is advisable to consider the degree of detail required to provide the necessary input to the following stages in the consequence calculations.

Important parameters for determining the consequences of accidents which may lead to explosions and fires:

1. *Flammable substances*: The quantity (rate and duration) directly determines the magnitude of the event, for example, the length of the jet fire or the peak overpressure of the explosion. It is usual to consider three leak sizes: 18, 50, and 100 mm. A maximum credible release size for process equipment is generally assumed to be 100 mm.
2. *Time to ignition*: A release that ignites at the time of rupture will give a jet fire whereas delayed ignition may result in an explosion. The size of the explosion will primarily depend on the quantity of the material that accumulates in the explosive range; this also determines the time to ignition. If the gas disperses, only toxic or secondary effects need to be considered.
3. *Mitigating factors*: The consequences of an accident event can be mitigated by the activation of protective systems.

The first level of mitigation is isolation following detection of a release. In the event of ignition the activation of the deluge system may prevent domino effects and the escalation of the accident. Passive protection provided by fire-walls or the fireproofing of structural members, piping, and pipe supports should also inhibit escalation of an accident event.

An assessment of the consequences of each accident event is achieved by analysing the full spectrum of all possible developments. For flammable substances, after determining the magnitude of the release and its duration the next step is to identify the form of the conflagration which may result, namely:

(a) jet fire
(b) flash fire
(c) pool fire
(d) unconfined explosion
(e) confined explosion.

Event trees are frequently employed to map the developments from the initiating event to the set of all possible outcomes. Figure 13.2 shows the event tree arising from 'gas release from an offshore pipeline' into the wellhead module of an offshore gas platform.

The paths from the initiating event 'gas release' are traced to the 21 possible outcomes depending on whether ignition is immediate or delayed, whether the detection systems (process, gas, or fire) operate to alert operators and trigger isolation of the source, and operation of the deluge system. Immediate ignition results in 13 possible jet fire scenarios, delayed ignition to seven explosion scenarios and one scenario when there is no ignition but with the possibility of casualties by inhalation and possible secondary effects such as ingestion of the released gas by an operating gas turbine. The consequences of each of the 21 outcomes need to be determined in terms of the effect of fire, explosion, and other possible secondary effects.

Gas and liquid jet fires

Two types of jet fire are considered viable. The first type arises from releases of flammable gas from process pipework and equipment, the second from releases of flammable liquids such as LNG/LPG which flash into vapour when released to the atmosphere. The analysis involves:

1. Calculating the size of the jet fire.
2. Calculating the distance to thermal radiation loads of 4 and 37.5 kW/m^2. The lower value is considered to be large enough to deter personnel from escaping through that level of radiation, the higher value capable of causing serious damage to process equipment and pipework if suitable protection is unavailable.

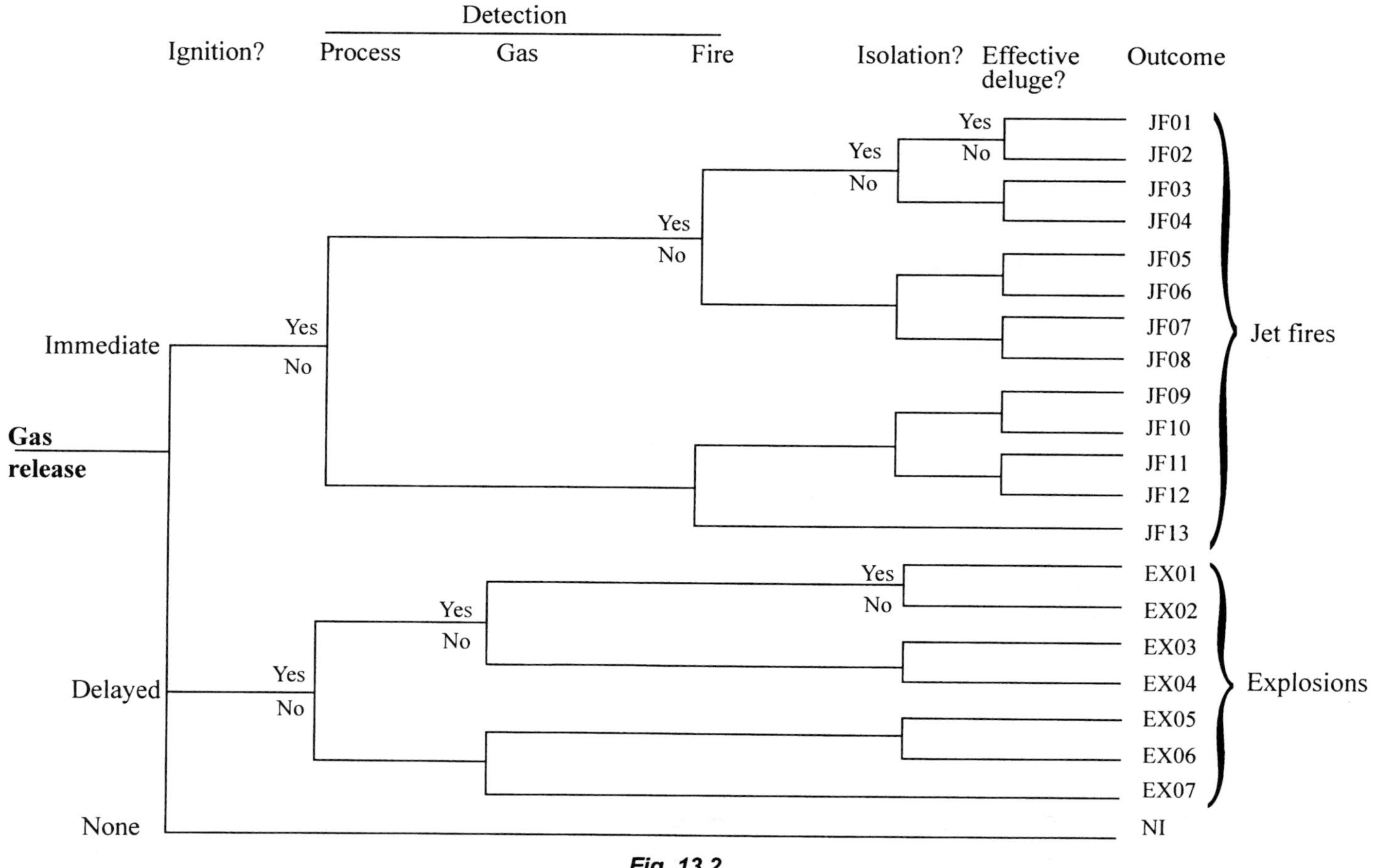
Detection
Ignition?
Process
Gas
Fire
Isolation?
Effective deluge?
Outcome
Yes
No
Gas release
Immediate
Delayed
None
JF01
JF02
JF03
JF04
JF05
JF06
JF07
JF08
JF09
JF10
JF11
JF12
JF13
EX01
EX02
EX03
EX04
EX05
EX06
EX07
NI
Jet fires
Explosions

Fig. 13.2

Liquid pool fires

Liquid pool fires result from releases of liquid hydrocarbons; this can be within an enclosed area, to the ground, or on water. In most cases the release quantities in enclosed areas are small and the consequences of an ignition are generally assumed to be contained within the enclosure. Releases from storage tanks on the other hand can be serious due to the quantity of material contained. This type of pool fire is analysed to:

(a) assess the size and characteristics of the pool fire
(b) critically appraise the consequences of the fire.

The vaporization model should take account of the various heat sources, such as heating from the ground or from water below the pool, whether the pool is evaporating or boiling, and also the possible increase in area as the pool spreads. The pool spreading rate generally depends on the depth of the pool. The viscosity of the liquid may also be important. Vaporization rates are a function of the pool area and the material properties; therefore, solution of the equations describing the simultaneous processes of spreading and vaporization requires significant computation, hence computer programs have been written to estimate this. If the spreading of the pool is contained by a bund then the calculations are easier and estimation of the peak evaporation rate can be carried out manually.

The modelling of pool fires and the effect of fire engulfment on process equipment and structural components can be involved and is generally achieved using proprietary software.

Confined and unconfined explosions

Explosions may arise where there is delayed ignition of a gas release. The analysis involves:

1. Calculating the expected quantities of material released within the explosive range for a number of different time intervals after initiation.
2. Calculating the overpressures generated by the different quantities of material in the explosive range.
3. Critically appraising the calculated overpressures to determine their impact on the immediate vicinity and on the installation as a whole.

There are a number of factors to be considered, such as the presence of obstacles which can increase flame speed and thus the overpressure. Materials such as propane can form refrigerated pools on land or water, due to the Joule–Thomson effect, which could escalate into an unconfined explosion or BLEVE. The area of venting also has an obvious effect on the explosion overpressure and will depend on the location at which the explosion occurs.

For all organic materials it is necessary to consider the different modes of combustion if the material is ignited at any stage below the limit of flammability. For example, LNG releases can burn in a number of different ways. A fireball (BLEVE) could be the result if a rapid release from pressurized storage is ignited immediately before significant pre-mixing with air. If a large vapour cloud is ignited some time after its formation a flash fire could result and the burning cloud could then burn back towards the source of release and eventually become a pool fire. The behaviour in vapour cloud explosions is complex and a number of sophisticated programs have been developed to model the shock wave. Most of the empirical models available use a TNT (trinitrotoluene) equivalent approach which is moderately satisfactory for the so-called 'far field' but less accurate for the 'near field'. Lees (**4**) gives examples of the calculation of TNT equivalents for detonations and chemical reactions. Baker *et al.* (**5**) give methods for estimating the intensity of the shock wave from bursting pressure vessels and other types of explosion. Failure of liquefied gas pressure vessels can also eject vessel fragments weighing several tonnes for considerable distances. This is a notable feature of BLEVEs although the principal threat to people is from the fireball.

Explosive failure of equipment in confined areas can arise, such as in gas turbine generators, emergency power equipment, or engine-driven fire pump modules. The quantities of material are generally small and, since the enclosures are usually fire and explosion overpressure rated, such accidents are unlikely to affect the external environment. Hydrogen explosions in battery rooms are potentially quite severe and could cause damage to adjacent modules. However, forced draft ventilation is generally provided so that the chance of such a hydrogen explosion causing a major accident is small.

The direct threat to people from explosion shock waves is limited. Casualties are mainly caused by missiles, being thrown against walls or other structural components, or by falling or flying debris. Structural integrity is clearly important and account must be taken of the effect of the peak overpressure of any potential shock wave.

However, it can generally be assumed that people caught within an exploding vapour cloud will be at greatest risk from the thermal or shock effects and that outside the cloud the effects can be related to the peak overpressure.

Toxic hazards

For toxic substances it is necessary to specify critical concentrations or concentration–time relationships to derive hazard ranges from a dispersion model. Defining these relationships can, however, be a particularly difficult area as data from human exposure in accident situations is extremely limited. Rather than assuming an LD_{50} criterion (the lethal dose that would kill 50 per cent of the exposed population) a threshold of exposure for fatalities is probably a more appropriate criterion for meeting the requirements of emergency planning and public information.

Calculating the dispersion characteristics of the gas cloud is complex because the effects of wind, rain, and other weather conditions need to be included as well as the characteristics of the installation and the material properties of the gas cloud. A number of computer programs have been written, many of which produce maps of the area surrounding the installation, with isobars for different concentrations of the toxic chemical.

13.4.3 Estimating event probabilities

For each of the postulated hazardous events considered in the consequence analysis it is then necessary to estimate the probability of occurrence. The process involves tracing the development of each potential accident scenario from the initiating event to the final outcome and combining the probabilities of all contributing events. Certain events, such as the possibility of an aircraft crashing on to a hazardous installation and collisions between ships and offshore structures, are relatively simple to model; for example, statistics regarding routes and frequency of flights will be available from sources such as the Civil Aviation Authority. Although these data may need modification to reflect the geographical location, plant layout, etc. it is generally a simple matter to estimate the probability of occurrence. In other cases, such as a fire following a major hydrocarbon leak where protection is installed to mitigate the impact of the accident, the progression will be more complex and may require modelling with techniques such as the event trees and fault trees discussed in Chapter 7.

Event trees are frequently employed for estimating event probabilities as well as for identifying the different possible outcomes for consequence analysis. For more complex scenarios fault trees are the usual method used in risk assessments for relating the various combinations of primary and secondary events which may lead to an accident. The methodology is well developed and computer programs are commercially available for calculating the probability of the top, unwanted event, and for ranking the primary events by importance. A simple example of a fault tree modelling the event combinations which could lead to a hydrocarbon gas release in the wellhead area of an offshore drilling platform is shown in Fig. 13.3.

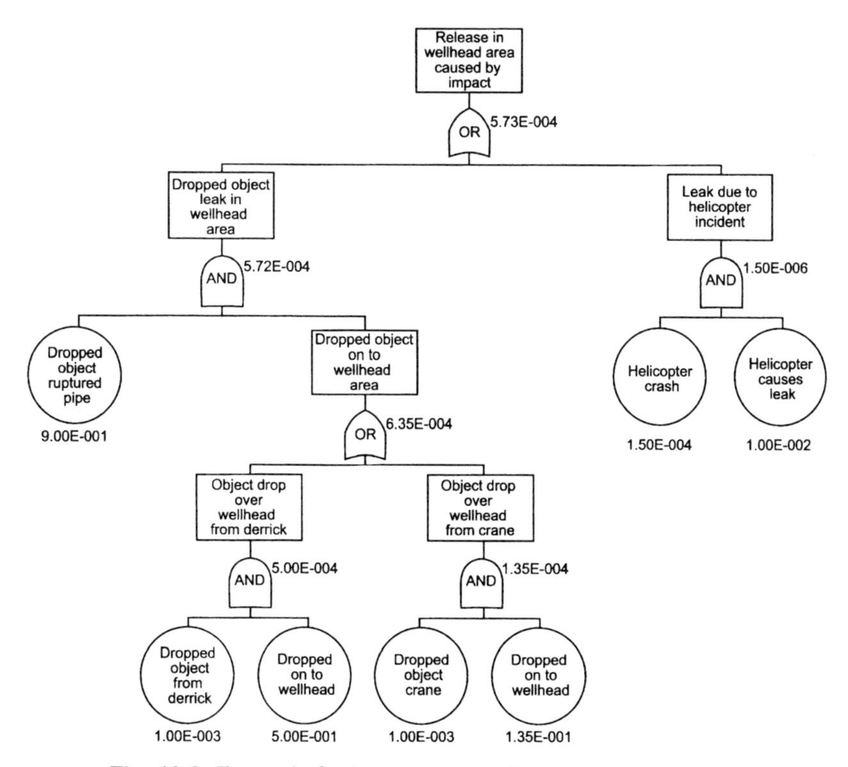

Fig. 13.3 Example fault tree – gas release in wellhead area

The AND gates require all input events feeding the gate to be present to generate an output event. An OR gate only requires one of the input events to be present to generate the output. Circles on the fault tree indicate primary events and rectangles combination events. The figures under the circles are the input probabilities (derived from statistics of similar failure events) which are entered into the fault tree program; the figures next to the gates show the probabilities of the combination events.

Cause–consequence diagrams incorporate features of the fault tree and event tree. They are constructed by defining a critical event and then developing consequence events and logical relationships which lead backwards to the initiating cause events. An example of a simple cause–consequence diagram is shown in Fig. 13.4.

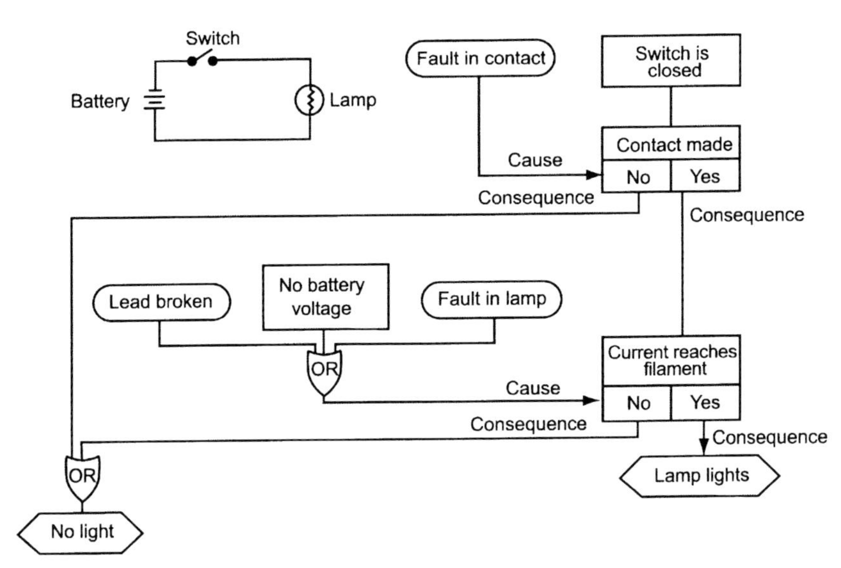

Fig. 13.4 A simple example showing the structure of a cause–consequence diagram

One advantage of the cause–consequence diagram is the scope for including time delays in the analysis.

13.4.4 Risk evaluation

As noted previously, risk is defined as the probability of a specified unwanted event or alternatively as the product of failure probability and consequence. The first definition is probably more applicable to safety, the latter to financial risk optimization. Deciding on a 'tolerable' level of risk has caused significant debate as evidenced by the 'Tolerability of risk from nuclear power stations' paper (**6**) which resulted from recommendations from the Sizewell B Public Enquiry (**7**). The Sizewell B report also asked the UK Health and Safety Executive to publish a paper on the relationship of safety standards in the nuclear industry with those in other industries; this was subsequently published in 1989 (**8**).

Summarizing some of the important conclusions from these papers it is notable that plant reliability and risk of plant failure need to be calculated in some way so that the area left for judgement is circumscribed. Nevertheless, it is concluded that a degree of engineering judgement will always be an indispensable feature of risk assessments. Clearly risk calculations will always tend to err on the cautious side so the actual risk of plant failure can be assumed to be generally lower than those calculated. However, great attention needs to be paid to the quality of the plant and the operation and maintenance procedures that are applied.

Risks need to be considered in three main areas, namely:

(a) risks to society
(b) risks to individuals
(c) economic risks.

The term 'societal risk' refers to the total harm suffered by a whole population and to the future of whole communities measured, for example, by the chance of a large accident causing a defined number of deaths or injuries. Individual risk applies to a member of the workforce or a member of the public, defined as anyone living within a defined radius of the plant. Economic risk is the financial loss that could result from an accident which destroys part or all of the plant. As well as the intrinsic value of the plant it will include lost production and possible loss of the organization's reputation as a responsible member of the industry.

Setting targets for individual and societal risks is a difficult and (sometimes) emotive problem. As long ago as 1971 Kletz (**9**) suggested that an acceptable target for the UK chemical industry was probably a fatal accident rate (FAR) of 0.4 in 10^{-8} h – a factor of 10 reduction on the average for the industry which equates to a risk of about 4×10^{-5}/man-year. For societal risks the UK Health and Safety Commission set up an Advisory Commission on Major Hazards which issued its first report in 1976 (**10**). With regard to accident frequencies it suggested that a 10^{-4} chance of a serious accident per year might perhaps be considered on the borderline of acceptability.

The approach in the UK has been based on the concept that the plant is a manmade hazard rather than a natural one and it is, therefore, reasonable to expect the manmade risk to be sufficiently low that *it does not make a significant difference* to the pre-existing comparable natural risk. Based on mortality statistics a man-made risk of about 1×10^{-6} per year would make about 0.22 per cent difference to the total risk of accidental death and about 0.01 per cent difference to the total risk from all causes. A risk of 1×10^{-6}/man-year would therefore appear to be an appropriate level of acceptable risk for the siting of potentially hazardous plant.

The various UK regulations for the control of major accident hazards, for example the COMAH regulations (**3**), require, *inter alia*, a demonstration that all hazards with the potential to cause a major accident have been identified, their risks evaluated, and measures taken to reduce risks to persons to 'as low as reasonably practicable' (ALARP). This recognizes that no industrial activity is entirely free from risk and that the benefits from the installation need to be balanced against these risks. Hence, the Farmer principle of reducing probability of occurrence with increasing consequence is generally applied to define three zones. Clearly the scaling of the zones will depend on a number of factors such as the location and public aversion to the perceived hazards of the installation. Floyd and Ball (**11**) give examples for different scenarios in a review of societal risk criteria. Figure 4 from their paper (Fig. 13.5 here) shows the ALARP diagram defining societal risk criteria originating from The Netherlands in the 1980s.

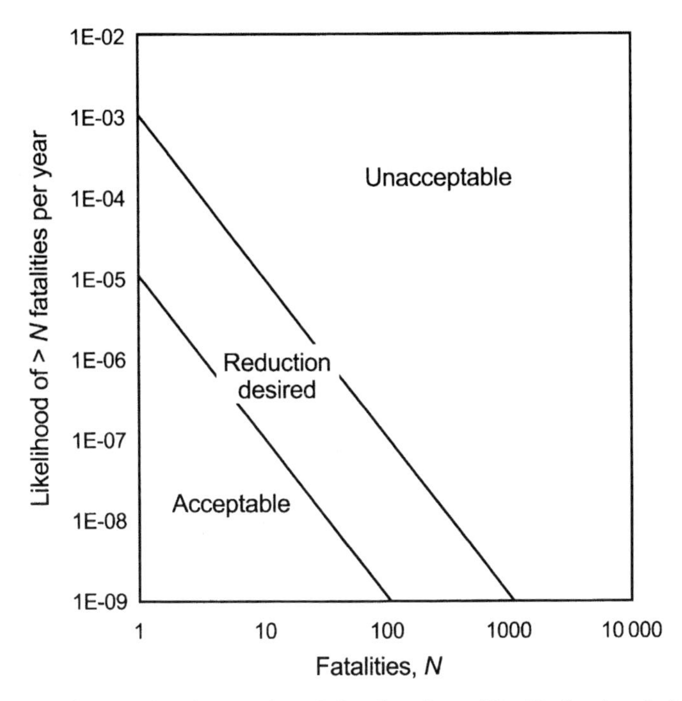

Fig. 13.5 Societal risk criteria originating from The Netherlands in the 1980s [from Floyd and Ball (11)]

Floyd and Ball define the three zones as:

(a) those that are so high as to be unacceptable or intolerable
(b) those that are so low as to be acceptable
(c) those in between where consideration needs to be given to the various trade-offs between the risks and the benefits.

In general terms the upper limit of risks regarded as acceptable tends to be set two or three orders of magnitude lower on the likelihood scale than those regarded as unacceptable, which tends to define a very broad band for ALARP risks. They conclude that given an 'anchor' and a gradient the resultant societal risk criteria can be prescribed. However, given the inherent uncertainties of risk assessment techniques – particularly in relation to low-frequency/high-consequence events – *FN* criteria should perhaps be considered as guidelines rather than rigid standards.

In conclusion, it can be seen that quantitative risk assessment offers a sound and systematic basis for evaluating potentially hazardous activities. Although it is clearly not legitimate to 'read across' risk figures from one type of hazard to another, to achieve some level which can be applied for the control of the whole set of risk estimates, there are benchmarks for ranking risks which can indicate the scale of resources to be applied to further reduce or mitigate them.

The methods of QRA are, however, specialized and frequently complex. An audit of the assessment is vital to ensure that a logical, consistent approach based on suitable consequence and probability models and relevant data has been adopted. Major accidents are rare events with return periods that will always be outside the combined experience of the design and operating teams. A systematic review of the potential hazards and assessment of their likely frequency and consequences will show where the most cost-effective measures can be introduced to improve safety. For the regulatory authorities it adds a further dimension to other technical procedures employed to evaluate the safety of specific proposals. The benefits of a probabilistic risk assessment to the operator can also be very significant.

13.5 Risk-based inspection and maintenance

13.5.1 General

The design of industrial plant is usually based on the assumption of a sequence of relatively long periods (typically 1 or 2 years) of high-availability, steady-state operation interspersed with short periods for planned maintenance and major overhauls. The length of the overhaul period will depend on the output requirements of the installation and, sometimes, on the season of the year. For example, the demand for electricity is lower in summer so onshore power station overhauls are generally more relaxed than, say, for an offshore power station serving an oil producing platform.

Statutory inspections are required on equipment where failure could pose a threat to plant personnel, society, or the environment. Planned maintenance is also required on equipment where loss of performance or degradation is likely to limit the objective of high-availability, steady-state operation. Because of the financial penalties of unscheduled maintenance, overhauls and planned maintenance strategies tend to be cautious. However, the combined effect on lost production can be significant and measures have recently been introduced in an attempt to rationalize the scheduling of inspections, planned maintenance, and renewals on sound, financial, risk-based procedures. These procedures are similar in concept to those employed in the QRA of potentially hazardous installations but include the aim to maximize productivity of the plant when the accidental risks are relatively low. Most of the developments of risk-based inspection has concerned in-service inspections for petrochemical and offshore plants since these are areas where commercial competition is most intense.

13.5.2 Risk-based inspection

Risk-based inspection (RBI) of in-service equipment can be considered as an alternative to the prescriptive approach currently employed in process plant for prioritizing and managing in-service inspection programmes. The aim of RBI is to simplify procedures and reduce costs without compromising safety. In the past, the prescriptive, calendar-based approach has been rigidly applied irrespective of the actual condition of the plant and the possible risk of equipment failure. An unrealistic scope for inspections can mean excessive downtime and unnecessary expense for plant operators leading to reduced production and profits.

13.5.3 Comparison of RBI and major accident hazard assessments

Although they are connected it is important to note the essential difference between RBI and major accident hazard (MAH) assessments since both are based on the concept of risk. RBI assessments are concerned mainly with **equipment**: the frequency and consequences of equipment failure due to one or more degradation mechanisms. The aim is to evaluate the risks associated with the different equipment failure modes and the likely impact on safety and productivity of extending (or reducing) the recommended inspection interval.

Major accident hazard assessments are concerned with **systems** and the potential major accident(s) which could develop from an initiating event (say, a pressure excursion) in combination with failure of the control and protection systems installed to mitigate the consequential effects on the plant, personnel, and the environment. The aim is to determine the probability of an unwanted event (major accident) and the consequential effects to ensure that adequate barriers and controls are installed to keep the risk below an acceptable threshold. The results from a safety-orientated QRA can sometimes be employed in the quantitative RBI assessment but in many cases the focus is different.

13.5.4 RBI assessment

RBI assessments range from subjective, qualitative methods to more complex evaluations involving quantification and ranking of the different risks. Assessments rely heavily on collaboration between designers, operators, maintenance personnel, and analysts with experience of risk and reliability techniques. The company engineers have in-depth knowledge and experience of the specific plant; the risk assessors have knowledge and experience of applying the RBI techniques. By carrying out a detailed, documented review of all relevant issues, including initial plant integrity, past inspection results, maintenance data, operating procedures, etc. the important failure modes and mechanisms can be identified and the risks estimated in terms of the likelihood of failure and the consequences.

A flowsheet of a typical plant RBI assessment is shown in Fig. 13.6. The different stages are:

1. *Definitions.* Clear specification of the objectives of the study, the operational boundaries of the plant, and the assumptions on which the assessment will be made.
2. *Functional analysis.* Breakdown of the plant into its major units, systems, and equipment to identify items of equipment at risk and interdependencies.
3. *Qualitative risk assessment.* Subjective analysis of equipment at risk and their broad classification on a consequence versus frequency criticality matrix partitioned into high-risk and low-risk bands – the area between covers the equipment which warrants further study.
4. *Quantitative risk assessment.* Formalized assessment for equipment which could pose threats to society, the environment, or the plant, to quantify likelihood and consequence of each equipment failure mode.
5. *Data analysis.* Review of past incident, inspection, and maintenance records and comparison with available generic data.
6. *ALARP.* Evaluation of qualitative and quantitative risk assessment and data analysis results to ensure risks in the high/moderately high risk bands are 'as low as reasonably practicable'.
7. *Risk optimization.* Studies to optimize inspection intervals based on safety and financial criteria.

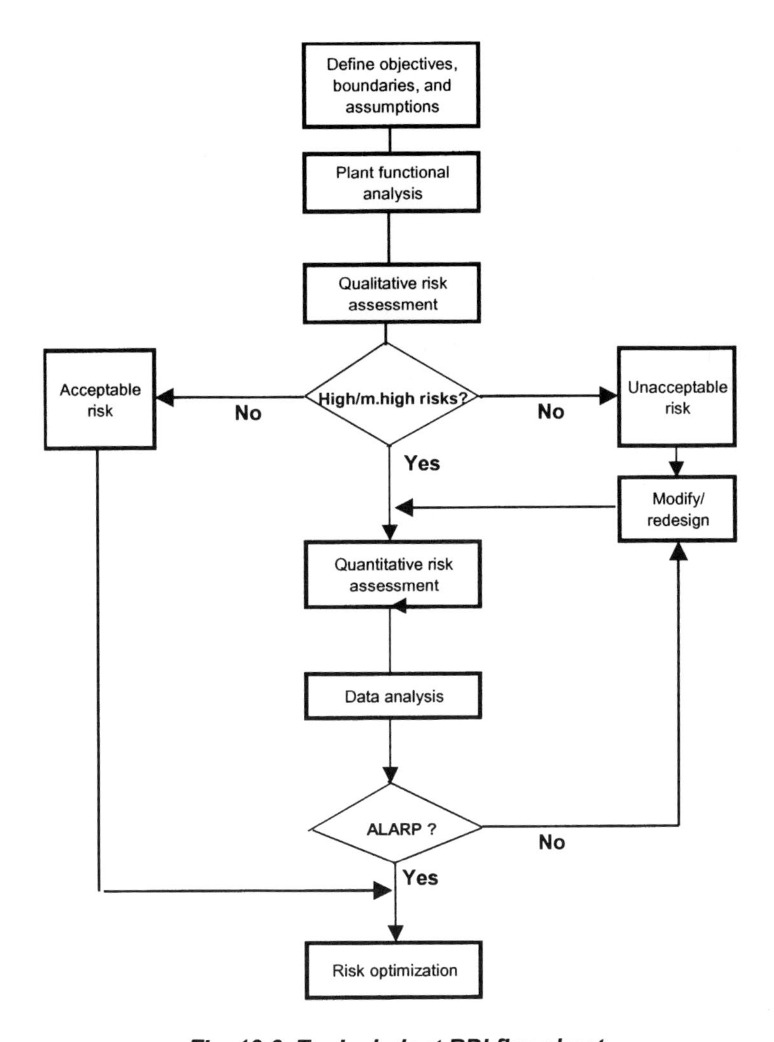

Fig. 13.6 Typical plant RBI flowsheet

13.5.5 API RBI assessment methodology

Under the auspices of the American Petroleum Institute (API) an assessment methodology, API RP 580 (**12**) has been developed for the petroleum and chemical process industries as a joint venture between 24 petrochemical companies and Det Norske Veritas (DNV) who created the support software. It is intended to be used in conjunction with existing industry codes covering the inspection of pressure vessels, piping systems, and above-ground storage tanks by organizations which maintain or have access to an authorized inspection agency, a repair organization, and technically qualified equipment engineers, inspectors, and examiners. It is specifically stated in API RP 580 that it shall not be used as a substitute for the original construction requirements governing pressure vessels, piping systems, or storage tanks before these items are to be placed in service; nor should it be used in conflict with any prevailing regulatory requirements.

Three approaches known as qualitative, semi-quantitative, and quantitative are proposed. Qualitative RBI is rule-based and is less precise than quantitative RBI but requires less effort. Likelihood and consequences are ranked by a number of factors which are plotted on a 5 × 5 risk matrix (Fig. 13.7) to give overall rankings for likelihood and consequences. The weighting of each factor for a qualitative assessment is derived from tables in the workbook which guide the user through the assessment.

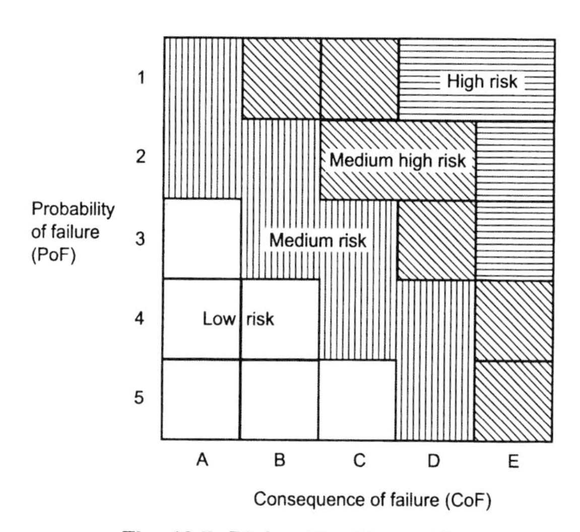

Fig. 13.7 Risk criticality matrix

The six factors that make up the overall likelihood factor are:

(a) amount of equipment (equipment factor)
(b) damage mechanisms (damage factor)
(c) appropriateness of inspection (inspection factor)
(d) current equipment condition (condition factor)
(e) nature of the process (process factor)
(f) equipment design (equipment design factor).

The consequence analysis determines a damage consequence factor for fire and explosion risks and a health consequence factor for toxic risks. Where these hazards are not involved the consequences are assumed to come in the low-risk area of the matrix.

The damage consequence factor is derived from a combination of seven sub-factors that determine the magnitude of a fire and/or explosion hazard. These are:

- inherent tendency to ignite (chemical factor)
- quantity that can be released (quantity factor)
- ability to flash to a vapour (state factor)
- possibility of auto-ignition (auto-ignition factor)
- effect of higher pressure operations (pressure factor)
- engineering safeguards (credit factor)
- degree of exposure to damage (damage potential factor).

The health consequence factor considers the following elements to express the degree of a potential toxic hazard:

- quantity and toxicity (toxic quantity factor)
- ability to disperse under typical process conditions (dispersibility factor)
- detection and mitigation systems (credit factor)
- population in vicinity of release (population factor).

The consequence categories (health and damage) are assigned letter scores and the one with the highest value is plotted on the horizontal axis of the 5×5 risk matrix to develop a risk rating for the unit.

The API quantitative RBI program is an equipment-level risk assessment tool that calculates the risk associated with each piece of operating equipment. It is based on performing a series of calculations to estimate the likelihood and consequences of failure of the pressure boundary of each piece of process equipment. The product of the likelihood and consequence of failure provides a measure of the risk associated with each item of equipment. These risk indices are used to rank and prioritize the equipment list by importance.

Semi-quantitative RBI is a simplified method of quantitative RBI for risk ranking on individual items of equipment in a process unit. It also uses a 5 × 5 risk matrix to establish priorities.

API RP 580 does not specifically cover rotating equipment, such as pumps and compressors, heaters, or furnaces but notes that these may be included as an option, provided means to consistently assess the risk have been developed.

RBI and RBM optimization

Time is an essential element of finance so optimization decisions are generally based on the time value of money. Depending on the assumed discount rate, the value of £1000 spent today can be significantly less than its value if the decision can be delayed by 1 or 2 years. A primary decision analysis tool in optimization problem solving is the influence diagram. It indicates a flow of information from decision alternatives on one side of the diagram to a single output node on the other side which represents 'net present value'. NPV is usually employed because it is a robust measure of value to the organization and retains its effectiveness in optimization as a function of time and across multiple-choice decisions. Quantities such as cost/benefit ratio, payback period, or internal rate of return can measure single project 'yes/no' decisions quite effectively. However, in addressing timing and multiple competing decisions with multi-year constraints, NPV is a more accurate, effective, and robust measure.

The influence diagram consists of connected nodes and is generally constructed moving from right to left starting with the node to be optimized. The net present value equals the difference in value between the benefit and the cost. Figure 13.8 shows a simple influence diagram for the decision 'inspect now' or 'delay inspection'.

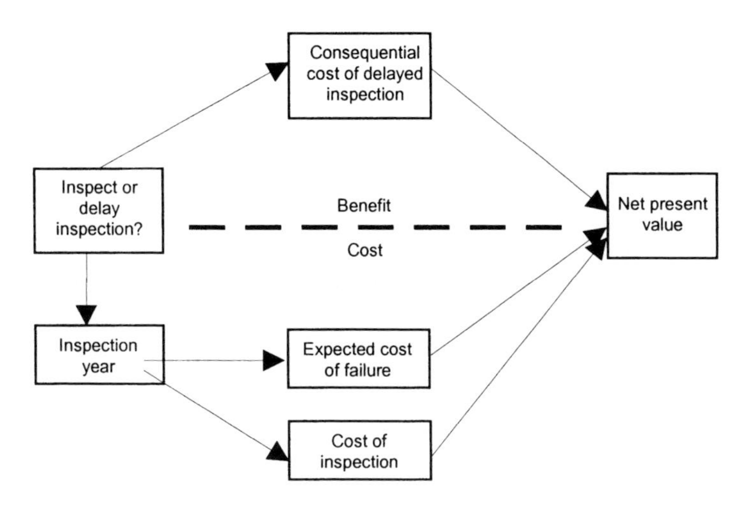

Fig. 13.8 Simplified inspection optimization influence diagram

Cash flow for the two options is the risk or expected cost of failure plus inspection cost in each case. It involves:

1. Calculation of the year-by-year net increment or decrement in expected value for that year.
2. Inflating or discounting the increment back to the initial year to create net present value. Plotting net present value against action year will show the cross-over point; that is, when the position changes from benefit to negative benefit as shown in Fig. 13.9.

Sensitivity analysis or simulation can be employed to ensure that the uncertainties inherent in the data and assumptions used for predicting the different alternatives are included in the model.

13.5.6 Experience with RBI

De Regt (**13**) describes the basic approach developed by DNV as part of the API RBI project and its later extension with major Norwegian oil companies to offshore topsides. Examples are given from the analysis of topside piping systems which isolated the small percentage of equipment that were categorized as very-high, high, and medium risk. More than 95 per cent of the piping items were allocated to the low and very low risk categories. Based on the results it was concluded that current inspection activities could be reduced by 25 per cent without compromising safety.

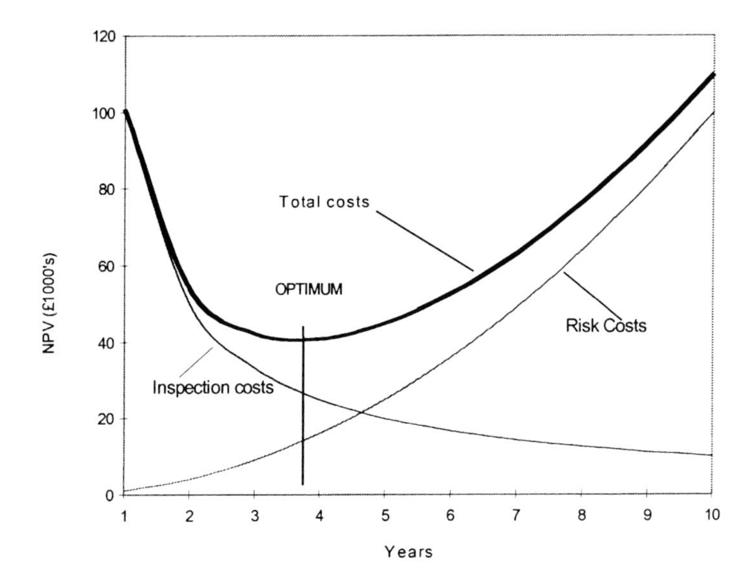

Fig. 13.9 Inspection interval optimization

RBI studies were also carried out to model the impact of changing feedstock in a refinery. It was shown that different crude compositions could be modelled in an RBI analysis to determine the increased risk as a result of higher corrosive feedstock composition. Cost–benefit analysis was employed to compare the cost of more corrosive crude. The increased cost of inspection to maintain the same level of risk could be balanced by the savings to purchase feed.

De Regt (**14**) also describes the use of the DNV RBI software tool as applied to pressure vessels and tanks as well as piping systems in the southern North Sea. The qualitative analysis isolated the 19 out of 29 and 8 out of 21 systems from the two complexes which were considered to require more detailed RBI analysis. The result indicated that about 10 per cent of the equipment items (vessels as well as piping) were above the safety risk criteria threshold. The assessment also showed that less than 10 per cent of the piping needed to be inspected over the next 5 years. The immediate cost saving of delayed inspection was estimated to be over £1 000 000 because of deferred production loss and inspection costs.

Koppen (**15**) discusses the development and application of RBI methodology and its effectiveness in improving plant reliability. The paper is based on studies of nearly 30 000 pieces of equipment worldwide. Results from a 3-week study of 180 pieces of equipment in a chemical plant supported the shortening of the planned annual overhaul to half the time originally allocated and identified a specific degradation risk associated with austenitic stainless steel equipment. Inspection frequencies were also examined for a refinery where the RBI study isolated the 'vital few' items of equipment which critically affect the risk to the unit. For the remainder the report recommended an increase of the inspection interval by 2 years with an overall 40-fold decrease in the risk of injury, business interruption, and equipment damage areas. Initial savings of the order of $100 000/year were estimated in addition to the greater long-term benefits of the more effective management of risk.

Perrollet (**16**) describes the RBI approach developed within Elf Exploration and Production. As a result of optimized inspection programmes significant OPEX savings have been achieved. The whole philosophy is geared by safety considerations with the first essential step to define the safety-critical components. This is effected by the conventional implementation of safety risk management using HAZOPs, QRA, etc. The methodology is basically complying with the API P 581 reference document (**17**); however, it has been significantly simplified in a number of areas in order to take full advantage of the lower hazard rating of specific process systems in oil and gas production. Since degradation modes are significantly less than in process refining systems, computerized degradation models were incorporated into the Elf software. This allowed the implementation of RBI analysis to new facilities in a simple manner and at minimum cost. At the same time the level of subjectivity when assessing critical parameters could be significantly reduced.

The benefits claimed include better control of risks associated with the integrity of installation, optimized inspection programmes with annual direct cost savings of $500 000 per installation, and indirect cost savings associated with improved availability of process systems.

Experience with risk-based maintenance

Less attention has been devoted to the use of risk-based maintenance (RBM) although there appears to be significantly more developments in this area than published to date. Smalley and Mauney (**18**) have applied risk-based methods to rotating machinery with the aim of reducing maintenance budgets. Methods of generating probability of failure data included:

- analysis and inference from historical data
- knowledgeable opinion
- inspection and simulation of flaw growth
- life prediction based on operating conditions
- operational monitoring and failure mode analysis.

Software has been developed to support the analysis which includes the capability for Bayesian combination of probability of failure information, from Weibull analysis of historical data, with knowledgeable opinion. The result is presented as a single probability curve from the combination of the two inputs.

Examples are given of applying the risk-based methodology to high vibration on a critical air blower in a refinery and the timing of gas turbine overhauls. In the former case the NPV is defined as the expected failure cost without balancing, minus the cost of balancing, minus the expected failure cost with balancing. The calculated NPV was positive, indicating that immediate balancing of the blower was the preferred option. For the gas turbine overhaul timing, the decision was also based on NPV to the company. The inputs to the model were the computed probabilities of failure and cost of repairs for the different degradation failure modes and the cost of lost revenue weighted by a utilization factor. The results are plotted as an NPV versus time curve for the gas turbine overhaul delay decision. The shape of the curve showed the sensitivity of the decision to timing. It particularly highlights when the delay is too long to add value to the corporation.

Christ *et al.* (**19**) also use NPV for optimizing inspection and overhauls of industrial steam turbines since insurance industry data identified steam turbines as a major machinery loss item with underwriters. The method followed was an adaptation of the ASME risk-based inspection guidelines (**20**, **21**). The process involved four steps:

(a) system definition
(b) qualitative risk assessment
(c) quantitative risk analysis – which included failure mode and effects criticality analysis (FMECA)
(d) inspection programme identification.

The result is incorporated in a computer model that permits different scenarios for individual turbines to be evaluated on a cost/risk benefit basis. Turbines are classified into five main groups according to design and operating characteristics because of the wide diversity of conditions which could be expected to generate different categories of likelihood and consequence. At the time the paper was presented (1997) the computer model had only been tested on utility turbine/ generators greater than 60 MW. The project did not consider lost production as part of its consequential impact but looked only at the cost to repair/replace the failed component. The results of analysis of other turbine classes were generally shown to agree with the data in their database except for a few cases. For example, supercritical pressure units were showing better risk for the liquid petroleum turbine than non-supercritical units. The constant reality checks being applied to the data should allow for continuing improvement of the model.

Labouchere (**22**) has considered the complex problem of delayed replacement of major capital assets using risk-based methods. A computer program has been developed and employed to consider a reactor vessel which had already exceeded its design life and was subject to degradation. The model uses discounted cash flow to calculate an equivalent annual cost of the reactor vessel over a range of possible lives. The decision hinges on the cost benefit of replacement by a new vessel as against retaining the existing vessel. The difference between the two options is expressed as a present-value lump sum difference. In this case the present-value lump sum value of replacing the reactor vessel now as compared with replacement 2 years hence was over £6 000 000.

However, under budgetary constraints a replacement cannot always be carried out at the optimum time. A replacement prioritization parameter (RPP) is therefore defined as:

$$\text{RPP} = \frac{\text{present value cost of delaying replacement by 1 year}}{\text{cost of replacement}}$$

By listing the replacements in a plant in order of their RPPs a priority list is generated. Replacements can then be scheduled as far down the list as the budget permits.

From the comprehensive range of case studies which now exist it is evident that a large percentage of the risk is associated with a small percentage of equipment in process plants. Risk-based inspection and maintenance has been shown to have a major impact on identifying and preventing mechanical failures resulting from corrosion, erosion, and other types of in-service deterioration. Even if the improvements obtained apply to only a proportion of the total losses in a plant it is clear that the impact can represent very significant savings. However, risk-based inspection and maintenance (RBIM) is only part of the overall discipline of risk management which is currently fragmented between safety, insurance, process, environment, and engineering. As the scope of RBIM becomes more widely known further advances in the methods and software available will obviously surface. Probably the most important advance currently required in the process industries is for improved rotating machinery RBIM techniques since these systems are responsible for a significant amount of plant downtime.

Other developments

As well as the American Society of Mechanical Engineers others have also applied risk assessment methods to scheduling overhauls, safety optimization, and safety system testing. The applications range from the use of Weibull analysis and simulation for equipment, such as aircraft gas turbines (where equipment populations are large), analysis and inference from historical data, indications from condition monitoring of equipment and knowledgeable opinion to obtain the input data required for RBIM studies.

With the API methodology primarily geared to the needs of US-based refineries DNV embarked on a further development to cater for the needs of the petrochemical industry worldwide. This culminated with the development of the supporting software which has now been applied to a wide range of equipment and piping systems and in support of onshore and offshore safety programmes. In the UK comprehensive software tools for asset management have been developed which include software for risk-based optimization of maintenance and inspection.

13.6 Summary

The use of risk assessment techniques for the analysis of major accident hazards is now well established. It involves the identification of the hazards associated with industrial installations and the evaluation of the consequences and likelihood of potential major accident scenarios to establish the level of risk.

Risk-based methods are also applied in industry to ensure that the scheduling of inspection and maintenance on critical equipment in the plant is based on robust, cost-effective criteria. Even if the improvements obtained apply to only a proportion of the total losses in a plant it is clear that the impact can represent very significant savings. However, RBIM is only part of the overall discipline of risk management, which is currently fragmented among safety, environmental protection, insurance, and engineering, so there is still significant scope for further development.

13.7 References

(1) **Farmer, F. R.** (1967) Siting criteria: a new approach. IAEA Symposium on *The Containment of Nuclear Power Reactors*, Vienna, April.

(2) USAEC 1974 – An assessment of accidental risk in US commercial nuclear power plants. WASH 1400 August.

(3) Health and Safety Executive (1999) A guide to the control of major accident hazards regulations. HSE Books.

(4) **Lees, F. P.** (1980) *Loss Prevention in the Process Industries*. Butterworth.

(5) **Baker, W. E.** *et al.* (1983) *Explosion Hazards and Evaluation*. Elsevier.

(6) Health and Safety Executive (1988) *The Tolerability of Risk from Nuclear Power Stations*. HMSO.

(7) Health and Safety Executive (1987) *Sizewell B Public Enquiry*, Report by Sir Frank Layfield. HMSO.

(8) Health and Safety Executive (1989) *Quantified Risk Assessment: Its Input to Decision Making*. HMSO.

(9) **Kletz, T. A.** (1971) Hazard analysis: a quantitative approach to safety, *Instn of Chem. Engs Symp. Ser.* **34**.

(10) Health and Safety Commission (1976) First report of the advisory committee on major hazards.

(11) **Floyd, P. J.** and **Ball, D. J.** (2000) Societal risk criteria, possible futures. In *ESREL 2000*, Edinburgh, June.

(12) API RP 580 (1996) *Recommended Practice for Risk Based Inspection* (Draft). American Petroleum Institute, New York.

(13) **de Regt, C.** (1999) Development of risk-based inspection. National NDT Conference, Aberdeen.

(14) **de Regt, C.** (1999) Risk-based inspection applied in the southern North Sea. National NDT Conference, Aberdeen.

(15) **Koppen, G.** (1999) Improving plant reliability by risk based inspection. GTF Reliability, Amsterdam.

(16) **Perrollet, C.** (1999) Risk based inspection: Elf experience in the North Sea. GTF Reliability, Amsterdam.

(17) API P 581 (1998) Base resource document on risk based inspection. American Petroleum Institute, New York.

(18) **Smalley, A. J.** and **Mauney, D. A.** (1997) Risk based maintenance of turbo-machinery. In Proceedings of the 26th Turbo-Machinery Symposium, Turbo-Machinery Laboratory, A. and M. University, Texas.

(19) **Christ, T. J.**, **Drosjack, M. J.** and **Tanner, G. M.** (1997) Steam turbine risk assessment: a tool to assist in optimizing inspection and overhaul of industrial steam turbines. In Proceedings of the 26th Turbo-Machinery Symposium, Turbo-Machinery Laboratory, A. and M. University, Texas.

(20) CRTD 20 (1) (1991) Risk based inspection, guidelines. ASME Research Report, Vol. 1, ASME, New York.

(21) CRTD 20 (3) (1994) Risk based inspection, guidelines. ASME Research Report, Vol. 3, Fossil Fuel-Fired Electric Power Applications. ASME, New York.

(22) **Labouchere, C.** (1999) Can we delay the replacement of this plant? Asset management in the utilities. ERA Conference, London.

Chapter 14

Case study 1 – Quantitative safety assessment of the ventilation recirculation system in an undersea mine

14.1 Introduction

A recirculation fan was added to the surface fan and the booster fans to provide a solution to the problem of ventilating the increasingly distant faces of an undersea mine. Conventional methods of maintaining air quantities were approaching practical limits with the undersea workings then 10 km from the shafts and the possibility of reaching 20 km in the future. Increased air quantities are required for a variety of reasons, these being primarily: to dilute mine gases, to dilute dust concentrations, and to maintain reasonable working temperatures. However, due to the safety implications the present Mines Acts prohibit recirculation of the primary ventilation in the mine. An exemption from these regulations has been granted by the Health and Safety Executive (H. M. Inspectorate of Mines) to allow, subject to imposed conditions, controlled recirculation to take place. The imposed conditions relate to the use of environmental monitoring and automatic control systems to provide an acceptable level of safety. The work described examines the ability of these systems to shut down the recirculation fan and prevent noxious gases, smoke, combustion products, or methane being transported in the intake stream.

14.2 Recirculation fan system description

The recirculation fan is located as shown in Fig. 14.1, taking air from the return roadway and feeding back into the intake. When adverse conditions exist in the mine this necessitates the stopping of the recirculation process. In order to perform this task effectively an automatic fan shutdown system has been installed. The system has been designed with the intent of stopping the recirculation fan when undesired environmental conditions are detected but to avoid unnecessary interruption of the mine ventilation due to spurious shutdown when conditions are satisfactory for recirculation to continue (1). This has been achieved by using a voting system in which two-out-of-three transducers are required to indicate an unhealthy condition to trip the system. This also enables the maintenance or replacement of transducers to take place with the system still operable. The controlling microprocessor used to process environmental data is installed locally to the fan site in an underground outstation to avoid the dependence on a data transmission line back to the surface. When a fan trip occurs in response to predetermined criteria a reset can only be carried out at the fan.

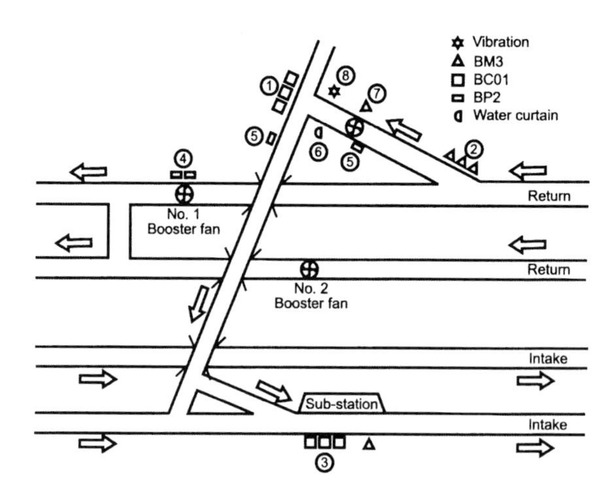

Fig. 14.1 Environmental monitoring systems at the 1570 level

14.3 Conditions for fan stoppage

14.3.1 Methane levels

A set of three monitors has been installed to measure the methane content in the main return air of the recirculation circuit (location 2, Fig. 14.1). The instruments installed have an operating range of 0–3 per cent and have a built-in rechargeable battery which provides approximately 24 h operation in the event of the primary 12V d.c. supply failing. A shutdown condition is produced when 1 per cent methane is detected by at least two of the three monitors.

14.3.2 Carbon monoxide levels

Carbon monoxide monitors have been installed as the most reliable means of detecting combustion products from a fire. The instrument has a similar battery back-up to that contained in the methane monitors. Carbon monoxide is measured at two locations on the recirculation system. The first set of three instruments is situated in the recirculated return air at the outlet side of the recirculation fan (location 1, Fig. 14.1). Instrument set number two is placed in the intake roadway of the recirculation circuit (location 3, Fig. 14.1). Both of these sets of monitors are designed to produce a shutdown condition when high carbon monoxide levels are indicated by at least two of them. Different trip levels are used for each set: the first set trips when 48 p.p.m. are detected; set two trips at the lower level of 20 p.p.m.

14.3.3 Recirculation factor

The exemption to the regulations permits a recirculation factor of up to 37 per cent. To monitor the air velocities which would enable this factor to be calculated requires the installation of velocity monitors and a computer to process the data. Unfortunately people or vehicles in the vicinity of the sampling heads would make this an unreliable means of controlling fan shutdown.

When booster and recirculation fans are running normally the recirculation factor is acceptable. Therefore a more reliable means of controlling the recirculation factor, although indirect, is by measuring the fan pressure at the recirculation fan and at the associated booster fans and interlocking the recirculation fan with the booster fan to prevent operation unless the booster fan is running at normal load. This is achieved by monitoring the pressure either side of the booster fan (location 4, Fig. 14.1) and comparing the reading with a preset normal. Fluctuation in the measured pressure by more than 15 per cent from the normal value automatically trips the recirculation fan. These arrangements are repeated at the recirculation fan (location 5, Fig. 14.1) and a similar system is installed at the No. 2 booster fan.

Two pressure monitors are used to measure the pressure at the two locations. No internal batteries are fitted in these monitors and to initiate an automatic shutdown of the recirculation fan both are required to indicate a trip condition.

14.3.4 Additional monitoring

In addition to the conditions described above, the recirculation fan is also required to stop in the event of excessive fan vibration, on the activation of the water curtain, or when the emergency stop button is depressed.

A monitoring and vibration indication system (MAVIS) monitor is installed at the fan site to detect any indication of a mechanical fault developing. When a preset limit is exceeded the fan is stopped automatically.

A fire control water curtain is installed at the outlet side of the recirculation fan. On experiencing a temperature of 57 °C a fusible link automatically activates the water curtain. Operation of the supply valve to the water curtain also automatically stops the recirculation fan.

14.4 Scope of the analysis

The environmental conditions under which the safety systems are designed to shut down the recirculated air flow can occur at any point in time. The greater the proportion of time the safety systems are available, that is, in the working state, the higher the chance of them correctly carrying out the required action. Probabilities have been predicted in this study to assess the chance of the protective systems being failed in an unrevealed or dormant mode and unable to trip the fan when a demand occurs. This is the most undesired type of failure of the ventilation recirculation protective systems, which could result in toxic or flammable gases passing into the ventilation intake.

A second way in which the protective monitoring systems can fail is in a revealed or spurious mode where trip conditions are indicated which do not exist. This results in an unplanned stoppage of the recirculation fan. While this event does not have the potential hazards involved in an unrevealed failure, if the resulting fan stoppage lasts for any length of time, the conditions at the coal face will deteriorate due to high dust and heat levels. Unplanned fan stoppages lasting 30 min or longer must be reported to H.M. Inspector of Mines. The recirculation system is required to provide ventilation to the coal face and although its spurious tripping is not as serious as an unrevealed system failure, its occurrence is inconvenient to the mine operations. As such, probabilities have also been predicted for a spurious fan shutdown.

Fault trees were constructed and then quantified to provide the probability assessment. Prior to the fault tree analysis a qualitative failure mode and effect analysis (FMEA) was also carried out. In order to facilitate the assessment, the protective system was considered in terms of its eight sub-systems. These sub-systems are those which provide the recirculation fan trip for each of the conditions described previously.

System 1. Carbon monoxide monitoring system located in the recirculated return air at the outlet side of the recirculation fan.

System 2. Methane monitoring system located at the recirculation fan intake from the return air.

System 3. Carbon monoxide monitoring system located in the recirculated air intake.

System 4. Pressure monitoring system at No.1 booster fan.

System 5. Pressure monitoring system at the recirculation fan site.

System 6. Water curtain system.

System 7. Emergency stop buttons.

System 8. Vibration monitoring system.

14.5 System description

The majority of components in each of the safety systems are common to all systems. Hence a description of only one of these systems is provided, this being the carbon monoxide monitoring system located in the recirculated return air at the outlet side of the recirculation fan.

This system is located at position 1 on the layout map shown in Fig. 14.1. The system features three units which monitor the carbon monoxide levels at the recirculation fan outlet as illustrated in the system schematic diagram shown in Fig. 14.2. When any two register a carbon monoxide concentration of 48 p.p.m. or greater then the recirculation fan will be tripped. Each instrument has both an analogue and a digital output channel. Analogue signals are relayed to the surface control room where an alarm will be raised when carbon monoxide levels are high. On the digital channel a 'flashing signal' output once every 15 s indicates a healthy state. This frequency increases to once every 2 s to signify the trip condition.

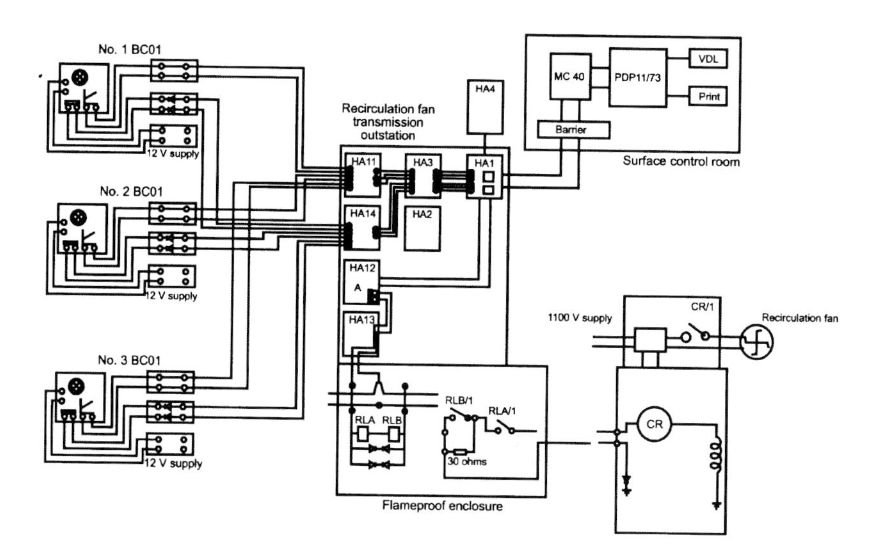

Fig. 14.2 BC01 monitoring sytem

From each carbon monoxide monitor the analogue signals travel via a junction box to the recirculation fan transmission outstation. Connection into the intrinsically safe cabinet is made by a connection board (HA11). Similarly, digital signals at the outstation are connected to the HA14. From the two connector boards both analogue and digital signals are passed to the HA3 board. Here the analogue signals are checked to ensure they are within bounds. In the event of the cable carrying the input signal being broken or any other failure which results in a voltage lower than 0.4 V the fault condition is set and relayed to the HA1 board. Digital signals are passed directly from HA3 to HA1. An open circuit on a digital channel is also interpreted as a fault condition.

The HA1 p.c.b. contains the control chips and it is this board which processes all the information received on each channel. Each analogue signal is used for continuous monitoring and alarm indication and is relayed directly to the surface MINOS computer system. The status of the digital channels is also relayed to the surface.

The configuration chip on the HA1 board checks the status of the digital inputs. When two out of three show a fault condition the chip de-energizes a relay located on the HA12 board, labelled relay A.

When a trip condition causes the relay located on HA12 to be de-energized this opens the relay contacts. The open circuit then de-energizes relay RLA which is located in the flameproof enclosure of the recirculation fan outstation. HA13 is a connector board between the intrinsically safe cabinet and the flameproof enclosure. When relay RLA is energized it completes the control circuit. The control circuit energizes the control relay CR which closes contacts to maintain the 1100 V supply to the recirculation fan. A trip condition will cause contacts RLA/1 to open which will in turn result in contacts CR/1 opening and the fan stopping.

Information relayed to the surface computer is carried on transmission lines from the recirculation fan outstation to the control room. These data lines pass through an intrinsically safe barrier into the MC40 interface connector. From the MC40s they are connected to the PDP11/73 computer system running the MINOS software. The analogue signals are used to inform the operator of high carbon monoxide levels prior to a fan trip by setting the alarm condition.

14.5.1 Section switch trip protection

Additional protection to ensure a trip is carried out when required is provided by the section switch trip. This monitors that the fan power isolation does result when instigated. If not then other relays are then de-energized to isolate power to the fan.

The methane and carbon monoxide trip systems are the most important of those installed on the recirculation system. Variations in these system designs to that described above only occurs for the carbon monoxide system located in the recirculated air intake. For this system the instrument outputs feed through two outstations rather than one. Initially the monitor outputs feed into the general-purpose monitoring outstation (GPMO). This processes the input signals and then transmits any fan trip condition into the recirculation fan transmission outstation. Using two outstations enables the system to cope with the large number of monitors installed.

The card layout within the GPMO is similar to that of the recirculation fan outstation. For the remaining five systems the types and numbers of instruments which monitor the environmental conditions are different. Pressure is monitored by two instruments for system 4 and 5 both of which must indicate an unacceptable pressure fluctuation to halt the fan. System 4 monitor inputs are accommodated by the GPMO. The water curtain, vibration monitor, and emergency stop buttons all provide single inputs to the out-station for a trip condition.

14.6 Fault tree construction

To quantify the probabilities of the undesired system outcomes, the method of fault tree analysis has been used. Fault tree diagrams have been developed for each of the eight safety systems which are associated with the ventilation recirculation. Two fault trees for each system were drawn. The first represents causes of a dormant failure where the system will fail to respond to an unhealthy condition and will not stop recirculation. The second fault tree develops causes of a spurious recirculation trip under normal, healthy environmental conditions. The construction of these tree diagrams for the carbon monoxide monitoring system at the recirculation fan outlet is described below.

14.6.1 Dormant or unrevealed system failure

For high carbon monoxide levels at the recirculation fan outlet not to result in a recirculation fan stop, the protection system must have failed. In developing causes of this event, the first gate below the top event identifies the two ways in which this can happen. Either the monitoring equipment and the control module fail to identify the trip condition OR the control module correctly identifies a trip condition but the fan remains energized. This logic is illustrated in the fault tree presented in Fig. 14.3. These two events are themselves then further developed.

Causes of the environmental conditions not being correctly identified occur in components located between the carbon monoxide monitors and the HA1 board in the outstation. The majority of failures of these components fail safe, i.e. indicate a trip/fault condition on the channel. Those which do not are the HA1 logic chip and the carbon monoxide monitors themselves drifting or being calibrated high. For the monitors themselves to fail the system, it would require at least two of the three units to fail in this mode, as indicated by the inclusion of a voting gate in the fault tree logic.

Development of the second branch below the top event traces the causes of failure to isolate the recirculation fan following the HA1 board command to do so. This part of the fault tree indicates one of the parallel features of the protection systems – the section switch trip. Since both the normal means of halting the fan and the section switch trip would need to be inoperable for this event in the tree to cause system failure, an AND gate is used to develop the failure logic. Causes for the first trip signal to stop the fan are the failure of the relay contacts opening to de-energize the required circuits which maintain power to the fan. Failure of the section switch trip facility contains both causes of the circuit remaining energized and causes of the monitoring circuit in the recirculation fan panel failing.

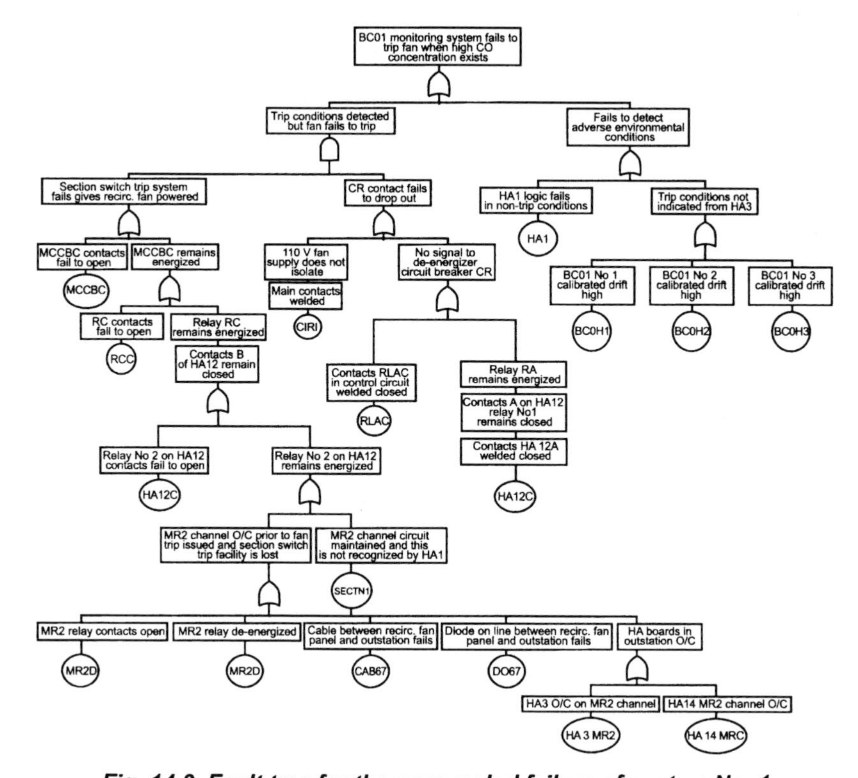

Fig. 14.3 Fault tree for the unrevealed failure of system No. 1

14.6.2 Spurious or revealed system trip

The fault tree constructed to represent the spurious trip of the recirculation fan due to the failure of components in the carbon monoxide monitoring system at the recirculation fan outlet is moderately large in size (of the order of 80 gates and 120 basic events). Hence it has not been reproduced. The development of this fault tree was carried out by starting at the recirculation fan itself and identifying causes for it to stop. If the fan motor is in a working condition then the power to the fan must have been removed. Power isolation can occur in two ways: either it is commanded by opening the CR relay contacts OR the 1100 V supply fails. Further development of this tree traces the fan power supply back to the surface supply, which is considered to be the system boundary, and also follows the de-energization of the fan back through the outstation HA boards to the carbon monoxide monitors.

Since the majority of logic gates in the fault tree are OR gates most component failures contained in the fault tree will individually trip the recirculation fan. It is only components of the system which provide inputs to the HA1 board which require more than one failure to isolate the fan.

14.7 Qualitative fault tree analysis of the system

The first stage in analysing the completed fault trees for the ventilation recirculation safety systems was to produce the minimal cut sets for particular system failure events. Minimal cut sets, sometimes known as system failure modes, are combinations of the basic events which are both necessary and sufficient to cause the top event of the fault tree to occur. The number of basic events or component failures appearing in each set is the order of the minimal cut set. In general, though not, as we will see here, on every occasion, the lower the order number of a minimal cut set the more likely it is to occur. That is, situations where single component failures will cause the system to fail are more likely to occur than system failures requiring combinations of two or more component failures.

14.7.1 Dormant or unrevealed system failure modes

Each of the trees which develop the causes of each system failing to respond to trip conditions were analysed in turn to yield the minimal cut sets. A summary of these qualitative results is given in Table 14.1. As can be seen from the table, each system failure resulted from either a single-component failure or a pair of components failing together. The total number of ways in which each system could be rendered inactive ranged from 32 for the water curtain and stop buttons to 43 for the methane monitoring system. Since there was only one parallel level within the system, provided by the section switch trip, minimal cut sets of greater order than two were not expected.

Table 14.1 Minimal cut sets for unrevealed system failure

		Number of minimal cut sets		
System	Monitoring	Order 1	Order 2	Order 3
1	Carbon monoxide	1	33	34
2	Methane	1	42	43
3	Carbon monoxide	3	33	36
4	Pressure	9	30	39
5	Pressure	8	30	38
6	Water curtain	2	30	32
7	Stop buttons	1	31	31
8	Vibration	4	30	34

In a qualitative analysis it is the lower order minimal cut sets which are deemed the most important. An examination of the minimal cut sets showed that the first-order sets for each system are consistent with the component failure modes identified during the FMEA as being possible to singly fail the system.

From the first-order minimal cut sets it is evident that one event appears in the list for each system. For systems 1, 2, and 7 this is the only single-event failure mode. This common event represents a failure of the configuration chip on the HA1 board in the recirculation fan outstation such that it fails to trip the recirculation fan. It would appear from this that the HA1 board is the weakest part of the system from a structure point of view. Since there is redundancy in most of the monitoring systems which feed into the HA1 board, and the section switch trip provides a check that the fan has tripped when required by the HA1, this result would seem to be a logical conclusion.

System number 6, the water curtain system, features two first-order minimal cut sets, the configuration chip failure, and failure of the water curtain switch. The water curtain switch failure appears singularly since it is a single detector unit; most other systems have more than one monitor.

System number 8, the vibration monitor, also has a single detector for the shutdown condition. For this system the additional first-order cut sets occur due to failures on the HA2 board which features a timer to permit higher vibration levels at the fan to occur during specific times, such as start-up.

The systems which monitor the pressure levels at the recirculation fan and the No.1 booster fan identified as system numbers 4 and 5 appear to be the weakest of the eight systems due to their high number of first-order cut sets. System 4, which monitors pressure at the No. 1 booster fan, can fail due to any of nine component failure modes. System 5, which monitors the recirculation fan output, has only one less minimal cut set.

There are two reasons which can be attributed to the causes of these increased number component failure modes for the pressure monitoring systems. The first is that these systems require both of the two pressure monitors to indicate a trip condition before the recirculation fan will be shut down; therefore there is no level of redundancy. Any failure which results in the inability of a particular channel to indicate a trip condition renders the safety system inactive. In addition to this, since the signals from the pressure monitors pass to the HA1 board via the HA2 board to allow out-of-range pressures at particular times, failures on either the HA2 board (or a particular HA2 channel time setting) or the monitors themselves all contribute single causes for the monitor channels, and hence the system, failing.

The final two-component failure modes which appear in the list of single-order cut sets occur in systems 3 and 4 and are contributions due to the use of the GPMO to feed the monitor signals into the recirculation fan outstation. The first of these failure events corresponds to a failure of the configuration chip on the HA1 board in the GPMO and is equivalent to the failure mode already described for the HA1 board in the recirculation fan outstation. The second failure event identifies a failure of the relay contacts to open on the HA12 board in the GPMO when a trip condition is required by systems 3 and 4. Both of these failure modes result in the trip condition not being communicated to HA1 in the recirculation fan outstation and the fan power being maintained.

The overall conclusion which can be drawn from these qualitative results is that the HA1 processor board in the outstations is structurally the weakest part of the system. Also, modification to individual systems can be identified which would improve their performance, if it were deemed necessary, such as the addition of redundancy for the pressure monitoring systems by adding an extra pressure transducer and using a voting system similar to that for the carbon monoxide and methane monitors. A monitoring system on the HA1 in the GPMO similar to the section switch trip facility in the recirculation fan outstation would also improve systems 3 and 4.

The extent that the single-component failure modes contribute to the system failures and the degree of improvement obtained by any design modifications can only be assessed by quantifying the fault trees.

14.7.2 Spurious or revealed system failure modes

The shutdown of the air recirculation system will result when any of the protective systems removes power to the fan. A spurious fan trip can therefore be the result of a spurious failure of any one of the protective systems. To assess the potential causes of this outcome, the fault trees constructed to represent the causes of a spurious fan trip from each system were combined. A combination of these fault trees has been analysed because of the large portions of the trip system which are common to each of the sub-systems. The power supply from the surface, the section switch trip facility, and the trip circuits which lie between the HA1 board and the fan are common in each system. Table 14.2 lists the number of cut sets resulting from this analysis. A spurious fan trip can result from any one of 4257 causes. Of these, 76 are single-component failures.

Table 14.2 Numbers of minimal cut sets which produce spurious fan trip

	Number of minimal cut sets				
Event description	Order 1	Order 2	Order 3	Order 4	Total
Spurious recirculation fan trip	76	2939	1135	107	4257
System 1	3	675	1105	41	1824
System 2	3	867	1251	33	2154
System 3	14	675	1106	34	1829
System 4	14	361	-	-	375
System 5	4	361	-	-	365
System 6	9	-	-	-	9
System 7	8	-	-	-	8
System 8	7	-	-	-	7
Components common to all systems	37	-	-	-	37

The remainder of Table 14.2 indicates the relative contributions to a spurious fan trip provided by each of the different protective systems and also the portions of the system which are common to all trip systems. Failure events indicated for each of these sub-systems cannot be made totally independent such that their sum over each sub-system will result in the overall numbers given on line 1 of the table. The reason for this is that several failures, such as GPMO components and the HA2 board, contribute to more than one system but not to all systems and so are accounted for more than once in the number of minimal cut sets. These systems are, however, considered to be independent to a degree which will enable the quantification of the events listed in the left-hand column of Table 14.2 to indicate where the most significant contributions to spurious fan trip are located.

14.8 Component failure and repair data

Quantification of the fault tree requires the probability of each basic event to be calculated. Two factors contribute to the proportion of time a component resides in a failed state: the component failure rate and the time to repair. The failure rate indicates how often a component makes a transition from the working to the failed state. An increase in failure rate increases the component unavailability. Once in the failed state the time to repair is the factor which governs how long it remains in this state and is made up from a number of contributions, depending on the mode of failure. The time taken to detect the failure, the time taken to repair or replace the component, and the time taken testing the system prior to bringing it back on-line all influence the overall time spent in the failed state. Mean time to repair is used to represent the average total restoration time in the quantitative analysis.

14.8.1 Component failure rate data

Obtaining accurate, representative component failure data is commonly one of the most difficult tasks in any quantitative system reliability study. This is particularly true when specialized components are used in the system construction. In this situation the rarity of the components means that good quality failure data are not usually available. Certain of the components used in the ventilation recirculation safety systems are specific to the mining industry and failure rates for these components cannot be obtained from databases of general component failures. In this category are the gas and pressure monitors and also the Transmitton electronic boards used in the outstations. Special efforts were made to estimate failure rates of these components. Other components, such as relays, diodes, transformers, and circuit breakers, are commonly used and as such their failure rates have been obtained from general component failure databases.

For the specialized components, neither the colliery nor the manufacturers collect data on component failures in a form which enables their failure rate to be extracted. Since the gas and pressure monitors are serviced by the mining company their failure rates can be estimated by inspecting the maintenance records at the colliery and the records at the service laboratories. Details of this information are given below for each type of monitor.

14.8.2 Carbon monoxide monitors

The BCO1 carbon monoxide monitors consist of a control unit with integral sensor. Records show that a decline in sensor sensitivity is the most likely cause of failure of this unit.

Failure modes and failure rates for the BCO1s have been derived from the testing and maintenance records of units located at 11 sites in the colliery. The 11 sites had totalled 12 055 days' operating time and had experienced four failures in this period (all resulting in the unit reading low). Such a small number of failures precludes any detailed statistical analysis and an estimate of the failure rate of 13.8 per million hours in the low mode has been made from these data (a mean time between failure of approximately 8.3 years). Some units were taken out of service due to such reasons as the LEDs failing. Since they would not affect the units producing a fan trip, this type of failure has been ignored when estimating the failure rates for this and other monitors.

14.8.3 Pressure monitors

BP2 pressure monitors consist of the control unit and a tube extruding from the monitor. The use of these units gives few problems once they are correctly installed. Problems may arise if fitted with an incorrect tube or if condensation enters the instrument.

Colliery data based on these units reveal five failures in a total of 4784 operating days. Of the five failures two resulted in the units reading high and three resulted in the units reading low. Again a detailed statistical analysis cannot be carried out on such a small number of failures the following failure rates have been estimated:

Unit fails low	26.1×10^{-6}/h (MTBF ~ 4.4 years)
Unit fails high	17.4×10^{-6}/h (MTBF ~ 6.6 years)

14.8.4 Methane monitors

Sieger BM3 monitors are used to measure the methane levels in the mine. These units comprise three modules: a control unit and a remotely located regulator/head assembly, which is connected to the control unit by a cable. The main problems experienced with this type of unit are with the pellistors in the detector heads. Environmental conditions can cause poisoning of the pellistors, which may lose sensitivity.

Data from the colliery were available for the maintenance test results for units in eight locations. Between the eight sites a total of approximately 7500 operational days had been accumulated. During this time the following number of failures were recorded.

BM3 control unit fails low	1
Regulator unit fails high	1
Regulator unit fails low	1
BM3 Head fails high	43
BM3 Head fails low	29

The following failure rates of the control and regulator elements have been estimated from these data:

BM3 control unit fails low	5.495×10^{-6}/h (MTTF ~ 20.8 years)
Regulator unit fails high	5.76×10^{-6}/h (MTTF ~ 19.8 years)
Regulator unit fails low	5.76×10^{-6}/h (MTTF ~ 19.8 years)

(Differences between these figures for the control units and regulators is due to the inability to gather all information required in their estimation from the maintenance records.)

The high number of failures recorded for the heads means that a more detailed analysis of these failures could be carried out. The censored data taken from the records has been plotted on Weibull probability paper. Two parameters for the Weibull distribution were then estimated. The β parameter is the most significant for these data since a value of one means the components have a constant failure rate, less than one indicates a decreasing failure rate, and greater than one an increasing failure rate. Data representing the times for the BM3 heads to fail both high and low provided reasonable fits to a Weibull distribution. For both failure modes β was estimated as greater than one (1.5 for high failures, 1.75 for low failures); this indicates an increasing deterioration with time.

Mean times to failure of heads obtained from this analysis are:

BM3 head fails high 2000 h (approximately 3 months)
BM3 head fails low 3600 h (approximately 5.4 months)

14.8.5 Component repair data

Component mean repair times were estimated by engineers at the colliery. For components whose failure is immediately revealed by either tripping the recirculation fan or by an annunciation system, the repair process can commence immediately and these times can be used directly in the analysis.

Components which feature unrevealed failure modes, such as relay contacts welding closed, have an additional time to include in the time it takes to restore the system: this is the time it takes to discover the fault. If component failure is unrevealed this will only be detected when either there is a demand on the safety systems or preventive maintenance takes place. For these components mean repair times are obtained by adding the repair times to half of the inspection interval used in the maintenance programme. Half the inspection period is used since components can fail at any time during the inspection interval; it will on average have failed mid-way between inspections.

For the BM3 and BCO1 monitors, checks on system functioning are carried out at 30-day intervals by introducing a gas sample to the units and ensuring that the recirculation fan trips. BP2 pressure monitors are also checked by ventilation staff each month.

14.9 Quantitative system analysis

14.9.1 System unavailability

Values for the unavailability of each of the safety protection systems (the proportion of time the systems reside in the failed state), are listed in column 3 of Table 14.3, headed 'Probability'. Column 5 gives the total number of hours a system will be unavailable in a 10-

year (87 600 h) period. Ordering the systems by their unavailabilities results in the rankings shown in the final column, where system 2, the methane detection system, has the highest unavailability of 3.41 per cent and others range between this and the lowest unavailability of 0.649 per cent. In addition to the methane detection system (with a failure probability of 3.41 per cent), the other systems with the most significant unavailabilities are system 4, the pressure monitoring system on the booster fan, and system 3, the carbon monoxide monitoring system at the recirculated air intake, with unavailabilities of 1.41 and 1.39 per cent respectively. These results emphasize that there is a large difference in the probabilities of failure for different components. If each component failure probability was of a similar order of magnitude, the largest unavailability would be associated with systems with the largest number of low-order minimal cut sets. In this case system 5, the pressure monitoring system, would be one of the poorer performers and the methane detection system would be one of the best performers. For these systems the variation in component probabilities means that minimal cut set order is not a good indication of system performance.

Table 14.3 Quantitative system analysis – unrevealed system failure

System	Monitoring	Probability (per cent)	Unconditional failure intensity (h^{-1})	Expected downtime in 10 years (h)	Expected number of failures in 10 years	Rank
1	Carbon monoxide	0.656	0.1835×10^{-4}	575	1.6	5
2	Methane	3.41	0.170×10^{-3}	2985	14.9	1
3	Carbon monoxide	1.39	0.3849×10^{-4}	1214	3.4	3
4	Pressure	1.41	0.1251×10^{-3}	1231	11.0	2
5	Pressure	0.677	0.105×10^{-3}	593	9.2	4
6	Water curtain	0.650	0.1876×10^{-4}	568	1.6	6
7	Stop buttons	0.649	0.1794×10^{-4}	568	1.6	7
8	Vibration	0.65	0.1921×10^{-4}	569	1.7	6

Importance measures (Fussel–Vesely) have been used to rank each component numerically according to the contribution to the failure of the system. As indicated by the qualitative results, the failure of the configuration chip on the HA1 board in the recirculation fan out-station provides a significant contribution to the failure of most of these systems. It is the main contributor for all except one of the systems. Only for system 2, the methane monitoring system, which has the highest unavailability, does it not have the highest importance measure. For this system the heads (PEL1, PEL2, PEL3) for the BM3s provide a higher contribution to the system failure due to their very high failure rate. It is the contribution of the BM3 heads and the HA1 board chip to the failure of system 2 which gives it the highest unavailability.

14.9.2 Unconditional failure intensity

The fourth column in Table 14.3 gives the frequency with which system failures occur. Multiplying this figure by a time period yields the expected number of system failures, which is given in the sixth column for a 10-year operational period.

14.9.3 Spurious recirculation fan stoppages

The causes of a spurious recirculation fan trip identified in Table 14.2 have also been quantified. Results for this analysis are contained in Table 14.4. The top line in the table considers the probability and frequency of a fan stoppage due to any of the safety systems. The overall probability of the system being unavailable due to a spurious stoppage is only 0.025 per cent. Very short repair times are responsible for the low value of this figure.

The remaining entries in the table show the contribution which each of the monitoring systems provides to the number of spurious system failures. These entries correspond to those in Table 14.2; the sum of the probability of each contributor gives the probability of the top line. The ranking of these events in the final column has been ordered according to the frequency with which each section causes a system trip. With such small repair times it is the number of times the system trips, rather than the total downtime, which is the most influential factor. From the rankings it can be seen that the majority of trips are due to components common to all systems.

14.10 Performance of the methane and carbon monoxide monitoring systems

Of the eight trip systems installed on the recirculation fan those which are the most important from a safety viewpoint are those which monitor the carbon monoxide and methane levels. It is the performance of these systems which is of most concern. Failure to trip the fan when methane is present in the recirculation system results from a failure of system 2 with the probability and performance characteristics given in Table 14.3.

In the event of high carbon monoxide levels existing in the mine, continuing recirculation would require both system 1 and system 3 to fail. System 1 monitors carbon monoxide levels in the recirculated return air, and system 3 in the recirculated air intake. If higher-than-acceptable carbon monoxide levels are present both of these systems should detect this and trip the recirculation process. Failure probabilities of 0.656 and 1.39 per cent for systems 1 and 3 respectively are given in Table 14.3. The fault trees for these two systems have been combined and re-analysed to predict the failure of both to detect carbon monoxide. The result of this analysis is presented in Table 14.5 together with the individual system results for comparison.

The probability that high carbon monoxide levels will not result in halting recirculation by either of the systems is 0.649 per cent, with an anticipated 568 h downtime in a 10-year operating period. These results provide only a marginal improvement in the performance of system 1 alone. It may be expected that requiring both systems to fail would produce an extra level of redundancy with a corresponding significant improvement in system performance. This is not the case since both systems are common in all but the monitors and their signal paths into the HA1 board. Failure of components between the HA1 board and the CR circuit breaker provide the majority of failure modes for both carbon monoxide monitoring systems. Hence failure in this section renders both system 1 and system 3 inactive.

14.11 Variations in system design and operation

Investigations into the sensitivity of the methane and carbon monoxide trip systems to potential variations in system design and maintenance procedures are described in the following section.

14.11.1 Design changes

The trip systems for the ventilation recirculation fan are already installed, hence comprehensive design changes would be prohibited on the grounds of expense. As such the effects of the design changes investigated have been restricted to minor modifications of the methane monitor configurations. The probabilities of unrevealed and revealed system failures are given for comparison with the installed system.

Table 14.4 Quantitative system analysis – revealed system failure

System	Monitoring	Probability (per cent)	Unconditional failure intensity (h^{-1})	Expected downtime in 10 years (h)	Expected number of failures in 10 years	Rank
Spurious fan trip		0.025	1.135×10^{-4}	21.58	10.00	
Contributors:						
1	Carbon monoxide	0.88×10^{-5}	0.23×10^{-7}	0.77×10^{-2}	0.2×10^{-2}	9
2	Methane	0.72×10^{-3}	0.47×10^{-5}	0.63	0.41	5
3	Carbon monoxide	0.46×10^{-2}	0.23×10^{-4}	4.04	2.00	3
4	Pressure	0.52×10^{-2}	0.25×10^{-4}	4.56	2.22	2
5	Pressure	0.59×10^{-3}	0.25×10^{-5}	0.52	0.22	6
6	Water curtain	0.25×10^{-2}	0.84×10^{-5}	2.23	0.73	4
7	Stop button	0.33×10^{-3}	0.83×10^{-6}	0.28	0.07	8
8	Vibration	0.11×10^{-2}	0.21×10^{-5}	0.97	0.18	7
	Common sections	0.01	0.50×10^{-4}	9.00	4.38	1

Table 14.5 Carbon monoxide monitoring systems – unrevealed failures

System failure description	Minimal cut sets				Probability (per cent)	Unconditional failure intensity (h^{-1})	Expected downtime in 10 years (h)	Expected number of failures in 10 years
	Order 1	Order 2	Order 3	Order 4				
System 1 fails	1	33	0	0	0.656	0.1835×10^{-4}	575	1.6
System 3 fails	3	33	0	0	1.39	0.3849×10^{-4}	1214	3.4
Systems 1 and 3 both fail	1	30	6	9	0.649	0.1794×10^{-4}	568	1.6

Three variations in the way that methane monitors can be used to trip the system have been considered. At present three BM3 monitors are installed and a fan trip will be initiated when two out of the three indicate an unhealthy condition. Table 14.6 lists the results achieved by variations in the usage of the monitors. From the first part of the table it can be seen that were it only required for one of the three monitors to indicate a methane presence to halt the recirculation then there is an improvement by a factor of about 4.6 in the chances of this being carried out successfully on demand. Only 2.2 failures would be expected in a 10-year operating period in comparison with 14.9 failures for the existing system.

Table 14.6 Effect of changes of the methane monitor configuration on the probability of system failure

	Probability (per cent)	Unconditional failure intensity (h^{-1})	Expected downtime in 10 years (h)	Expected number of failures in 10 years
Failure on demand:				
2 out of 3	3.41	0.170×10^{-3}	2985	14.9
1 out of 3	0.736	0.252×10^{-4}	646	2.2
1 monitor	10.24	0.281×10^{-3}	8968	24.67
Spurious trip:				
2 out of 3	0.72×10^{-3}	0.47×10^{-3}	0.65	0.41
1 out of 3	0.476	0.159×10^{-3}	417	140.0
1 monitor	0.169	0.58×10^{-3}	148	50.9

The reason that the one-from-three voting option was not initially chosen was due to the concern that this would result in a large number of unnecessary shutdowns. This argument is shown to be valid from the results for spurious trips shown in the second part of Table 14.6. A one-in-three monitor system produces an increased probability of the system failing due to a spurious trip by a factor of approximately 650. The expected number of spurious trips in 10 years also increases from 0.41 to 140, more than one spurious trip per month. So while the option does improve the probability of failure on demand, the number of spurious trips which would result may not be an acceptable trade-off.

The third option listed in Table 14.6 shows the results obtained with only a single BM3 monitor installed. As expected this option is the poorest in terms of safety, with a failure to trip on demand of around 10 per cent. It would also result in over 50 expected spurious failures in 10 years' operation. It is outperformed on both demand failures and spurious fan halts by the present installed system.

14.11.2 Inspection interval changes

Under the maintenance policy currently in operation at the colliery the BCO1 and BM3 monitors on the trip system are checked with known gas samples every 30 days. Unrevealed failures of the system can occur at any point in time between inspections. The less frequently checks are carried out the longer these dormant failures can exist, therefore increasing the fraction of time that the system is unable to detect high gas levels. Since the inspection intervals are much larger than the repair times once faults are identified they are the dominating influence on the mean time to restoring a failed system.

The results of varying the inspection interval on the probability of system failure are shown in Tables 14.7 and 14.8 for the methane and carbon monoxide detection systems respectively. Inspection intervals of 1 week, 2 weeks, and 2 months were investigated in addition to the 1-month interval currently operated.

Table 14.7 Effect of changes of the maintenance inspection interval on the unrevealed system failure probability for the methane monitoring system

Inspection interval	Probability (per cent)	Unconditional failure intensity (h^{-1})	Expected downtime in 10 years (h)	Expected number of failures in 10 years
1 week	0.332	0.5923×10^{-4}	291	5.12
2 weeks	0.969	0.9606×10^{-4}	849	8.41
1 month	3.41	0.170×10^{-3}	2985	14.9
2 months	10.50	0.273×10^{-3}	9196	23.9

Table 14.8 Effect of changes on the maintenance inspection interval on the unrevealed system failure probability for the carbon monoxide monitoring system

Inspection interval	Probability (per cent)	Unconditional failure intensity (h^{-1})	Expected downtime in 10 years (h)	Expected number of failures in 10 years
System 1:				
1 week	0.155	0.1808×10^{-4}	136	1.6
2 weeks	0.3069	0.1816×10^{-4}	269	1.6
1 month	0.656	0.1835×10^{-4}	575	1.6
2 months	1.315	0.1869×10^{-4}	1152	1.6
System 3:				
1 week	0.33	0.3832×10^{-4}	288	3.4
2 weeks	0.65	0.3837×10^{-4}	570	3.4
1 month	1.39	0.3849×10^{-4}	1214	3.4
2 months	2.76	0.3871×10^{-4}	2419	3.4

14.11.3 Methane detection system

For the methane detection system, the probability of system failure when high gas levels exist is shown to be very sensitive to variation in the inspection interval. To double the inspection interval gives a deterioration factor of 3 in the failure probability and increases the number of failures expected in a 10-year operating period from 14.9 to 23.9. This result indicates that relaxing the current testing policy would have a considerable detrimental effect on the performance of the system.

If the inspection frequency increased to once per week the system performance would be improved by an order of magnitude from a failure probability of 3.41 to 0.33 per cent, reducing the downtime in a 10-year period from 2985 days to only 291 days. Testing each week would also reduce the number of expected failures by a factor of 3. Since much of the trip system is common to all sub-systems, checking the operation of the common elements would have a beneficial effect on the probability of successful operation of them all.

14.11.4 Carbon monoxide detection system

The effects on varying the inspection interval on the system per formance of the two carbon monoxide detecting systems are shown in Table 14.8. Though not quite as dramatic as for the methane system these results show a substantial sensitivity to changes in inspection frequency. The probability of unrevealed system failure can be improved by a factor of 2.1 for both systems by carrying out an inspection every 2 weeks. Further improvement by a factor of 4.2 is attained for weekly inspection. Results indicate that the number of times the BCO1 systems would fail is unaffected by changes in testing frequency, which means that the improvement in probability of successful operation on demand can be totally attributed to reducing the time it takes to rectify the system failure.

14.12 Conclusions

The results of the fault tree study carried out on the eight safety systems installed on the ventilation recirculation system predicts probabilities of failure to shutdown on demand between 0.649 and 3.41 per cent. System 2, the methane detection system, has the highest of these probabilities. These results are in the range which could be anticipated for these types of trip system and are comparable with results of studies carried out in other industries. The acceptability of these systems as performing adequately must be judged in terms of their overall risk by considering the consequences of their failure when undesired conditions occur.

For the methane and carbon monoxide detection systems an improvement in performance could be achieved by several means should it be necessary. Change in system design, shorter component repair times, or more frequent maintenance inspection are some of the options available. Since the system is installed redesign would not be a very cost-effective option. The priority given by the colliery staff to the repair of failures in these systems means that repair times are already short and there is not an obvious way that these times could be reduced to the level which would result in significant reliability improvements. The third option, reducing the time between system tests, seems to be the most promising. Very good improvements in the probability of the systems to function on demand are obtained by doubling the inspection frequency to every 2 weeks. These improvements are dramatic if this is further increased to every week. For the methane trip system the improvement in the failure probability is an order of magnitude.

To achieve a recirculation fan trip all components in the safety systems beyond the HA1 board are common in all eight sub-systems. More frequent testing of the methane and carbon monoxide sub-systems would also improve the chances of all other safety systems functioning on demand.

Case study 2 – Failure mode and effects criticality analysis of gas turbine system

14.13 Introduction

The problem with most generic equipment reliability data is that information on the application and mode of operation are seldom included in the source documentation. In some cases, therefore, methods such as the well-tried technique of failure mode and effects criticality analysis (FMECA) may need to be employed to generate representative data for the important equipment featured in a systems reliability model. Expert opinion can be effective for identifying the component failure modes and frequencies required for input to the FMECA.

In this case study a DELPHI exercise was employed to generate best-estimates of the input data for the FMECA of a gas turbine used to drive an offshore electricity generating system. It benefited from the combined experience of experts in gas turbine manufacturing and operating companies who were members of the Loughborough University Rotating Machinery Reliability Group. The results were subsequently compared with generic data on similar equipment recorded in the *OREDA 92 Offshore Reliability Data Handbook* (**6**).

14.14 Gas turbine FMECA

The FMECA was based on the methods discussed in Chapter 4. Employing these techniques a preliminary FMEA (Fig. 14.4) was developed which listed possible failure modes and their expected effect on sub-system and system availabilities.

A DELPHI exercise was then performed to obtain consensus estimates of failure mode severity, failure rate, detectability, and

Unit	Sub-unit	Failure mode	Local failure effect	System failure effect	Comment
Gas turbine/generator	Air inlet	Filter fouling	Reduced output/efficiency	Degraded output	External atmospheric effect
		Filter blockage	Filter implosion	Potential catastrophic failure	Ditto
	Compressor/ turbine	Overspeed	Major damage to rotor and casing	Potential catastrophic failure	Sudden load change
		Rotor out of balance	High vibration and trip	Forced outage	Fouling/degradation
		Rotor bend distortion	High vibration and trip	Forced outage	Component degradation
		High temperature	Distortion/material damage and HT trip	Forced outage	Sudden load change
		High vibration	High vibration trip	Forced outage	Fouling/alignment change
		Tip rub	Reduced power output	Degraded output	Component degradation
		Blade failure or inlet guide vane failure	Damage to other blades and high vibration trip	Forced outage	Component degradation
		Thrust bearing failure	Major damage to rotor and stator	Forced outage	Component degradation
		Radial bearing failure	Major damage to rotor and stator	Forced outage	Ditto
		Foreign object damage	Possible blade/combustion chamber damage	Forced outage	External event
	Bleed air system	Blockage of external cooler	Over-heating leading to blade or vane failure	Forced outage	Hostile local environment
		Valve stuck shut	Start system inhibited	Failure to start	No effect
		Valve stuck open	Turbine over-temperature, gas generator overspeed	Forced outage	Corrosion/oxidation
	Combustion chamber	Explosion in combustion chamber	Possible fire, distortion of hot gas path and debris in turbine	Forced outage	Change in op. conditions
		Failure of refractory lining	Debris in turbine with consequential blade damage	Forced outage	Component degradation
	Burners	Combustion instability (pre-mix)	Pressure pulse humming and possible fatigue damage to turbine	Forced outage	Component malfunction
		Flame out	Engine rundown and stop	Forced outage	Change in op. conditions
		High temperature	Possible blade damage and HT trip	Forced outage	Ditto
	Fuel system	Sensor/actuator failure leading to malscheduling of fuel flow	High flow – turbine over temperature, gas generator overspeed, shaft over torque	Forced outage	Component malfunction
			Low flow – failure to achieve power setting	Degraded output	Ditto
	Fuel pipework	Fuel leak	Possible explosion or fire	Forced outage	Component degradation

Fig. 14.4 Gas turbine system FMEA

Unit	Sub-unit	Failure mode	Local failure effect	System failure effect	Comment
Gas turbine/generator, cont.	Inlet guide vane system	Sensor/actuator problems	Vanes malscheduling open – possible surge and compressor damage	Forced outage	Component malfunction
			Vanes malscheduling closed – possible gas generator overspeed and turbine over temerature	Forced outage	Ditto
	Casing	Air leak	Reduced output/efficiency	Degraded output	Component degradation
		Seal failure	Leakage of hot gases to atmosphere with possible fire	Forced outage	Ditto
		Valve/pipework failure	Ditto	Forced outage	Component degradation
		Exhaust flexible joint failure	Ditto	Forced outage	Ditto
Power turbine	Rotating assembly	Blade failure	Damage to other blades	Forced outage	Component degradation
		Rotor bend	Vibration and major damage to rotating assembly and stator	Forced outage	Ditto
		Tip rub	Reduced power/vibration	Degraded output	Ditto
		Thrust bearing collapse	Major damage to rotor and stator	Forced outage	Ditto
		Radial bearing failure	High vibration and trip	Forced outage	Ditto
Control and monitoring	Sensors	Spurious instrument fault	Spurious shutdown	Spurious trip	Component degradation
		Instrument fault	Fail to detect off spec. condition	Potential catastrophic failure	Ditto
	Control unit	Control hardware fault	Possible overspeed or reduced output	Forced outage/degraded output	Ditto
		Control software fault	Ditto	Forced outage/degraded output	Ditto
Lubrication system	Oil supply	Loss of supply or contamination	Bearing damage and vibration	Forced outage	External malfunction
		Fail to scavenge	Bearing chamber flooding. Rubs and damage due to out of balance forces. Internal oil fire and disc burst if leakage occurs into turbine.	Forced outage	Component degradation

Fig 14.4 Continued

active repair time for the different components in the system. This involved obtaining independent estimates from each contributor, combining these estimates into an anonymized table and circulating this table for members to confirm or modify their original estimates. The process was then repeated until general consensus was obtained.

It became evident after the first pass of the DELPHI exercise that members were significantly more comfortable with range rather than point estimates so a method based on the risk priority number (RPN) approach was adopted for the FMECA. The first meeting held to discuss the preliminary FMEA also concluded that the initial exercise was not sufficiently focused – for example the fail-to-start failure modes (which were included in the preliminary FMEA) are clearly irrelevant to a continuously running gas turbine. The FMEA, culled of all such failure modes, was then expanded into the list used in the DELPHI table. A typical initial return from one contributor to the DELPHI exercise is shown in Fig. 14.5.

For the FMECA, two risk priority numbers were defined, as follows:

$$\text{RPN1} = \text{severity} \times \text{failure rate} \times (11 - \text{detectability})$$

Similar to the definitions in Chapter 4 except that failure rate range was scored 1–4 (corresponding to the linguistic descriptors low, moderate, high, and very high) rather than 1–10. Detectability was also redefined as the probability of detection prior to failure in operation rather than detection at the design stage.

$$\text{RPN2} = \text{RPN1} \times \text{repair hours}$$

Sub-unit	Failure mode	Severity (1–10)	Failure rate (1–4)	% Critical- (forced outage rate)	Detectability (1–10)	Active repair time
Gas turbine	Overspeed to destruction	10	1	100	10	Swap
Air inlet	Filter fouling	2	3	5	8	8 h on-line
Air inlet	Filter blockage	3	2	100	10	8 h
Compressor	Blade/inlet guide-vane failure	5	2	100	10	Swap
Compressor	Surge	8	1	60	10	Swap
Combustion chambers	Explosion	8	1	100	10	Swap
Combustion chambers	Failure of refractory lining	1–5	2	100	5	24 h to change can or swap
Burners	Combustion instability	3	3	20	10	No stoppage
Fuel pipework	Fuel leak	2	3	100	5	8 h
Power turbine	Blade failure	5	1	100	10	4 wks/ 672 h
Power turbine	Rotor bend	8	1	80	10	24 h
Thrust bearing	Collapse	8	1	100	10	72 h
Control and monitoring	Instrument spurious fault	1	4	10	5	2 h
Control and monitoring	Instrument off spec.	3	4	10	2	8 h
Lubrication system	Loss of oil supply/contami nation	5	1	100	5	48 h
External cooler (for bled cooling air)	Blockage	3	2	100	5	48 h
Lubriction system	Fail to scavenge	3	2	100	8	8 h
Fuel system	Sensor or actuator problem	3	3	80	5	8 h
IGV system	Sensor or actuator problem	3	3	80	5	8 h
Compressor bleed air system	Valve stuck shut	5	2	100	10	72 h
Compressor bleed air system	Valve fails open	3	2	100	10	72 h

Note: Swap = 4 days

Fig. 14.5 Example of DELPHI response

Unit	Sub-unit	Failure mode	Local failure effect	System failure effect	Comment	Severity range (1–10)	Failure rate range (1–4)	Detectability (1–10)	Active repair time (h)	RPN1	RPN2	RPN1 rank	RPN2 rank
Gas turbine/generator	Air inlet	Filter fouling	Reduced output/efficiency	Degraded output	External atmospheric effect	3	3	8	8	27	216		
		Filter blockage	Filter implosion	Potential catastrophic failure	Ditto	3	3	8	6	27	162		
	Compressor/turbine	Overspeed	Major damage to rotor and casing	Potential catastrophic failure	Sudden load change	10	1	10	1000	10	10000		2
		Rotor out of balance	High vibration and trip	Forced outage	Fouling/degradation	4	1	7	200	16	3200		
		Rotor bend distortion	High vibration and trip	Forced outage	Component degradation	6	2	4	200	84	16800	5	1
		High temperature	Distortion/material damage and HT trip	Forced outage	Sudden load change	3	3	9	8	18	144		
		High vibration	High vibration trip	Forced outage	Fouling/alignment change	3	3	9	8	18	144		
		Tip rub	Reduced power output	Degraded output	Component degradation	3	1	5	100	18	1800		
		Blade failure or inlet guide vane failure	Damage to other blades and high vibration trip	Forced outage	Component degradation	7	2	9	100	28	2800		
		Thrust bearing failure	Major damage to rotor and stator	Forced outage	Component degradation	8	2	10	72	16	1152		
		Radial bearing failure	Major damage to rotor and stator	Forced outage	Ditto	3	3	9	8	18	144		
		Foreign object damage	Possible blade/combustion chamber damage	Forced outage	External event	4	1	3	100	32	3200		
	Bleed air system	Blockage of external cooler	Over-heating leading to blade or vane failure	Forced outage	Hostile local environment	4	3	5	24	72	1728		
		Valve stuck shut	Start system inhibited	Failure to start	No effect	5	2	5	8	60	480		
		Valve stuck open	Turbine over-temperature, gas generator overspeed	Forced outage	Corrosion/oxidation	3	2	4	8	42	336		
	Combustion chamber	Explosion in combustion chamber	Possible fire, distortion of hot gas path and debris in turbine	Forced outage	Change in op. conditions	6	2	9	100	24	2400		
		Failure of refractory lining	Debris in turbine with consequential blade damage	Forced outage	Component degradation	6	3	7	30	72	2160		
	Burners	Combustion instability (pre-mix)	Pressure pulse humming and possible fatigue damage to turbine	Forced outage	Component malfunction	4	3	8	10	36	380		
		Flame out	Engine rundown and stop	Forced outage	Change in op. conditions	4	3	9	8	24	192		
		High temperature	Possible blade damage and HT trip	Forced outage	Ditto	3	4	9	8	24	192		
	Fuel system	Sensor/actuator failure leading to malscheduling of fuel flow	High flow – turbine over temperature, gas generator overspeed, shaft over torque	Forced outage	Component malfunction	3	3	4	8	63	504		
			Low flow – failure to achieve power setting	Degraded output	Ditto	3	3	4	8	63	504		
	Fuel pipework	Fuel leak	Possible explosion or fire	Forced outage	Component degradation	5	3	7	8	60	480		

Fig. 14.6 Gas turbine system FMECA

Unit	Sub-unit	Failure mode	Local failure effect	System failure effect	Comment	Severity range (1–10)	Failure rate range (1–4)	Detectability (1–10)	Active repair time (h)	RPN1	RPN2	RPN1 rank	RPN2 rank
Gas turbine/generator, cont.	Inlet guide vane system	Sensor/actuator problems	Vanes malscheduling open – possible surge and compressor damage	Forced outage	Component malfunction	3	4	5	8	72	576		
			Vanes malscheduling closed – possible gas generator overspeed and turbine over temerature	Forced outage	Ditto	8	4	5	24	192	4608	1	4
	Casing	Air leak	Reduced output/efficiency	Degraded output	Component degradation	4	3	5	48	72	3456		5
		Seal failure	Leakage of hot gases to atmosphere with possible fire	Forced outage	Ditto	5	2	8	8	30	240		
		Valve/pipework failure	Ditto	Forced outage	Component degradation	5	2	8	8	30	240		
		Exhaust flexible joint failure	Ditto	Forced outage	Ditto	2	1	6	5	10	50		
Power turbine	Rotating assembly	Blade failure	Damage to other blades	Forced outage	Component degradation	8	2	9	100	32	3200		
		Rotor bend	Vibration and major damage to rotating assembly and stator	Forced outage	Ditto	8	2	8	100	48	4800		3
		Tip rub	Reduced power/vibration	Degraded output	Ditto	3	1	5	100	18	1800		
		Thrust bearing collapse	Major damage to rotor and stator	Forced outage	Ditto	8	2	10	82	16	1152		
		Radial bearing failure	High vibration and trip	Forced outage	Ditto	3	3	9	8	18	144		
Control and monitoring	Sensors	Spurious instrument fault	Spurious shutdown	Spurious trip	Component degradation	3	4	5	2	72	144		
		Instrument fault	Fail to detect off spec. condition	Potential catastrophic failure	Ditto	3	4	3	6	96	576	2	
	Control unit	Control hardware fault	Possible overspeed or reduced output	Forced outage/degraded output	Ditto	3	1	7	5	12	60		
		Control software fault	Ditto	Forced outage/degraded output	Ditto	4	2	6	5	40	200		
Lubrication system	Oil supply	Loss of supply or contamination	Bearing damage and vibration	Forced outage	External malfunction	5	3	5	4	90	360	3	
		Fail to scavenge	Bearing chamber flooding. Rubs and damage due to out of balance forces. Internal oil fire and disc burst if leakage occurs into turbine.	Forced outage	Component degradation	4	3	4	8	84	672	4	

Fig 14.6 Continued

RPN2 was an attempt to reflect the importance of the repair process to system availability. These definitions were employed in the FMECA shown in Fig. 14.6, from which the top-five important failure modes can be ranked by criticality with respect to expected system availability performance. The failure modes and their RPN rankings are shown in Table 14.9. It can be seen that the effect of including repair time (RPN2) changes the ranking of the top-five failure modes and highlights the importance of failures in the rotating assemblies. In the view of the experts RPN2 was significantly more effective in identifying and ranking the failure modes critical to system availability.

Table 14.9

Sub-unit	Failure mode	RPN1 rank	Sub-unit	Failure mode	RPN2 rank
Inlet guide vane system	Sensor/actuator problems	1	Compressor/ turbine	Rotor blend distortion	1
Control and instrumentation	Instrument fault	2	Compressor/ turbine	Overspeed	2
Lube oil system	Loss of supply or contamination	3	Power turbine	Rotor bend distortion	3
Lube oil system	Fail to scavenge	4	Inlet guide vane system	Sensor/actuator problems	4
Compressor/ turbine	Rotor blend distortion	5	Casing	Air leak	5

The use of range estimates of the different parameters also allowed the construction of a 1629A-type criticality matrix relating failure mode frequencies to repair times. The figures in the different cells of the matrix shown in Fig. 14.7 indicate the number of component failure modes in each category.

Failure rate range		< 10 h (1)	10–99 h (2)	100–999 h (3)	1000+ h (4)
4	> 1 f/year	1	1		
3	0.1–1.0 f/year	11	3		
2	0.01–0.1 f/year	4	6	1	
1	< 0.01 f/year	1	1	1	1

Repair time range

Fig. 14.7 Gas turbine criticality matrix

The critical failure modes are in the repair time ranges 4 and 3, namely compressor/turbine (CT) overspeed, CT rotor bend distortion, and CT rotor out-of-balance, with similar – but less extreme – consequences for the power turbine failure modes. Thus it can be seen that a 1629 FMECA can also be effective in identifying the failure modes likely to result in the long-outage events which impact on system availability performance if the criticality matrix is configured in this way.

From the criticality matrix it is also possible to see the likely distribution of modal failure rates and repair times. The two histograms, shown in Fig. 14.8 with regression lines, indicate a log-normal distribution of component failure rates and a positively skewed repair time distribution.

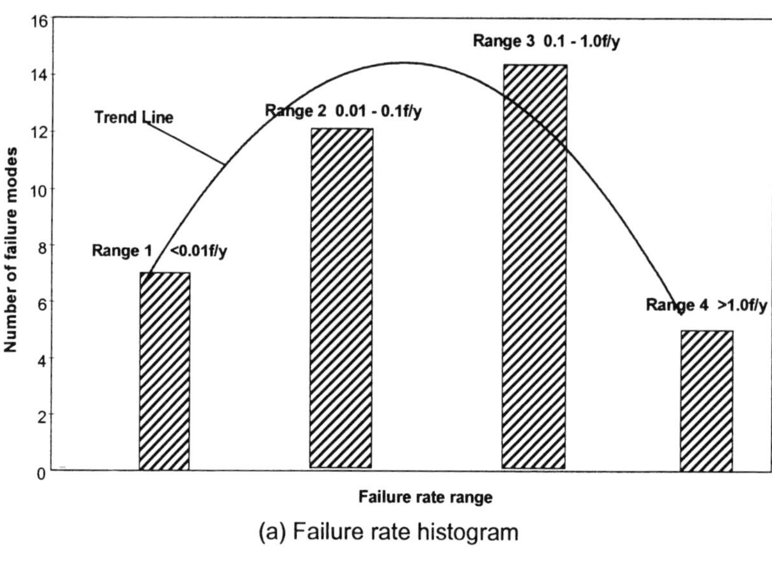

(a) Failure rate histogram

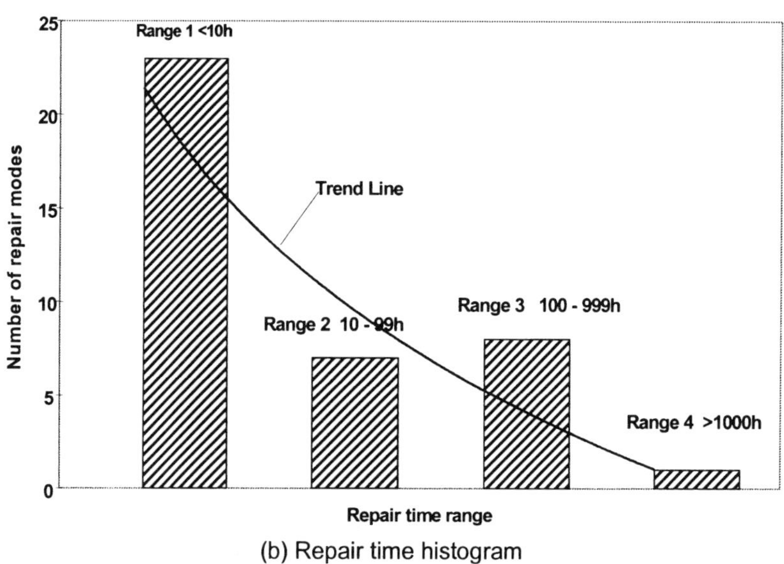

(b) Repair time histogram

Fig. 14.8

Since the scaling for failure rates is logarithmic it would seem reasonable to assume the geometric mean as a best-estimate of the average failure rate within each failure rate range. An estimate of the forced-outage rate for a gas turbine system can then be obtained by multiplying these average failure rates by the number of failure modes in each range. Table 14.10 shows the calculations.

Table 14.10 Gas turbine system – estimate of mean failure rate

Failure rate range (f/year)	Geometric mean of range (f/year)	No. of failure modes in range	Range failure rate (f/year)
1. <0.01	0.003	4	0.012
2. 0.01–0.1	0.03	11	0.33
3. 0.1–1.0	0.3	14	4.2
4. >1.0	3.0	2	6.0
		TOTAL	10.5

From this table an estimate of the overall failure rate for a gas turbine system is 10.5 f/year (1203 f/Mh), cf. 1163 f/Mh for gas turbines given in *OREDA 92* (**6**).

A similar approach can be adopted for estimating the mean repair time, see Table 14.11.

Table 14.11 Gas turbine system – estimate of mean repair time

Repair time range (h)	Geometric mean of range (h)	No. of failure modes in range	% of failure modes in range
1. < 10	3	17	55
2. 10–100	30	11	35
3. 100–1000	300	2	7
4. > 1000	3000	1	3

Ignoring the very high outages in range 4, since this will probably involve replacement rather than repair, then an estimate of the mean active repair time (MART) is

$$\text{MART} = \frac{(17 \times 3 \text{ h}) + (11 \times 30 \text{ h}) + (2 \times 300 \text{ h})}{30} = 33 \text{ h}$$

Estimates generated by OREDA data are typically about 23 h, with the minimum and maximum of the range given as 1.0 and 4280 h respectively. Because the distribution of failure mode repair times is so strongly skewed it is perhaps politic to calculate gas turbine availability from the estimated proportion of failures in each repair time range.

For range 1 with 55 per cent of failures:

$$\mathrm{MTBF}_1 = \frac{8760}{0.55 \times 10.5} = 1516\ \mathrm{h}$$

Hence

$$Q_1 = \frac{\mathrm{MTTR}_1}{\mathrm{MTBF}_1 + \mathrm{MTTR}_1} = \frac{3}{1516 + 3} = 0.002$$

Similarly $Q_2 = 0.012$, $Q_3 = 0.025$, and $Q_4 = 0.097$. Hence an approximate steady-state unavailability estimate for the system is

$$Q_s = 0.002 + 0.012 + 0.025 + 0.097 = 0.136$$

The expected availability A_s of an aero-derivative gas turbine system operated continuously onshore is therefore about 86 per cent. The importance of the single high-impact, low-probability (HILP) failure mode in repair time range 4 is particularly worthy of note since it is responsible for over 70 per cent of the total estimated downtime.

14.15 Discussion

The most effective use of FMEA/FMECA has always been in the detailed studies carried out at the design stage. This is how it was originally conceived and developed for equipment where failure during a mission could be catastrophic. More recently FMEA has been intelligently adapted to other applications, such as reliability-centred maintenance, where consideration is usually limited to the functional failure modes of the different sub-units rather than the component failure modes of particular importance to the designer of one-off devices. This exercise also demonstrates its value for repairable systems where it can supplement the information obtained from field data analysis both in identifying the critical failure modes (that is, those failure modes which affect the operational reliability of the equipment), the failure mechanisms which are likely to cause component degradation, and the high-impact, low-probability (HILP) events which can result in very long outage times. What is less apparent is the importance of including and quantifying the impact of the repair process on the operational reliability of the equipment, in addition to the identification and quantification of failures modes which can be hazardous to both personnel and plant.

It is notable that the use of range estimates for failure frequency and consequence of failure is much easier for engineers to accept than making point estimates of the likely mean failure rate and (to a lesser degree) the mean time to repair for each failure mode in the FMECA. Engineers work every day with exact measurement of size, pressure, temperature, etc. When there is considerable uncertainty, as for example, in guessing how frequently something will fail, there is often a mind block, particularly when remote from the operating equipment. Component failure rates are needed in FMECAs since (with the assumption of constant failure rates) the failure rate of the equipment is the aggregated total of the component failure rates. The assumption of exponentially distributed failure times is clearly acceptable for one-off devices, particularly when many of the components are electronic, and is probably the best that can be done currently for mechanical items given the paucity of the available data and the uncertainties associated with operational, environmental, and maintenance conditions. Here it is evident that robust estimates of both mean failure rate and mean repair time can be derived from an FMECA by using ranges for both failure frequencies and repair times within the horizontal and vertical divisions of the criticality matrix. These figures may, however, need modifying to reflect the expected operational and environmental conditions associated with a specific application.

14.16 Summary

An FMECA provides a complementary analysis of failure characteristics of complex rotating machinery systems such as gas turbines, particularly for identifying the failure mechanism likely to be associated with the critical sub-units and the infrequent high-impact, low-probability events which can result in very long outage times.

The benefit of the RPN method for FMECA is limited for repairable systems unless a factor is included for repair time in the RPN definition. On the other hand a criticality matrix employing range estimates for both failure mode frequency and repair time appears to offer significantly more information on the likely failure and repair characteristics of a complex system such as a gas turbine.

This can be obtained from a MIL HBK 1629A assessment when severity is defined in terms of repair time.

The criticality matrix can also provide robust estimates of mean failure rate and repair time for a complex system. If there is significant skewness in either distribution it becomes evident from the histograms of modal failure rates and repair times. This can indicate whether the assumption of a single repair mode is valid or whether a number of different categories of failure/repair mode need to be considered.

In this FMECA estimates of MTBF and MTTR were obtained which are evidently in good agreement with field data analyses and generic sources for gas turbines.

Case study 3 – In-service inspection of structural components (application to conditional maintenance of steam generators)

Henri Procaccia: ESReDA (European Safety Reliability & Data Association) co-founder and Honorary Member, Electricité de France/ Direction des Etudes et Recherches (EDF/DER), Scientific advisor of French Dependability Institute (ISDF).
André Lannoy: Electricité de France/Direction des Etudes et Recherches (EDF/DER). Vice-President of the French Dependability Institute (ISDF) and present ESReDA President-Elect.

14.17 Introduction

In the field of nuclear plant reliability, probabilistic methods are used to evaluate the effects of deterioration due to ageing, particularly on major passive structural components such as steam generators, pressure vessels, and primary piping. The remaining life of these components will generally depend on different types of degradation mechanisms, such as corrosion, fatigue, erosion-corrosion, stress-corrosion cracking, and embrittlement under irradiation.

Probabilistic methods provide an alternative to a purely deterministic approach in assessing the effect of these different degradation mechanisms. The results can offer a more comprehensive evaluation of safety margins as a function of component age with the scope for applying sensitivity analysis to determine the effect of uncertainties in the model and the data employed. It is also possible to identify and quantify the impact of key factors affecting safety, such as the systems, procedures, and preventive maintenance actions provided to mitigate the consequences of failure. As noted by Flesch *et al.* (7) the information obtained can also be applied to support realistic safety assessments and component lifetime predictions and to optimize in-service inspection (ISI) programmes and maintenance strategies.

Fortunately the rupture of a structure is an extremely rare event in industry [Procaccia *et al.* (**8**)] and data used for safety and risk estimation in the decision analysis process are rarely failure data but essentially results from the in-service inspections which periodically give the status of the structure. These collected data include the detection of minor cracks from which the risk of primary water leakage (affecting plant availability and maintenance) or the risk of failure (affecting safety) can be derived from a deterministic degradation model in which all uncertain parameters are probabilized.

Critical structures affected by ageing in nuclear plant include the steam generators (SG) and worldwide experience has indicated significant stress-corrosion cracking (SCC) problems in steam generator designs employing Inconel 600 tube bundles (**9**). Initiation and propagation of longitudinal cracks (more than 80 per cent of detected cracks) have been observed in the kiss-rolled zone (Fig. 14.9) of the 3500 tubes of a specific design.

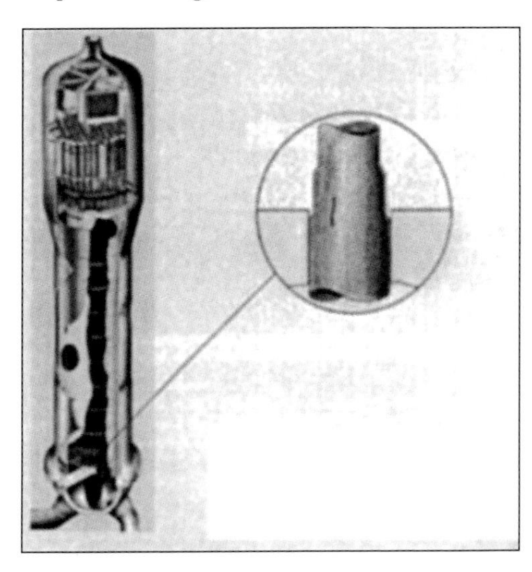

Fig. 14.9 Steam generator and tube kiss-rolling zone

Primary water leakage can occur when the size of a crack reaches 6 mm. This type of incident affects plant availability because a shutdown of the installation is necessary to perform the corrective maintenance action (plugging the affected tube). A tube rupture could follow when the size of the crack becomes greater than 25 mm under accidental steam generator overpressure, although plant safety may not be significantly affected. However, the subsequent operational transient can be very severe and difficult to manage.

For plant operators a primary water leak is considered to be a maintenance-related incident. Tube rupture is always associated with plant safety. Safety and availability maintenance objectives for steam generators are, therefore, as follows.

1. To limit the probability of tube rupture to a reasonable level. The accepted probabilistic risk criterion of tube rupture affecting plant safety is 10^{-2}/reactor-year. To ensure this criterion can be met with a conservative margin, tubes with crack sizes $\geq$ 13 mm are plugged. (This crack size was determined from laboratory tests which showed tube rupture at 25 mm crack size under maximum SG overpressure of 155 bar, and 45 mm crack size at the operating pressure of 70 bar.)
2. To avoid any primary water leaks > 5 l/h in operation. Leaks of this size trigger an automatic shutdown of the plant. This criterion allows plant operation to continue with SG tube cracks $\leq$ 6 mm.

To ensure these objectives are met, Electricité de France (EDF) has defined a preventive base programme of steam generator tube control. This periodic base programme is applied at each plant shutdown for refuelling (about once every operating year). Both safety and availability criteria and the state of ageing of the steam generator define the periodicity of systematic control in particular zones of the tube bundle, and also the rate of tube plugging in these zones. Depending on the observed inspection results a helium test can be applied to detect micro-leaks. A regulatory hydraulic test is also applied every 10 years to verify the overall status and resistance of the tube bundle.

14.18 Data needed for safety and maintenance objectives

The French OMF method (optimisation de la maintenance par la fiabilitie, or reliability-centred maintenance) is suitable for degradation mechanisms with slow kinetics such as encountered in steam generators. When such a degradation is observed a corrective maintenance action can be carried out before the occurrence of a failure.

Crack growth mechanisms affecting the tube bundle can be divided into two sub-groups:

(a) the initiation phase
(b) the propagation phase.

A probabilistic extension of the deterministic degradation and tube rupture model, based on laboratory tests and the analysis of the results of systematic steam generator controls, has been developed by Pitner *et al.* (**9**). This model predicts the initiation of a crack which can propagate until there is a primary water leak, and its subsequent development if not detected before tube failure. The difficulty is to define failure criteria and their formalization into mathematical language using easily measurable or calculated physical parameters. For cracking mechanisms a critical defect size is usually defined. For steam generator tubes this critical size is a random variable derived from inspection data which can be characterized in the following ways.

1. As a function of the maintenance targets. This includes all the detected crack sizes potentially leading to a primary water leak of > 5 l/h – that is, all cracks predicted to grow to more than 6 mm during the next operating cycle of the plant. The maintenance-critical crack size criterion of 6 mm is designed to be conservative.
2. As a function of the safety aspects. This includes all cracks potentially leading to a tube rupture in the next operating cycle of the plant. To be conservative the safety-critical crack size criterion is defined as 13 mm.

Detection of new initiated cracks and propagation of existing cracks in the SG tube bundle are measured with eddy-current probes. The crack size detection limit of the probes is 1 mm and the probability of detection is 0.9999 for crack sizes > 3 mm. The steam generator preventive maintenance actions are based on the results of inspections made on a fraction of the tube bundle.

14.19 The steam generator maintenance programme

Initially the SG base in-service inspection (ISI) control programme aimed to inspect one-eigth of the tube bundle at each refuelling of the plant. When a crack was predicted to reach a critical size relative to rupture or a primary water leak > 5 l/h (13 and 6 mm crack sizes respectively) during the next plant operating period, a conditional maintenance action, consisting of the systematic plugging of the affected tubes, was carried out. When 15 per cent of tubes were plugged the affected steam generator was replaced.

This base inspection and maintenance programme was subsequently optimized by using a structural reliability approach (SRA) model to predict the probability of occurrence of new cracks and the propagation of existing cracks during future plant operating cycles. The optimization programme determines the periodicity of tube bundle in-service inspections, the number of tubes to inspect in different zones, the size of cracks relevant to the maintenance and safety criteria, and an estimation of the remaining lifetime of the steam generator.

14.20 Expected benefits of the probabilistic ISI base programme

Deterministic methods traditionally used in nuclear plant at the design stage are mainly based on design codes and regulatory requirements. They aim to guarantee the integrity of components by including a number of conservative assumptions on inspection data in the degradation model to predict, typically, a very conservative performance attribute through safety margins. The conservatism introduced is acceptable as long as the margins are large, but with ageing the pessimistic assumptions can give results too far from reality to determine realistic lifetime strategies.

In order to deal more precisely with the various uncertainties in the measured data and the degradation model and to avoid accumulating pessimistic (and costly) assumptions, a probabilistic approach is used to provide more realistic and consistent estimates of the reliability of components. Also it is obvious that where extremely rare events are concerned, such as SG tube rupture, the classical or statistical methods are inappropriate. For these situations a Bayesian structural reliability approach (SRA) has been developed. This methodology is based on an understanding of the failure mechanisms, the deterministic degradation model, and the introduction of influential uncertainties into this 'deterministic' model in the form of random variables. In addition to estimating the probability of failure (tube leak or rupture) the SRA model can include sensitivity studies to evaluate the impact of various sources of uncertainty to identify the key parameters of the model. This provides the necessary data to optimize the in-service inspection and maintenance programmes of passive structural components.

The Bayesian SRA model developed by Madsen *et al.* (**10**) clearly provides a more realistic and detailed image of component condition (including the prediction of failure probability) than the deterministic analysis which only determines if the current situation is acceptable or not. The additional information included in the Bayesian SRA model is shown in Fig. 14.10.

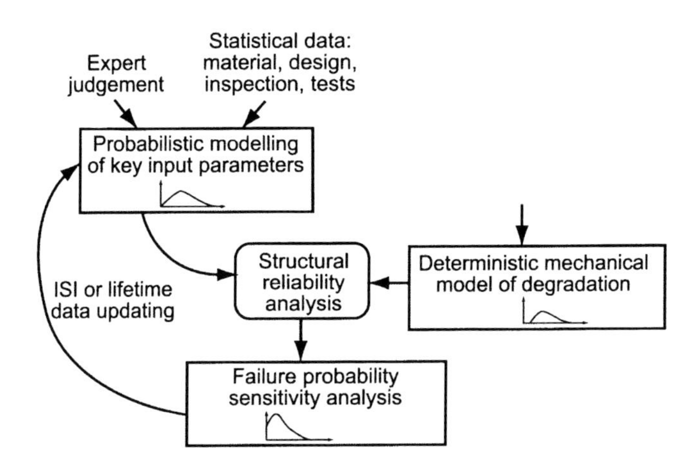

Fig. 14.10 Simplified Bayesian structural reliability analysis diagram

The model includes consideration of the uncertainties in the basic data. These uncertainties are derived from various sources.

1. *The variability.* Associated with the properties of materials used in manufacturing but also with previous experience of the tube manufacturers' and suppliers' past performance.
2. *The sizing geometry.*
3. *The degradation mechanism.* Stress-corrosion model, load variations, local stress intensity factor, presence of initial defects, etc.

4. *Performance characteristics of in-service inspection equipment*: probability of detection, accuracy of measurement, NDE and operator errors.
5. *The potential consequences of preventive maintenance*. To mitigate the impact of potential failures – repairs, tube plugging, shot-peening of the SG tubes for local stress release, etc.

These uncertainties are included in the analysis in the form of random variables. The objective is to translate knowledge concerning the uncertainties as accurately as possible by probability distributions (histograms or parametric distributions). This feature of the model is an important step in the analysis which requires particular care.

Different inspection and operation scenarios can be tested to quantify their effect on the risk and remaining lifetime of the steam generator. Another advantage of this analysis is that it provides a quantification of the influence of the uncertainties on the probability of failure; thus it is possible to concentrate solely on the scope for reducing the failure probability. If the uncertainty associated with a model variable has a significant impact on the final result, Bayesian techniques are used to balance the weakness of observation or statistical knowledge with expert opinion or, alternatively, for updating the prior knowledge of the variable as new data are collected (**11**, **12**).

14.21 Data for safety and data for maintenance

The data collected during a specific in-service inspection include information on defects (internal longitudinal cracks of different sizes) detected on the inspected tubes in the control sample. The actual steam generator preventive maintenance programme is then defined from the predicted status of the tube bundle derived from these observed data. This prediction is based on the results of more than 200 laboratory tests performed on SG extracted tubes to determine the crack detection performance, the kinetics of the defects, the tube rupture characteristics, and particularly:

- the performance of eddy-current probes: detection capability and accuracy
- the kinetics of crack propagation
- the size of crack associated with a leak
- the kinetics of the leak flow rate
- the size of crack leading to a tube rupture.

The first four test statistics are mainly maintenance-orientated; the last one concerns plant safety.
These tests have shown that:

(a) A crack initiation can be detected when its size reaches 1 mm;
(b) When a crack is initiated the crack propagates with a mean velocity < 1 mm per operating year;
(c) A small leak always occurs when the crack size (a) reaches about 6 mm (leak before break).

This means there is a long time between tube leakage and tube rupture. In particular, even in cases of accidental overpressure of the SG (155 bar; normal operating pressure is 70 bar) a detectable leak would occur during operation before the occurrence of a tube rupture. (Note that tube rupture only occurs when the crack size is greater than 25 mm.)

The EDF base maintenance programme can accept operation of steam generators with cracked tubes if the potential leak rate of primary water is less than 5 l/h and the risk of tube rupture remains at an acceptable level.

With these results and with numerous in-service observations made on all inspected steam generators, a probabilistic fracture mechanics code has been written (COMPROMIS: code de mecanique probabiliste pour la maintenance et l'inspection en service; Fig. 14.11) by Pitner *et al.* (**13**).

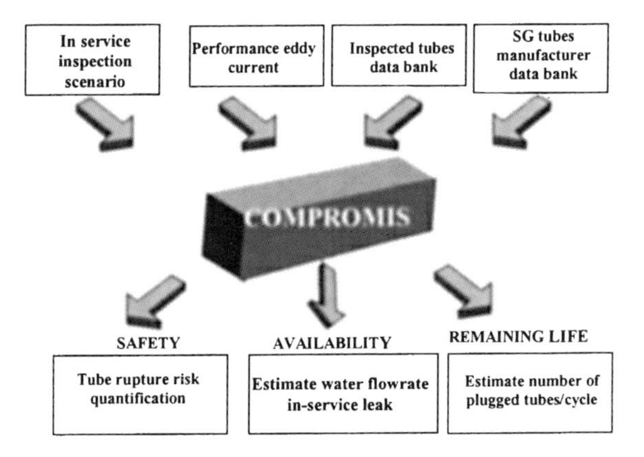

Fig. 14.11 Architecture of COMPROMIS code

Taking into account the performance of eddy-current probes (probability of crack detection and accuracy of crack size measurement) and the number of tubes inspected, the structural reliability model evaluates the expected number of affected tubes in the tube bundle, the number of tubes which must be plugged (conditional maintenance for all tubes with cracks ≥ 5 mm), and the risk of a tube rupture during the next operating cycle of the plant. The code also optimizes the in-service inspection strategy and preventive maintenance programme with respect to time, as a function of the safety and availability of a specific steam generator.

The code input data are derived from the results obtained from eddy-current probes during a specific in-service inspection, the tube bundle design data, the plant operating data, the performance of the eddy-current probes, and the fracture mechanics model calibrated with laboratory test data.

14.22 The probabilistic fracture mechanics model

The probabilistic fracture mechanics analysis is based on an understanding of the deterministic model which, as shown by Procaccia and Morilhat (**14**), can be used to simulate, *a priori*, the behaviour of the tube bundle by taking into account all uncertain parameters in the form of probability distributions. Schematically, if a tube failure is caused by a crack-like flaw the failure condition is

$$a > a_{\mathrm{c}}$$

where a is the crack size at a given time and a_{c} is the critical size potentially causing either a leak (a_{c} > 6 mm) or a tube failure (a_{c} > 13 mm). In fact a and a_{c} are random variables, as noted previously. There are many factors contributing to the uncertainties as indicated in the COMPROMIS flowchart (Fig. 14.12). Examples are as follows.

1. The statistical distribution of crack sizes found at the last inspection. This is characterized by a log-normal law with parameters derived from observed data.
2. The performance of the in-service inspection equipment (detection and measurement of flaws). This is characterized by a Weibull law derived from eddy-current performance data tests:

$$P_{\mathrm{D}} = 1 - \exp(-a/\eta)^{\beta}(1-\varepsilon)$$

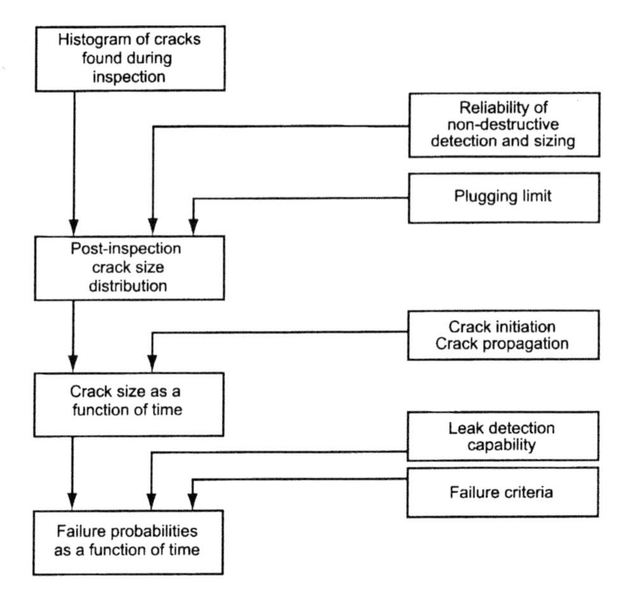

Fig. 14.12 General flowchart of COMPROMIS code

Obviously the sensitivity of the results has been tested against other appropriate statistical distributions.

3. The probability of isolating flaws (tube plugging).
4. Variability associated with the distributions. Modelling the time to crack initiation (t_f) and the crack propagation kinetics which are, respectively

$$t_f = k\sigma^{-4} \exp(-Q / RT)$$

$$da / dt = C[K(a)]^m$$

with

$$K(a) = x \log(a) + ya + z$$

Where

σ = local stress
T = temperature in degrees Kelvin
Q = activation energy in J/mol
k = an empirical coefficient characterized by material data

R = gas constant
C = Paris law constant (log-normal law)

The constants m, x, y, and z are derived from field data ($1 < m < 4$). The constant k is difficult to estimate because of differences in the local stress field (particularly complex in the kiss-rolled zone of the tube), the operational stresses, and the materials data. In practice k varies between a k(threshold) value (0 μm/h between 10 and 20 $MPa.m^{1/2}$ for different categories of Inconel 600) and a k(asymptotic) value (0.8 to 8 μm/h above 20 $MPa.m^{1/2}$).

5. The leakage flowrates. These are determined from an empirical law deduced from numerous laboratory tests and the probability of detection of leaks in-service.
6. The variability of parameters used in calculating the failure criteria and the safety margin

$$M^* \sigma_\theta^* = \sigma_f$$

determined from an empirical law deduced from numerous laboratory tests. M^* is the geometric shape factor of the cracked tube which takes into account the local effect of the SG tubesheet, σ_θ^* is the circumferential stress in the tube due to the applied pressure (particularly on the lips of the flaw), and σ_f is the characteristic stress of the tube material depending on its elastic and rupture limits.

Figure 14.13 shows the results of the laboratory tube rupture tests, the conservatism of the critical crack size (13 mm), and the deterministic worst-case calculation (17 mm) compared with the real test results.

The parameters of the distribution laws used in the code are deduced from data banks on material properties, laboratory test results on tube leak flow rates, tube rupture conditions, and from field data (measured flaws) taking into account all potential sources of error. Table 14.12 summarizes the main calculated and observed data.

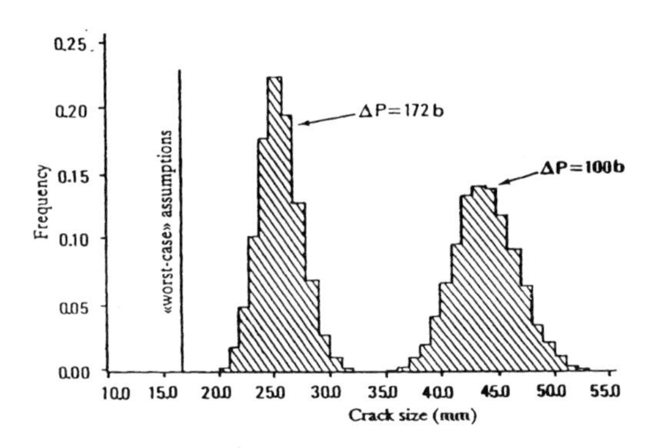

Fig. 14.13 Tube rupture laboratory test results in normal or accidental situation

Table 14.12 Probability distributions and parameters used in the model

Parameter		Statistical distribution	Expectation or scale parameter	Standard variation	Confidence interval
Tube thickness	e (mm)	Normal	1.27	0.04	$-3\sigma, +3\sigma$
Diameter	D (mm)	Normal	22.22	0.06	$-3\sigma, +2.2\sigma$
Crack initiation probability	δ (MPa)	Weibull	46 000	1.2	
Elastic and critical stress	MPa	Normal	956	50.3	$-3\sigma, +3\sigma$
Stress	K (Mpa.m$^{1/2}$)	Normal	0.58	0.01	$-3\sigma, +3\sigma$
Initiated flaw	$2a$ (mm)	Normal	3	1.5	
Paris coefficient	C	Log-normal	−12.6	0.84	
Crack probability detection	P_D	Weibull	2.37	2.25	Residual error $\varepsilon = 10 \times 10^{-4}$

14.23 Safety and maintenance-orientated results

As noted previously the COMPROMIS code calculates the probability of occurrence of a steam generator leak (leading to immediate plant shutdown and immediate corrective maintenance if the water flowrate exceeds 5 l/h) and the probability of plant rupture (degradation of plant safety) by combining the laws of probability of the various input data of the model. There are four calculation steps performed by the code, starting from the observed in-service inspection data to the final prediction of the risk of tube failure during the next plant operating cycle.

14.23.1 Developing the preventive maintenance strategy

Estimation of the general state of the steam generator tube bundle is based on a sample of the inspection data. Starting from a specific inspection strategy (rate of inspected tubes in different zones of the tube bundle) and the histogram of cracks detected with eddy-current probes, the code determines the probability distribution of all cracks inside the tube bundle and predicts the distribution of small initiated cracks which cannot be detected with eddy-current probes (Fig. 14.14).

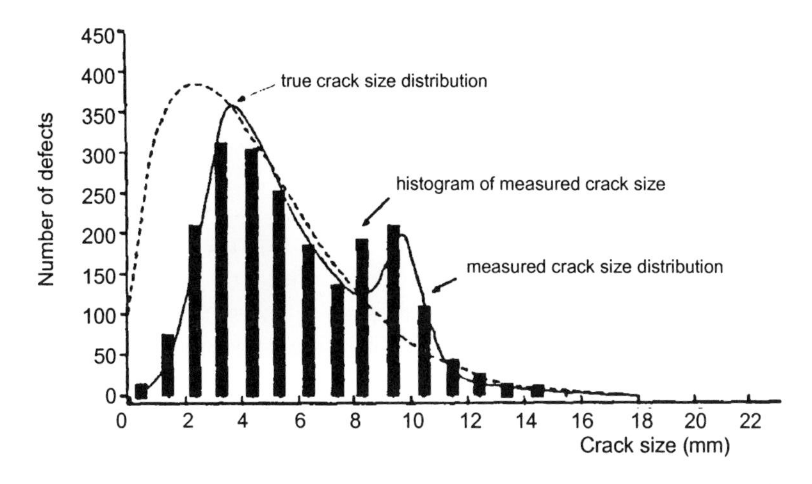

Fig. 14.14 Observed and computed distributions of crack sizes

Note that the histogram of the results indicates a second mode for the measured size of defects. This is due to a misinterpretation of eddy-current probe results when modifying the (analogue) calibration of the probe. This error was detected by the probabilistic model and was corrected by a subsequent modification to the probe electronics. The maintenance programme is derived from the observed data and the computed probable evolution of these defects with time.

14.23.2 Evolution of the crack size distribution with time

During the next plant cycle new cracks will be initiated and existing cracks will propagate. For these conditions the COMPROMIS code computes the future risk of tube leaks and the risk of tube rupture during plant operation (in normal and accident situations) from the probable evolution of the crack size distribution (Fig. 14.15).

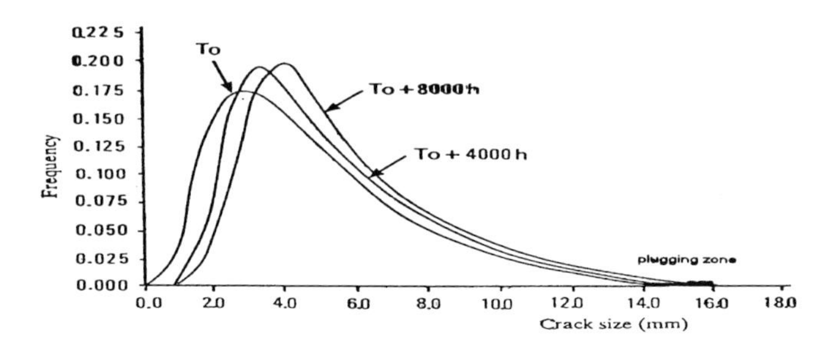

Fig. 14.15 Evolution of crack size versus time

The final predicted crack distribution takes into account the inspection strategy, information on the observed cracks, the probable distribution of all cracks in the steam generator, the plugging strategy, and the results of the eventual helium test.

14.23.3 Determination of future leak and rupture risks

To determine the optimum plant availability and safety strategy before plant start-up, the code determines the number of future potential leaks (among the cracked tubes), the corresponding primary water flowrate, the risk of plant shutdown, and the risk of tube rupture during the next and following plant operating cycles. The maintenance programme aims to plug all cracked tubes which could develop into a leak greater than 5 l/h during the next operating cycle and, obviously, all tubes that could reach the safety-critical size during this time.

14.23.4 Determination of steam generator residual life – strategy for SG replacement

When 15 per cent of the tubes in the steam generator tube bundle are affected (about 500 tubes) the performance of the heat exchanger decreases so much that the steam generator needs replacing. Obviously the SG remaining life strongly depends on the recommended plugging criterion, as demonstrated by the simulation for a particular steam generator shown in Table 14.13.

Table 14.13 Remaining lifetime calculated for different plugging criteria. At the beginning of the simulation N_o = 72 985 h and 127 tubes are already plugged

	Prediction of the number of tubes plugged as a function of the plugging criteria		
Operating cycle	11 mm	13 mm	15 mm
N_o	124	47	20
$N_o + 1$	89	38	22
$N_o + 2$	110	53	33
$N_o + 3$	139	73	49
$N_o + 4$	175	98	69

In this table the remaining life is calculated for different plugging criteria. At the beginning of the simulation the steam generator operating hours value was 72 985 and 127 tubes were already plugged. If the SG replacement criterion is 500 plugged tubes (15 per cent) this operation is required:

(a) after two further plant operating cycles for 11 mm cracks
(b) after four cycles for 13 mm cracks – the accepted safety criterion
(c) after more than six cycles if a 15 mm crack size criterion is adopted.

It can be seen that choosing a 15 mm criterion will substantially increase the life duration of the steam generator while keeping within an acceptable level of risk (1000–10 000 times less than the tube rupture safety criterion).

14.24 Sensitivity analysis

One of the main advantages of the probabilistic fracture mechanics model is the scope for sensitivity analysis. For example the simulated data in Table 14.13 show the impact on steam generator remaining life based on different tube plugging criteria.

14.24.1 Comparison between probabilistic and deterministic models

Figure 14.16 shows the difference for tube rupture risk determined probabilistically compared with the result from a conservative deterministic model.

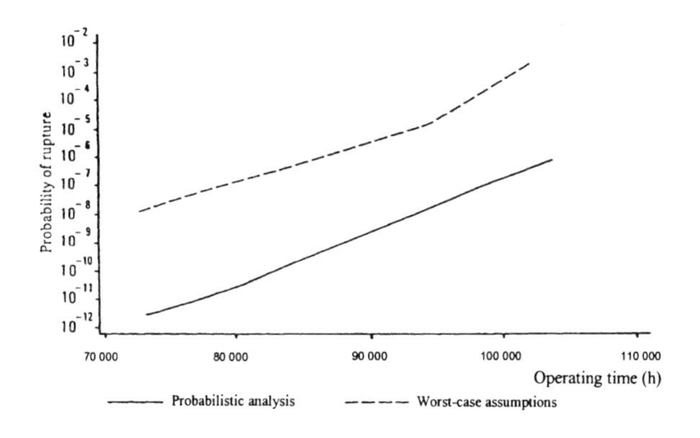

Fig. 14.16 Probabilistic and deterministic SG safety level versus time (without maintenance)

The simulation does not take into account the next planned in-service inspection which would be carried out after about a further 10 000 hours of operation. This frequency of in-service inspections maintains the probabilistic risk of tube failure at an approximately constant level of 10^{-6}/plant-year.

14.24.2 Impact of plugging criteria – data for plant safety strategy

Figure 14.17 shows the results of a sensitivity analysis based on critical crack size criteria – the reference criterion derived from laboratory tests is 13 mm. It is worth noting that 13 mm is very conservative compared with 25 mm for the accident case and 45 mm in normal operation, hence choosing an 11 mm criterion will lead to a decreasing risk at each new inspection.

The probability of missing a crack decreases with the number of inspections; however, it is notable that the residual risk always stays very small and is not comparable with other plant accepted risks (10^{-10} versus acceptable risk of 10^{-2}).

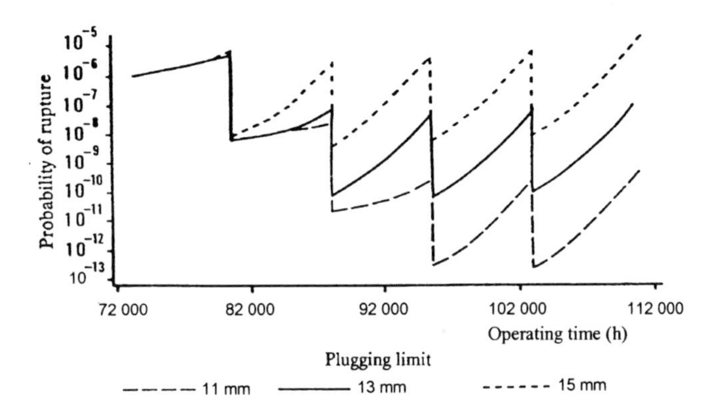

Fig. 14.17 Influence of the plugging criteria on the SG safety level (accidental situation, annual maintenance)

14.24.3 Influence of the rate of controlled tube inspections – data for maintenance strategy

Inspection of 100 per cent of the tube bundle after 73 000 operating hours equates to negligible risk even after three further operating cycles without any new inspection. However, inspection of a control sample of only one in eight tubes is still acceptable after 8 to 10 calendar years of steam generator operation compared with the accepted risk criterion (10^{-2}/year) as shown in Fig. 14.18.

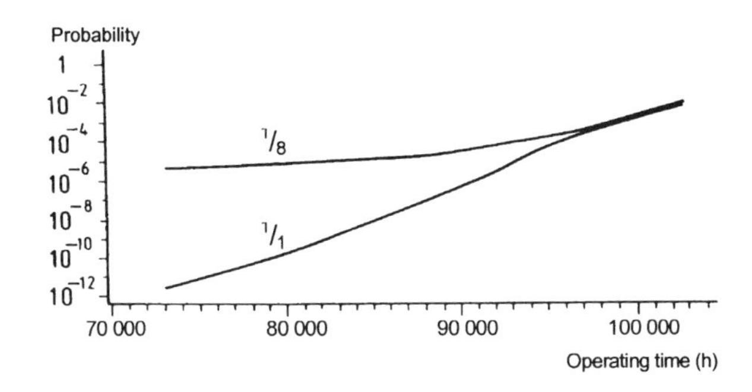

Fig. 14.18 Influence of the rate of inspected tubes versus time

14.25 Conclusions

The COMPROMIS code produces significant quantitative data for operators of pressurized water reactor (PWR) steam generators concerned with expected current and future safety status as a function of inspection/maintenance strategies. It supports decisions aimed at optimizing monitoring policy (periodicity of inspections, rate for controlled tube as a function of safety, and maintenance objectives) and the prediction of expected life duration of the steam generator. The code is currently used for the following functions.

1. To test and compare various maintenance scenarios with a view to cost/safety optimization (type of inspection, inspection periodicity in different zones of the tube bundle, sampling strategy, and plugging criterion).
2. To predict the residual life of a steam generator by simulating the number of tubes plugged versus time (Table 14.13).
3. To validate the consistency of the extensive input data used in the model by simulation (e.g. crack propagation model or leakage flowrate model).

The software has been operational since the end of 1990 and has been progressively applied to determine optimal replacement dates for stress-corrosion affected steam generators in about 20 plants. It is capable of dealing with longitudinal and circumferential cracking (the majority of the stress-corrosion cracking maladies of Inconel 600 material) and can cope with other types of tube material, or other shapes of the tube kiss-rolled zone (integral rolling for instance).

Input to the code is user-friendly and flexible, permitting the definition of steam generator inspection and maintenance strategies as a function of accepted safety risk on specified steam generators of Electricité de France.

Case study 4 – Business-interruption risk analysis

14.26 Introduction

Unreliability of major items of equipment can have a material effect on the profitability of process plant. Many modern production systems are single stream with relatively little redundancy so that unplanned repair or replacement of equipment can substantially reduce the yearly output and profitability of the installation. A comprehensive availability assessment during the conceptual design phase can ensure that the critical systems and equipment are identified and their inspection and maintenance strategies optimized to maximize product availability. For existing plant the scope for modifying the main equipment configurations is limited; however, a simpler approach concentrating on the reliability-critical systems and equipment can be applied as shown in this case study.

Identifying and quantifying the likely frequency and consequences of major outages fosters a systematic review of the needs for spares holding, inspection/maintenance frequencies, and business-interruption insurance. Developing an availability model of the production process goes a long way towards understanding the effect of different damage mechanisms and the uncertainties inherent in the operation of the process. The method is illustrated here with an assessment of a combined-cycle electricity generating plant which was carried out to determine the likely frequency and consequences of plant outages and the need for business-interruption insurance. Because long outages are almost invariably associated with deterioration of large equipment the model was based on the relatively few critical systems. These were represented on a set of integrated spreadsheets using simple analytical models. Simulation was subsequently employed to review the implications of different assumptions and the effect of uncertainties in the input data.

14.27 Risk assessment

In the process industries risk is generally defined as the product of likelihood and consequence – the frequency of a specified unwanted event (say, a high-energy, toxic, or flammable release) and its consequences in terms of harm to personnel, the plant, or the environment. The Control of Major Accident Hazards (COMAH) regulations have been developed over many years for this type of situation. Probabilistic risk assessment (PRA) may well be applied. Fundamentally it is concerned with the combination of events linked to the initiating failure mode and the impact of the different safety systems and barriers installed to limit the extent of the damage.

More recently interest in the process industries has focused on risk-based inspection (RBI) and in particular on the pressurized equipment covered by statutory inspection regulations. The objective in this case is to assess the likely effect of increasing (or decreasing) the inspection interval on the safety and integrity of the equipment. RBI assessments range from subjective, qualitative methods to more complex evaluation involving quantification and ranking of the different risks.

Both of the above are primarily concerned with safety – PRA with **system** safety and RBI with **equipment** safety. Rotating machinery reliability is not a specific issue except in terms of its accident potential – say, from rupture of its containment. Business-interruption risk analysis (BIRA), however, is not exclusively concerned with safety, and the availability of rotating machinery – which can be responsible for up to 50 per cent of plant downtime – must be included although it seldom involves safety issues. The primary objective of business-interruption risk analysis is to ensure that the chance of a major outage which could materially affect the commercial viability of the plant is minimized. Production risks are then defined in terms of product loss. An availability model can be a significant asset in BIRA assessments and was the basis of the study discussed here.

Since the major impact will clearly be from the relatively infrequent long outages this study concentrated on the high-cost/high-energy equipment in the plant. It employed analytical models which have a number of advantages, particularly the speed with which a study can be completed. Also, if the data and models are developed on computerized spreadsheets, there is scope for exercising the models in real time with respect to the uncertainties in the input data, increases or decreases in equipment redundancy, the impact of different maintenance strategies, and the optimum level of important spares.

14.28 Combined-cycle plant assessment

Models for this study were developed on a set of integrated EXCEL spreadsheets in which the necessary input data and calculations were included. This facilitates the study of possible changes in the basic data and equipment configurations which can be immediately reflected in the output from the model.

The preliminary appraisal identified the major items of equipment in the plant. These and their interconnections are shown in the functional block diagram of Fig. 14.19. Gas and steam turbines drive electric generators, each giving one-third of the total generation capacity. The hot exhaust gases from the gas turbines are used to raise steam in the heat recovery steam boilers. The residual heat from the steam turbine exhaust is employed in the feed-water heater and returned through the condenser which provides warm water for input, via the condensate and boiler feed pumps, to the heat recovery steam boilers. The cooling water pump supplies the condenser heat sink. Bypasses exhausting the hot gases from the gas turbine to atmosphere can be enabled to allow each gas turbine generator to continue operating in the event of failure of steam side components.

To facilitate analysis two basic functional groups (trains) were defined, comprising different items of equipment in series, namely:

(a) gas turbine generation train comprising a gas turbine and electric generator and transformer;
(b) steam turbine generation train comprising the steam turbine, electric generator and transformer, and all other equipment concerned with steam raising.

As well as the bypasses between the gas turbines and heat recovery steam boilers other bypasses are installed (for example, around the feed-water heater). These are not included in the model because their impact was considered to be relatively small. The gas turbine bypasses are assumed to be 100 per cent available.

The simplification which groups all steam side components in the steam turbine train as a series system leads to a relatively small number of defined output states, as follows.

(a) All three trains available – 100 per cent output;
(b) Steam train unavailable – 66 per cent output;
(c) One-out-of-two gas turbine trains unavailable and steam turbine train available – 50 per cent output;
(d) One-out-of-two gas turbine trains unavailable and steam turbine train unavailable – 33 per cent output;
(e) Other combinations – assumed 0 per cent output.

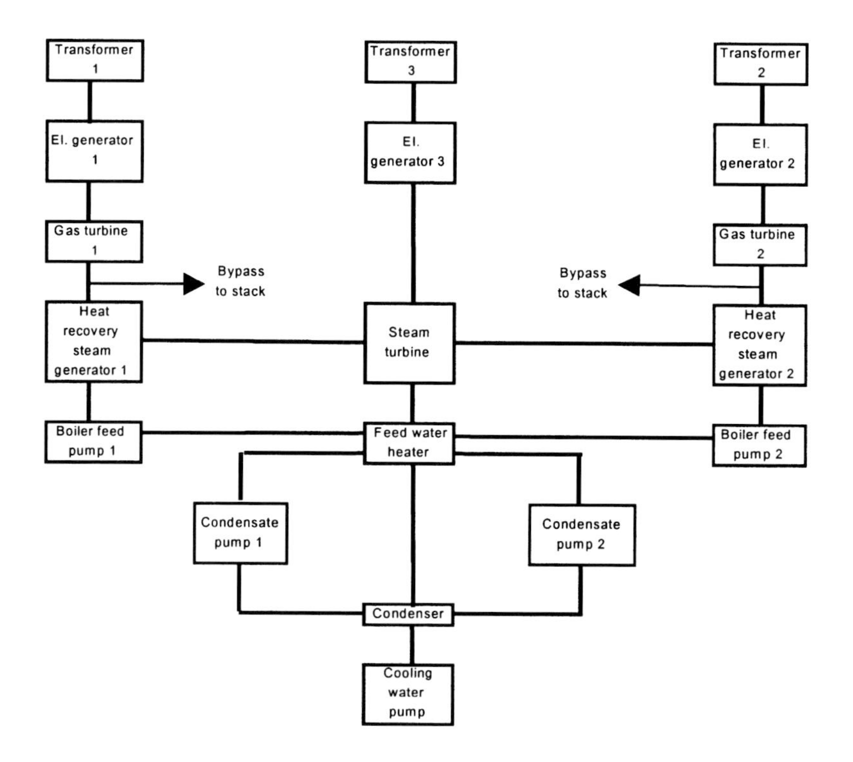

Fig. 14.19 Functional block diagram – combined-cycle plant

14.29 Data and basic assumptions

A literature survey identified two relevant papers concerned with combined-cycle plant, by Kirchsteiger *et al.* (**15**) and Lofe *et al.* (**16**). The data from these papers were combined with other relevant information to derive best-estimates of equipment failure rates and repair times shown in Table 14.14. These best-estimates were used initially to calculate the probabilities of residence in the different output states.

Table 14.14 Equipment reliability data

Equipment	Failure rate (f/year)	Failure mode	Failure mode rate	MTBF	MTTR	Equipment availability	Equipment unavailability
Transformer	0.03	Forced outage	0.03	292 000	200	0.9993	0.0007
El. generator	0.3	Forced outage	0.3	29 200	500	0.9832	0.0168
Gas turbine	6	Forced outage	6	1460	30	0.9799	0.0201
Heat recovery steam boiler	1.5	Forced outage	1.5	5840	30	0.9949	0.0051
Boiler feed pump	1.4	Forced outage	2.8	3129	54	0.9830	0.0170
Condensate pump	0.4	Forced outage	0.8	10 950	30	0.9973	0.0027
Circulating water pump	0.2	Forced outage	0.2	43 800	25	0.9994	0.0006
Condenser	0.4	Forced outage	0.4	21 900	36	0.9984	0.0016
Feed-water heater	0.3	Forced outage	0.3	29 200	50	0.9983	0.0017
Steam turbine generator	0.6	Forced outage	0.6	14 600	40	0.9973	0.0027

The assumption made was that all failure and repair times are independent, identically distributed (IID) events from exponential distributions and that gas turbine and steam turbine trains comprise equipment in series. The steady-state availability for items of equipment is then obtained from the expression:

$$A = \frac{\text{MTBF}}{\text{MTBF} + \text{MTTR}}$$

where MTBF is the reciprocal of the equipment failure rate and MTTR is the mean time to repair. Here MTBF is defined as the mean operating time between failures. Train availability (A_T) is the product of the relevant equipment availabilities. From these data it is also possible to compute train MTBF and MTTR, i.e.

$$\text{MTBF} = \frac{8760 \text{ h}}{\Sigma\lambda_i}$$

and

$$\text{MTTR} = \frac{(1 - A_{\text{T}}) \times 8760}{A_{\text{T}} \Sigma \lambda_i}$$

where

λ_i = failure rate of the ith item of equipment in the train, in failures/year
A_{T} = availability of a train

14.30 Plant availability prediction

The availability logic and calculations are shown in Fig. 14.20. The aggregated output state probabilities sum to unity so the probability of no output from the system is effectively zero. However, since several items of equipment in the different trains are identical the impact of common cause outages cannot be ignored. Applying the β factor method, as described in Chapter 8, and assuming a β factor of 0.1 the common cause failure rate for the gas turbine trains λ_{cc} = 0.63 f/year, MTBF_{cc} = 13 840 h, and common cause availability A_{cc} = 0.9978. The result of including common cause availability in the model would decrease the state 1 and state 2 probabilities to 0.8940 and 0.0288 respectively. This increases the estimated probability of no output from zero to about 20 h/year. At this stage, however, common cause failures are not included in the availability model.

Output state product availability (PA) is obtained by multiplication of the output percentage by the probability of being in the defined state. Product availability (defined as the percentage of achievable output assuming continuous operation) is the sum of the PAs in the different output states. As shown in Fig. 14.20, plant product availability

$$\begin{aligned} \text{PA} &= 0.8960 \times 100\% + 0.0289 \times 66\% + 0.0709 \times 50\% \\ &\quad + 0.0023 \times 33\% = 95.36\% \end{aligned}$$

Table 14.15 shows the different equipment types ranked by failure frequency. Gas turbine failures will clearly dominate, being over 30 per cent higher than the failure rates of all other equipment combined. The data used to derive best-estimates for the initial availability analysis are shown in Table 14.16 from which it is evident that there is

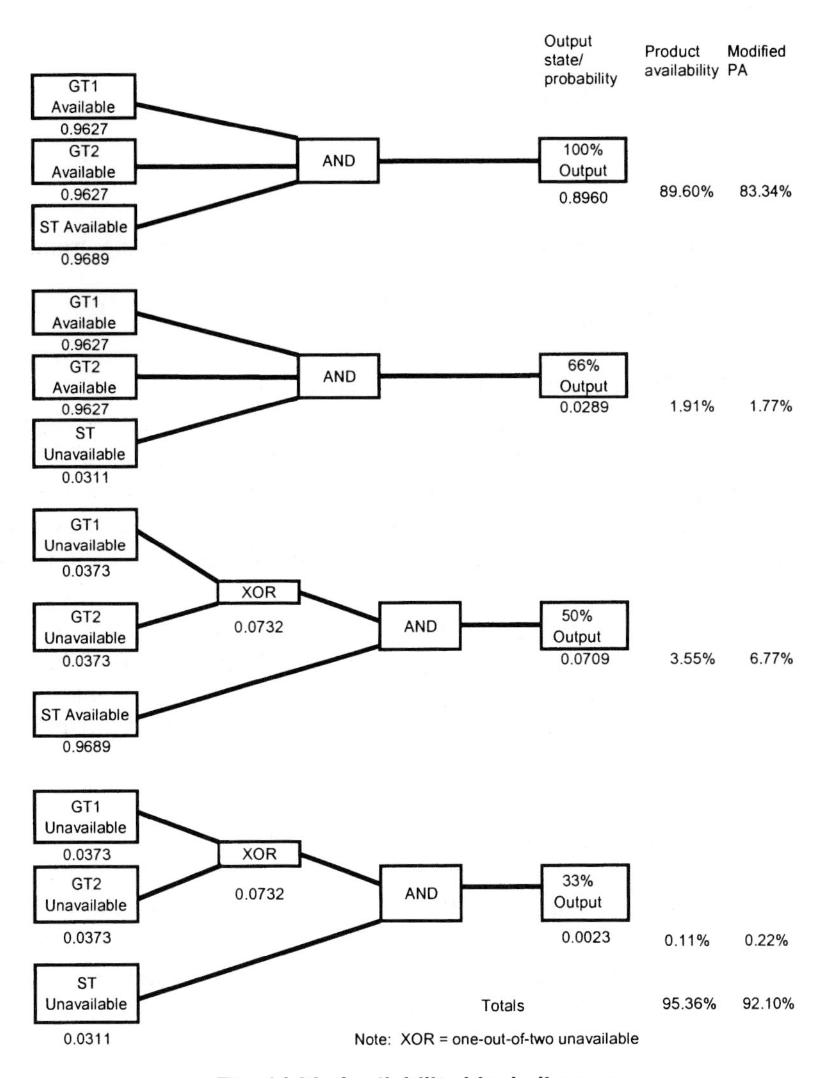

Fig. 14.20 Availability block diagrams

significant uncertainty in the failure rate data. The impact of doubling the failure rates in Table 14.14 for the gas turbines and electrical generators [in line with the data used by Lofe *et al.* (**16**)] is shown in the 'Modified PA' column of Fig. 14.20. With this increase product availability is reduced from 95.36 to 92.10 per cent, mainly due to the transfer of 300 h from output state 1 (100 per cent output) to state 3 (50 per cent output). Assuming gas turbine and steam trains each have equal generation capacity of 100 MW then the difference of 3 per cent

Table 14.15 Equipment ranked by failure rate

Equipment	Failure rate (f/year)	No. installed	Total failure rate (f/year)	Rank
Gas turbines	6	2	12	1
HRS boilers	1.5	2	3	2
Boiler feed pumps	1.4	2	2.8	3
Condensate pumps	0.4	2	0.8	4
El. generators	0.3	2	0.6	5
Steam turbine	0.6	1	0.6	6
Condenser	0.4	1	0.4	7
Feed-water heater	0.3	1	0.3	8
Circulating water pump	0.2	1	0.2	9
Transformers	0.03	2	0.06	10

Table 14.16 Source data and best-estimates

Equipment	Source 1 (**15**)	Source 2 (**16**)	TWP database	Best-estimate (f/year and MTTR)
Transformer	0.03/230 h		0.02/150 h	0.03/200 h
El. generator	0.09/531 h	14.6/24	0.5/450 h	0.3/500 h
Gas turbine	4.0/37 h	(El. gen and GT)	7.8/20 h	6.0/30 h
Heat recovery steam boiler	1.2/30 h	6.6/25 h	1.7/30 h	1.5/30 h
Boiler feed pump		0.9/54 h	2	1.4/54 h
Condensate pump		0.3/31 h	0.4	0.4/30 h
Circulating water pump		0.1/29 h	0.2/20 h	0.2/25 h
Condenser		0.4/36 h	0.3	0.4/36 h
Feed-water heater		0.3/51 h	0.3	0.3/50 h
Steam turbine generator	0.9/37 h	78.7/71 h	0.3/40 h	0.6/40 h

is equivalent to about 80 000 MW h lost output per year. The effect of doubling the failure rate of a transformer on the other hand has little effect on the predicted product availability.

At this stage the estimates were based on the assumption of a single failure/repair mode for all equipment. Recent research by Moss (**17**), however, shows that this assumption is probably too simplistic. For complex equipment such as gas turbines it was evident that repair times are strongly skewed with short outages dominating the

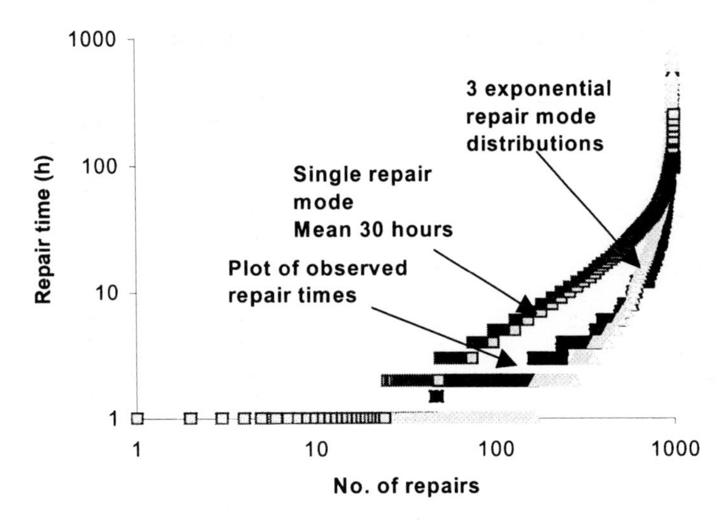

Fig. 14.21 Distribution of gas turbine repair times

downtime distribution. Fig. 14.21 shows the distribution of 1000 ordered repair times recorded over a number of years from gas turbines used in power generation. None of the standard distributions provide a satisfactory fit to these data. The alternative of three exponential repair time distributions with 50 per cent of repairs in the range 1–10 h, 40 per cent between 10 and 100 h, and 10 per cent between 100 and 1000 h is considered to provide a satisfactory match to the observed data. The result of simulating 1000 repair times based on this assumption is also shown in Fig. 14.21 together with the ordered repair times generated for a single repair mode with mean of 30 h as assumed in the original prediction.

The availability predictions can now be revisited to see the impact of this alternative model. Figure 14.22 shows the effect of using the three-failure/repair mode model for estimating state and product availabilities. This indicates an increase in the system product availability to 95.96 per cent. Including common cause failures increases the no output downtime from zero to 26 h/year. Doubling the gas turbine failure rate reduces the system product availability to 93.22 per cent.

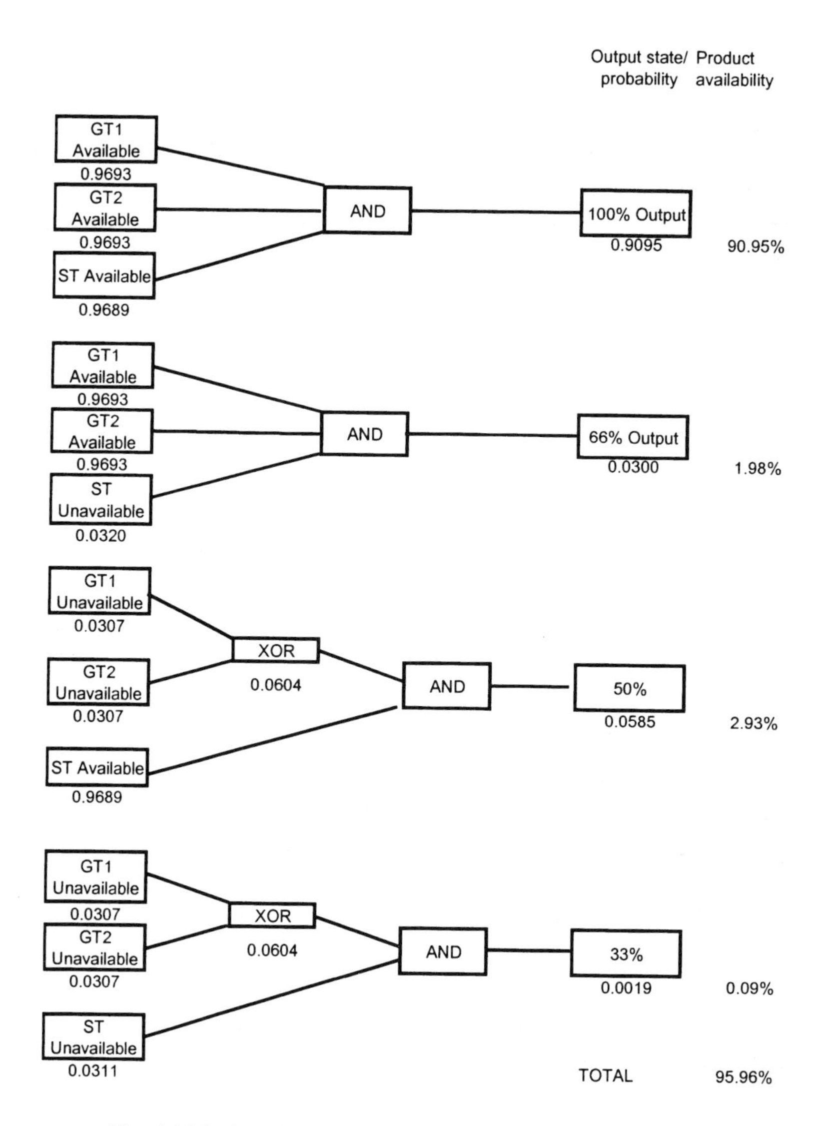

Fig. 14.22 Availability logic (three failure/repair modes)

14.31 Risk estimation

From the availability assessments it is evident that the majority of operating time will be spent in the 100 per cent output state. For this scenario all three trains (and therefore all equipment) can be considered as components in a series system. This considerably simplifies the risk calculations. However, since there are a number of identical components common cause outages must also be included. Based on the discussion in the previous section, risk estimates can now be made either on the assumption of a single failure/repair mode or alternatively on three failure/repair modes. From the best-estimates listed in Table 14.15 it is evident that the system failure rate is approximately 21 failures/year, including common cause failures. Using the expression in Section 14.29 a single failure/repair mode system MTBF and MTTR can now be calculated, i.e.

$$\mathrm{MTBF}_1 = \frac{8760}{\Sigma\lambda_i} = \frac{8760}{21} = 417\ \mathrm{h}$$

$$\mathrm{MTTR}_1 = \frac{(1 - A_\mathrm{T}) \times 8760}{A_\mathrm{T}\Sigma\lambda_i} = \frac{(1 - 0.8960) \times 8760}{0.8980 \times 21} = 47\ \mathrm{h}$$

The alternative – assuming three failure/repair modes – has the following characteristics:

Mode 1: $\lambda_{31} = 0.5 \times \Sigma\lambda_i = 10.5$ f/year MTBF = 834 h
MTTR = 3 h (log mean of range 1–10 h)

Mode 2: $\lambda_{32} = 0.4 \times \Sigma\lambda_i = 8.4$ f/ year MTBF = 1043 h
MTTR = 30 h (log mean of range 10–100 h)

Mode 3: $\lambda_{33} = 0.1 \times \Sigma\lambda_i = 2.1$ f/year MTBF = 4171 h
MTTR = 300 h (log mean of range 100–1000 h)

With these data the probability of an outage greater than t = 1000 h can now be calculated:

$$P(\mathrm{outage} > t\ \mathrm{h}) = \exp(-t/\mathrm{MTTR})$$

For the single-mode model this probability is extremely low ($\sim 6 \times 10^{-10}$) but significantly higher (~ 0.04) for the three-mode model.

The *f–n* (frequency versus number) curve shown in Fig. 14.23 is the regression line for the single- and three-repair mode data. From the curve it is evident that the probability of an outage exceeding 1000 h is about 0.01. This would be a mode-3-type outage for which the frequency (including the common cause failure rate) is about 2.1/year (sum of mode-3 failure rates for gas turbine and steam turbine trains). The risk is given by

$$\text{Risk} = \text{frequency} \times \text{consequence} = \lambda_{\text{mode 3}} \times P(t > 1000\ \text{h})$$
$$= 2.1 \times 0.01 = 2 \times 10^{-2}/\text{year}$$

$$\text{Probabilistic cost} = \text{risk} \times (\text{lost product cost} + \text{manhour cost} + \text{cost of spares})$$

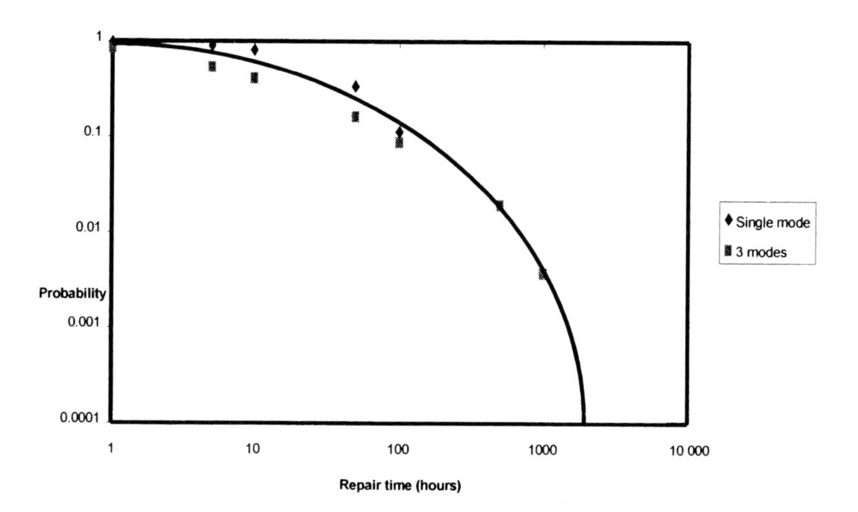

Fig. 14.23 F–n curve for combined cycle plant [P(outage > t h)]

Assuming manhour cost and cost of spares to be 20 per cent of lost product cost and a nominal production-loss cost of £10/MW h, then the probabilistic cost (PC) of an outage greater than 1000 h is not negligible (as might be assumed from a single mode model) but, in fact, is very significant:

$$\text{PC} = \lambda_{\text{mode 3}} \times P(t > 1000\ \text{h}) \times 300\ \text{MW} \times £10 \times 1.2 \times 1000\ \text{h}$$
$$= £72\,000/\text{year}$$

These cost estimates are clearly approximations and could be refined by simulation; however, they provide the basis for consideration of business-interruption insurance requirements which, of course, are also dependent on the risk aversion of the company. Because all of the logic and calculations are on linked EXCEL spreadsheets the overall model provides scope for 'what-if' analyses – failure rates, repair mode proportions, and cost parameters for plant and equipment can be quickly varied in line with the company's experience to see the impact of changes in the basic assumptions.

The EXCEL model can also be used in conjunction with more sophisticated tools – for example, the software developed in the MACRO project for plant and equipment asset management. It provides the basis for subsequent plant performance monitoring for the high-risk sub-systems and equipment. These are the areas where the more detailed methods, such as reliability-centred maintenance, root-cause analysis, and reliability-based inspection can be profitably applied.

In summary, best-estimates for the expected performance of this plant are:

Product availability = 95.5 per cent
(based on unscheduled plant outages – scheduled maintenance and inspection will reduce this figure but could be included in the model).

Top-five equipment at risk:	percentage of outages
Gas turbines	60
HRS boilers	15
Boiler feed pumps	14
Condensate pumps	4
Electric generators	3
Total	96

Risk of an outage > 1000 h = 2×10^{-2}/year

14.32 Conclusions

The objective of this exercise was to demonstrate the power of relatively simple analytical models for assessing the engineering and financial risks associated with process plant. The models were executed on an integrated set of EXCEL spreadsheets which allowed a variety of different design, operation, and maintenance scenarios to be rapidly evaluated.

Clearly this analytical model is a simplified version of the actual combined-cycle system. It can easily be extended into more detail if required; however, given the uncertainties which always exist in the input data and the model, it is considered adequate for identifying and ranking the risks in the first instance and has the advantage that it can be completed within a relatively short time scale.

Using three repair modes simplifies the process of estimating mean times to repair, which is frequently the most difficult task in an availability assessment. It is evident that most engineers are significantly more comfortable with allocating repair times for specified failure modes into bands such as 1–10 h, 10–100 h, and 100–1000 h rather than making point estimates of the mean time to repair. The proportions of the equipment failure rate in each mode can then be adjusted to reflect the plant engineer's best-estimate of the likely downtime distribution.

14.33 References

(1) **Mitchell, D.** (1989) Safety considerations in the development of recirculation techniques in primary mine ventilation systems. Presented at the North of England Branch of the Institution of Mining Engineers, Newcastle, 16 March.

(2) *RMC Reliability Data Handbook* (1985) RM Consultants Ltd.

(3) **Smith, D. J.** (1988) *Reliability and Maintainability in Perspective*. Macmillan.

(4) **Green, A. E.** and **Bourne, A. J.** (1972) *Reliability Technology*. John Wiley.

(5) MIL HDBK 217E, US Military Standard.

(6) *OREDA 92 Offshore Reliability Data Handbook*, 2nd edition. Det Norske Veritas, Hovik, Norway.

(7) **Flesch, B.**, **Pitner, P.**, **Procaccia, H.**, **Riffard, T.**, and **Granger, B.** (1993) Etudes mécaniques probabilistes appliquées à l'optimisation de la maintenance des faisceaux tubulaires des générateurs de vapeur. *Revue Française de Mécanique*, **1**.

(8) **Procaccia, H.**, **Arsenis, S.**, and **Aufort, P.** (1998) EIReDA (European Industry Reliability Data Bank). Edition CUP, Heraklion, Greece.

(9) **Pitner, P.**, **Procaccia, H.**, **Riffard, T.**, **Granger, B.**, and **Flesch, B.** (1993) Optimisation du controle et de la maintenance des faisceaux tubulaires des générateurs de vapeur. *Revue Générale Nucléaire (RGN)*, **3**.

(10) **Madsen, H. O.**, **Krenk, S.**, and **Lind, N. C.** (1986) *Methods of Structural Safety*. Prentice-Hall Inc., Englewood Cliffs, New Jersey.

(11) **Procaccia, H., Piepszownik, L.,** and **Clarotti, C. A.** (1992) Fiabilité des equipements et théorie de la décision statistique fréquentielle et Bayésienne. *Collection de la Direction des Etudes et Recherches d'Electricité de France*, **81**, Edition EYROLLES.

(12) **Clarotti, C. A., Procaccia., H.,** and **Villain, B.** (1994) The ARCS code: a tool for decisions on the basis of field data. In Proceedings of the 6th ESREDA Seminar on *Maintenance and System Effectiveness*, Chamonix, France.

(13) **Pitner, P., Procaccia, H.,** and **Riffard, T.** (1999) COMPROMIS, pour un contrôle et une maintenance intelligente des générateurs de vapeur. *Revue EPURE*, **32**.

(14) **Procaccia, H.** and **Morilhat, P.** (1996) Fiabilité des structures des installations industrielles. Théorie et applications de la mécanique probabiliste. *Collection de la direction des etudes et recherches d'Electricité de France*, **94**, Edition EYROLLES.

(15) **Kirchsteiger, C., Teichmann, T.,** and **Balling, L.** (1995) Probabilistic outage analysis of a combined-cycle power plant. *Power Engng Jl*, June.

(16) **Lofe, J. J., Pouriau, J. L.,** and **Richwine, R. R.** (1997) Availability predictions for combustion turbines, combined-cycle and gasification combined-cycle power plants. ASME paper, 87-JPGR-Pwr-64.

(17) **Moss, T. R.** (1997) Rotating machinery reliability. In ESREL 97, ESReDA session.

Appendix A

Expectation, Variance, and Standard Deviation

A.1 Definition of mathematical expectation

The expectation of a discrete random variable X having possible values $x_1, x_2, x_3, \ldots, x_n$ is defined as

$$E(X) = x_1 P(X = x_1) + \cdots + x_n P(X = x_n) = \sum_{j=1}^{n} x_j P(X = x_j)$$

or equivalently, if $P(X = x_j) = f(x_j)$,

$$E(X) = x_1 f(x_1) + \cdots + x_n f(x_n) = \sum_{j=1}^{n} x_j f(x_j) = \sum x f(x)$$

where the last summation is taken over all appropriate values of x. As a special case of this where the probabilities are all equal we have

$$E(X) = \frac{x_1 + x_2 + \cdots + x_n}{n}$$

which is called the **arithmetic mean** or simply the **mean** of $x_1, x_2, \ldots, x_n$.

For a continuous random variable X having density function $f(x)$ the expectation of X is defined as

$$E(X) = \int_{-\infty}^{\infty} xf(x)\mathrm{d}x$$

The mean or expectation of X gives a single value which acts as an average of the values of X and for this reason it is often called a **measure of central tendency**.

A.2 Definitions of variance and standard deviation

Another quantity of great importance in probability and statistics is called the variance and is defined by

$$\mathrm{Var}(X) = E\left[(X-\mu)^2\right]$$

where μ is the mean.

The variance is a non-negative number. The positive square root of the variance is called the **standard deviation** and is given by

$$\sigma_x = \sqrt{\mathrm{Var}(X)} = \sqrt{\left\{E\left[(X-\mu)^2\right]\right\}}$$

The standard deviation is often denoted by σ and the variance in such a case is σ^2.

If X is a discrete random variable having probability function $f(x)$, then the variance is given by

$$\sigma_x^2 = E\left[(X-\mu)^2\right] = \sum_{j=1}^{n}(x_j-\mu)^2 f(x_j) = \sum(x-\mu)^2 f(x)$$

In the special case where the probabilities are all equal we have

$$\sigma^2 = \left[(x_1-\mu)^2 + (x_2-\mu)^2 + \cdots + (x_n-\mu)^2\right]/n$$

which is the variance for a set of n numbers $x_1, \ldots, x_n$.

If X is a continuous random variable having density function $f(x)$, then the variance is given by

$$\sigma_x^2 = E\left[(X-\mu)^2\right] = \int_{-\infty}^{\infty}(x-\mu)^2 f(x)\,\mathrm{d}x$$

The variance (or the standard deviation) is a measure of the **dispersion** or **scatter** of the values of the random variable about the mean. If the values tend to be distributed far from the mean, the variance is large (Fig. A1).

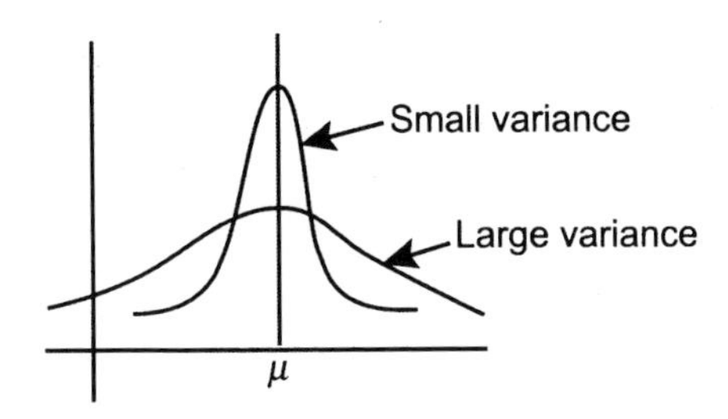

Fig. A1

Appendix B

Standard Laplace Transforms

1. The Laplace transform of $F(t)$ is

$$\mathcal{L}\left[F(t)\right] = \int_0^\infty F(t)\, \mathrm{e}^{-st} \mathrm{d}t$$

2. Table of transforms:

$F(t)$	$\mathcal{L}\{F(t)\} = f(s)$
a	$\frac{a}{s}$
e^{at}	$\frac{1}{s-a}$
$\sin at$	$\frac{a}{s^2+a^2}$
$\cos at$	$\frac{s}{s^2+a^2}$
t^n	$\frac{n!}{s^{n+1}}$
(n is a positive integer)	

3. Laplace transforms of derivatives:

$$\mathcal{L}\left[\frac{\mathrm{d}F}{\mathrm{d}t}\right] = s\,\mathcal{L}[f] - F(0)$$

where $F(0)$ is the value of F when $t = 0$:

$$\mathcal{L}\left[\frac{d^2F}{dt^2}\right] = s^2\,\mathcal{L}[F] - SF(0) - F'(0)$$

4. First shift theorem:

$$\text{if } \mathcal{L}[F(t)] = f(s) \quad \text{then} \quad \mathcal{L}\left[e^{-at}F(t)\right] = f(s+a)$$

5. Convolution theorem:

$$\text{if } \mathcal{L}[g(t)] = G(s) \quad \text{and} \quad \mathcal{L}[f(t)] = F(s) \quad \text{then}$$

$$\mathcal{L}\left[\int_0^t g(u)f(t-u)\,du\right] = G(s)\cdot F(s)$$

6. Inverse transform:

$$\text{if } \mathcal{L}\{F(t)\} = f(s) \quad \text{then} \quad \mathcal{L}^{-1}\{f(s)\} = F(t)$$

7. Table of inverse transforms:

$f(s)$	$F(t)$
$\frac{a}{s}$	a
$\frac{1}{s+a}$	e^{-at}
$\frac{n!}{s^{n+1}}$ (n is a positive integer)	t^n
$\frac{1}{s^n}$	$\frac{t^{n-1}}{(n-1)!}$
$\frac{a}{s^2+a^2}$	$\sin at$
$\frac{s}{s^2+a^2}$	$\cos at$

Glossary of Reliability and Risk Assessment Terms

Active repair time The time required to repair and return an item to a state where it is ready to resume its functions.

Average individual risk The average chance of any individual in a defined population sustaining a given level of harm from incidents which are considered to be limited to that population.

Boundary The interface between an item and its surroundings.

Calendar time The interval of time between the start and end of data surveillance for a particular item, selected input events or states.

Checklist A method for hazard identification by comparison with experience in the form of a list of failure modes and hazardous situations.

Command fault A component is in the incorrect state due to improper control signals or noise.

Common cause failure The failure of more than one component, item, or system due to the same cause.

Confidence limits End points of the confidence interval that is believed to include the population parameter with a specified degree of confidence.

Cut set A list of components such that if they fail then the system is also in the failed state.

Demand A condition which requires a protective system to operate.

Diversity The performance of the same function by a number of independent and different means.

Downtime The total time an item has been out of service due to a failure since the failure occurred, including, but not restricted to, the time to detect the failure, any delays and waiting time, active repair time, and time for testing and start-up after repair.

Environmental stress The stress to which an item is exposed due soley to its presence in an environment.

Event tree analysis A method for illustrating the sequence of outcomes which may arise after the occurrence of a selected initial event.

Fail-safe fault A fault which results in no deterioration of safety.

Fail-to-danger fault A fault which moves a plant towards a dangerous condition or limits the ability of a protective system to respond to a dangerous condition.

Failure The termination or the degradation of the ability of an item to perform its required function.

Failure cause The original cause of the failure; the circumstances during design, manufacture, assembly, installation, or use that have led to the failure.

Failure mode The effect by which a failure is observed on the failed system.

Failure mode and effect analysis (FMEA) The study of the potential failures that might occur in any part of a system, to determine the probable effect of each on all other parts of the system and on probable operational success.

Failure mode and effect criticality analysis (FMECA) An FMEA, the results of which are ranked in order of seriousness.

Fatal accident rate (FAR) The number of deaths that have occurred or are predicted to occur in a defined group, in a given environment, during 10^8 hours of total exposure.

Fault tree analysis A method of representing the logical combinations of various system states which lead to a particular failure outcome (top event).

F–N **curve** A plot showing, for a specified hazard, the frequency of all events causing a stated degree of harm to *N* or more people, against *N*.

Fractional dead time The mean fraction of time in which a component or system is unable to operate on demand.

Frequency The number of occurrences per unit of time.

Gate A symbol in a logic diagram which specifies the logical combination of inputs required for an output to be propagated.

Guide words A list of words applied to system items or functions in a hazard study to identify undesired deviations.

Hazard A physical situation with the potential for human injury, damage to property, damage to the environment, or some combination of these.

Hazard analysis The identification of undesired events that lead to the materialization of a hazard, the analysis of the mechanisms by which these undesired events could occur and usually the estimation of the extent, magnitude, and likelihood of any harmful effects.

Hazard indices A checklist method of hazard identification which provides a comparative ranking of the degree of hazard posed by particular design conditions.

Hazard and operability (HAZOP) study A study carried out by the application of guide words to identify all deviations from design intent with undesirable effects for safety or operability.

Hazardous substance A substance which by virtue of its chemical properties constitutes a hazard.

Implicant set A set of component states, failure or working, which cause the specific system failure mode.

Individual risk The frequency at which an individual may be expected to sustain a given level of harm from the realization of specified hazards.

Individual risk criteria Criteria relating to the likelihood with which an individual may be expected to sustain a given level of harm from the realization of specified hazards.

Initiating event A postulated occurrence capable of leading to the realization of a hazard.

Item A common term used to denote any level of hardware assembly, i.e. system, sub-system, maintainable items, and parts.

Logical diagram A representation of the logical combination of sequence of events leading to or from a specified state.

Loss prevention A systematic approach to preventing accidents or minimizing their effects. The activities may be associated with financial loss or safety issues.

Maintainable item An assembly of parts that are normally the lowest indenture level during maintenance.

Major hazard An imprecise term for a large-scale chemical hazard, especially one which may be realized through an acute event. Or, a popular term for an installation which has on its premises a significant quantity of a dangerous substance.

Major incident criterion Criterion (expressed as a frequency) for incidents falling within a defined category of consequences.

Minimal cut set A cut set such that if any term is removed from the list the system will no longer fail.

Minimal path set A path set such that if any item is removed the system will no longer function.

Operating mode The normal, or most frequent, duty of operation (continuously running, standby, alternating, etc.).

Operating time The period of time during which a particular item performs its required function(s), between the start and end of data surveillance.

Operation (or rating) stress The stress to which an item is exposed due solely to the performance of its function.

Path set A list of components such that if they all work then the system is also in the working state.

Perceived risk The risk thought by an individual or group to be present in a given situation.

Population The total number of items of one particular type in service during the period of event data surveillance.

Prime implicant set An implicant set such that if any event is removed it no longer causes the specific system failure mode.

Primary fault Any fault of a component that occurs under conditions for which the component was designed to work or that occurs from natural ageing.

Probability A number in a scale from 0 to 1 which expresses the likelihood of an event.

Redundancy The performance of the same function by a number of identical but independent means.

Reliability The probability that an item is able to perform a required function under stated conditions for a stated period of time.

Residual risk The risk remaining after all proposed improvements to the facility under study have been made.

Risk The combined effect of the probability of occurrence of an undesired event and the magnitude of the event.

Risk assessment The quantitative evaluation of the likelihood of undesired events and the likelihood of harm or damage being caused together with judgements made concerning the significance of the results.

Safety The freedom from unacceptable risk or personal harm.

Sample The group of items of one particular type in service during the period of event data surveillance.

Secondary fault Any fault of a component that occurs as a result of the component being under a condition for which it was not designed, either in the past or at present, perhaps due to the failure of other components in the system.

Societal risk The relationship between frequency and the number of people suffering from a specified level of harm in a given population from the realization of specified hazards.

Societal risk criteria Criteria relating to the likelihood of a number of people suffering a specified level of harm in a given population from the realization of specified hazards.

Standby redundancy That redundancy wherein the alternative means for performing a given function are inoperative until needed.

Sub-system An assembly of units that provide a specific function that is required for the system to achieve its intended performance.

System The highest indenture level including sub-systems and small entities belonging to that system.

Taxonomy A systematic classification of items into generic groups based on factors possibly common to several of the items, e.g. function, type, medium handled, etc.

Top event The selected system outcome whose possible causes are analysed in a fault tree.

Trip Unexpected shutdown of equipment. Trips are either spurious or real.

Waiting time The time to isolate the equipment from the process, repair, delay, and waiting for spare parts or tools, and any time after the repair has been completed if the item is not put into service immediately. Time for testing is included when this is an integral part of the repair activity.

Index